Landscape: from knowledge to action

Martine Berlan-Darqué,
Yves Luginbühl,
Daniel Terrasson

Éditions Quæ
c/o Inra, RD 10, 78026 Versailles Cedex

Collection *Update Sciences & Technologies*

Conceptual Approach to the Study of Snow Avalanches.
Maurice Meunier, Christophe Ancey, Didier Richard,
2005, 262 p.

Qualité de l'eau en milieu rural.
Savoirs et pratiques dans les bassins versants,
Philippe Mérot, coordinateur
2006, 352 p.

Biodiversity and Domestication of Yams in West Africa.
Traditional Practices Leading to *Dioscorea rotundata* Poir,
Alexandre Dansi, Roland Dumont, Philippe Vernier, Jeanne Zoundjihèkpon,
2006, 104 p.

Génétiquement indéterminé.
Le vivant auto-organisé
Sylvie Pouteau, coordinatrice,
2007, 172 p.

L'éthique en friche.
Dominique Vermersch,
2007, 116 p.

Agriculture de précision.
Martine Guérif, Dominique King, coordinateurs,
2007, 292 p.

Territoires et enjeux du développement régional.
Amédée Mollard, Emmanuelle Sauboua, Maud Hirczak, coordinateurs,
2007, 240 p.

Estimation de la crue centennale pour les plans de prévention des risques d'inondations.
Michel lang, Jacques Lavabre, coordinateurs
2007, 232 p.

Paysages : de la connaissance à l'action.
Martine Berlan-Darqué, Yves Luginbühl, Daniel Terrasson
2007, 316 p.

Translations credits
French original papers were translated into English by Gail Wagman, Sauve (30, France)
and Mary Shaffer, Paris (75, France).
English language reviewed by Valérie P. Howe (V.P.Hiwe@greenwich.ac.uk).

© Éditions Quæ, 2008   ISBN: 978-2-7592-0060-3   ISSN: 1773-7923

All rights reserved.
No part of this book may be reproduced in any form or by any means without permission in writing from
the publisher.

Table of contents

Chapter 4. Health, identity and sense of place: the importance of local landscapes

Chapter 5. Where does grandmother live? An experience through the landscape of Veneto's 'città diffusa'

Section 2
Public spaces in the cityscape

Chapter 1. What role does plant landscape play in urban policy?

Chapter 2. Green cityscapes and social inclusion in three major metropolitan areas of Switzerland

Section 3
Landscape policies:
from conception to implementation

Chapter 3. From tree-lined banks to hedge landscapes: the dynamics underlying the ideas about the landscape that have inspired public policies

Chapter 4. New hedgerows in replanting programmes: assessment of their ecological quality and maintenance on farms

Chapter 5. A typology of intercommunal actions related to the landscape

Chapter 6. The mayors' polyphonic discourse during a landscape intervention

Section 4
Public participation in landscaping action

Chapter 1. From the landscape perception until landscaping action. How long is the way?

Chapter 2. The incorporation of public participation processes in three landscape planning projects in the Murcia region of Spain

Chapter 3. Landscape at the crossroads of local development choices: What knowledge for which issues? Which tools for action?

Chapter 4. Landscape: a window of opportunity for regional governance? Landscape scenario workshops as a participatory planning tool

Foreword

As a result of the growing interest in the landscape, landscape issues became involved at every level of public action. Legislation explicitly concerning the landscape was enacted. Some of these laws aim at protecting specific sectors (those related to coastal and mountain regions), or at planning economic development, particularly in rural areas.

More recently, the European Landscape Convention, which went into effect in France on 1 July 2006, is the first international treaty specifically devoted to the landscape.

This convention, known as the Florence Treaty, provides a precise definition of the landscape and also defines notions of 'landscape policy' and 'landscape quality objectives'. It promotes the simultaneous development of landscape policies at three different levels: protection, management and land use. Moreover, the Florence Treaty sees the landscape as a guiding principle for the improvement of the quality of life of concerned populations, encouraging contracting countries to implement public policies in which the citizenry has had a say.

The aim of the research programme, 'Landscapes and Public Policies', launched in 1998 by the French Ministry of the Environment, was to evaluate the effects of these different public policies on the landscape. This research programme was innovative because even if scientific communities had already been mobilised on landscape issues, no research programme actually existed whose prime objective was to contribute scientific knowledge on this theme to public policy. Within the framework of this programme, 24 research projects were thus initiated between 1999 and 2001. In order to make the findings of this programme available to all those concerned (e.g. governments, elected officials and professionals, users and citizens), different ways of disseminating scientific knowledge (symposia, articles and training) were implemented and encouraged.

In support of the implementation of the European Landscape Convention, and in order to promote the role of landscape in European research and to strengthen the role of French research teams, the French Ministry of Ecology and Sustainable Development organized a European conference in Bordeaux, in partnership with Cemagref, to provide an opportunity to present the major results of this research, the aim of which was to understand the role of public action on the landscape.

We feel confident that the synthesis of the different points of view presented in this book will strengthen the action implemented by the Ministry of Ecology and Sustainable Development and, in particular, that of our two departments. We also hope that these contributions will give impetus to the emergence of landscape research with a specifically European character.

We would like to take this opportunity to thank all of those who contributed to the success of this research programme: the programme's entire scientific advisory board and, in particular, its two successive presidents, Georges Bertrand and Yves Luginbühl, as well as Daniel Terrasson who was responsible for the scientific co-ordination of this programme, and Martine Berlan-Darqué, who spearheaded it alongside Jean-François Seguin, head of the landscape office.

Guillaume SAINTENY
Director of Economical Studies Landscapes
and Environmental Evaluation

Jean-Marc MICHEL
Director of Nature
and Landscapes

Introduction

Daniel TERRASSON

Within the global context of the rapidly increasing concern for the environment, the landscape has progressively become a social issue, particularly in Western countries. However, the conditions that led to this growing awareness were very specific. The landscape was not the focus of urgent warnings from the scientific community, nor did it provoke major controversies like those brought about by global warming, natural or technological risks, pollution, health issues, erosion of biodiversity or water shortages. The landscape was not the rallying point around which major environmental organizations challenged our forms of development. Rather, it evolved on its own impetus, an issue whose importance became increasingly obvious as a result of the convergence of two dynamics. On the one hand, an elite, initially made up of several isolated personalities, became interested in the landscape, especially when it revealed a distinct cultural or outstanding aspect. This elite progressively acquired a stronghold at the operational level, as well as in the domain of research. On the other hand, ordinary citizens became concerned with a degradation of their living environment that was becoming increasingly evident. A phenomenon of society in the beginning, this concern then spread to the political and scientific arenas, and evolved from the extraordinary to the ordinary. The landscape issue has invaded the media today where it is now a recurring theme. We no longer count the number of books, exhibitions and TV shows devoted to the landscape. It is a vehicle used to promote travel and local products; it is adopted by multiple associations that take responsibility for its protection, its transformations, etc.

Nevertheless, the landscape was not a priority issue in policy discussions within international fora. Most Western countries developed a wide range of regulations and public action policies aimed at protecting or managing the landscape. These measures differ considerably, depending on the cultural and political context of each country. The European Landscape Convention (ELC), adopted by the Council of Europe in Florence on 20 October 2000, and which came into effect on 1 March 2004 after its ratification by ten member states, provides new momentum. It endows the landscape with a value of general interest and emphasizes the necessity of looking for a higher degree of consistency in public action between the different European countries. It also implicitly recognizes the inherently innovative character of public action by recommending the exchange of research results and experiences.

Initiatives also increased within the scientific domain. With its multiplicity of meanings, the landscape touches on a variety of fields, both in the social sciences and the natural sciences. It is the focus of organizations, as well as research programmes, geared exclusively to the landscape. In the first case, we can mention the International Association for Landscape Ecology (IALE), the Permanent European Conference for the Study of the Rural Landscape (PECSRL), Landscape Europe, Landscape Tomorrow, the Nordic Landscape Research Network (NLRN) in the Scandinavian countries, the Landscape Research Group (LRG) in UK, the European Council of Landscape Architecture School (ECLAS), etc. In the area of research, national programmes exist or have been recently completed – in Austria ('Forschungsprogramm Kulturlandschaft'), in Switzerland (PNR 48: 'Landscape Development in Mountain Regions') and in France ('Landscape and Public Policies', and since 2005, 'Landscape and Sustainable Development'). Finally, several European projects of the Fifth and Sixth Framework Programmes for Research and Development deal extensively with the landscape or are exclusively devoted to it: ATLAS, ELCAI, FORAM, REGALP, SENSOR, etc.

Many books intended for the general public and much academic literature on the landscape have been published in recent years. We would, therefore, like to explain the underlying motives and the originality of this volume by taking a closer look at the series of events that led to its publication.

This book is the outcome of the wish to compare the ideas discussed within the framework of the national research programme, 'Landscape and Public Policies', implemented in France at the initiative of the Ministry of Ecology and Sustainable Development, to research carried out in other countries. It is also an attempt to evaluate its results from the perspective of the European Landscape Convention. The French research programme, which took place from 1998 to 2004, had two major objectives (www.ecologie.gouv. fr/article.php3?id_article=5665). First, it set out to address the government's concerns about the real effectiveness of a recently implemented policy concerning an issue for which it was difficult to define an administrative and a standardization framework and, second, it attempted to give impetus and structure to a scientific activity that seemed to be in need of new dynamics. Several observations could be made upon completion of this programme. First, a dynamic scientific community was established in France as a result of this programme, and original results were produced at different levels: theoretical reflections on concepts linked to the landscape, the design and implementation of public action, the role of stakeholders, etc. On the other hand, this research, published in French, was rarely distributed abroad. Generally speaking, there appeared to be a lack of dialogue at the international level between different research communities, reinforced by disciplinary or thematic barriers and affiliations with schools of thought limited by geographical boundaries. Finally, from a general point of view, the scientific community only partially satisfied the needs of public action. Knowledge had obviously advanced in the analysis of landscape transformation dynamics and the explanation of the determinants of these dynamics, the functioning of ecosystems and consequences in terms of the erosion of biodiversity, social representations in relation to cultural contexts, landscape characterization and inventory, and data management and mapping tools. Nevertheless, gaps remained in linking these different approaches, in the knowledge of the long-term impacts of landscape policies, particularly at the economic and social levels, in the methodologies for evaluating these policies and in the dialogue between the scientific and operational arenas.

These observations led to the organization of a conference that was held in Bordeaux (France) in December 2004. This book is not the proceedings of this symposium, which is available on CD-Rom and the website, *SYMPOscience* (www.symposcience.org), nor an assessment of the results of the 'Landscape and Public Policies' programme, but instead, an additional effort to compare and find a common ground between similar experiences that took place and are taking place in Europe today. Its aim is to bring together a certain number of research projects devoted to public action on the landscape for the purpose of improving it. This concern with the relationship to action led us to give preference to actual case histories, chosen for their exemplariness and mutual resonance, instead of attempting to cover the entire diversity of public action forms that exist at this time. These examples were primarily chosen within the European context to address the concerns of the Council of Europe.

The first section includes five papers that deal with the way the landscape issue interacts with ecological and social priorities. In some countries and particularly in France, theoretical debates and power struggles have sometimes led to separating and assigning priorities to these issues in heated debates: man and nature, nature and culture, the elite and the ordinary, the subjectivity of the landscape and the reality of the erosion of biodiversity, etc. The authors remind us just how much the idea of landscape leads to the overlapping of these priorities without exclusion. They also show, as emphasized in the European Landscape Convention, that this issue is above all related to the relationship of ordinary people to their daily living environment, and that the areas within proximity of this environment, whether they be urban, forest or other, play a very important role.

The second part deals with the relationship between landscape and public space in an urban environment from two different perspectives: open spaces and the role of vegetation in the city. Parks and gardens have played an historical role in the emergence of landscape theory. We must go beyond aesthetic considerations in this case in order to develop the social priorities inherent in their role as public spaces and find a meeting ground between the inhabitants' individual practices and city planning. The issues involved in public space and its opening are also applicable to rural environments, as suggested in several of the papers in the first section. Therefore, comparative analysis of the subjects covered in these two sections reveals that the question of landscape involves considerations that transcend the traditional boundaries of the rural and the urban.

The third section describes examples of public actions in favour of the landscape stressing the principles that led to their design and the conditions for their implementation. It demonstrates the necessity, on the one hand, of taking stock of them, characterizing them, developing instruments for action and evaluating them and, on the other hand, of coming to terms with the time frames of public action, landscapes and perceptions. These aspects are illustrated in the range of papers presented here.

The last section deals with the contribution of citizens to public action as recommended by the European Landscape Convention. It reveals the considerable differences in practices and conditions under which initiatives were implemented. The idea here is not to analyse these differences but to show, in terms of the landscape, several examples of applications where scientists and those responsible for public action interact. The reader's attention is drawn to the experimental nature of this new form of governance.

Finally, the conclusion, by integrating the findings of actual local experiences presented here, interprets them within the framework of landscape research today.

The landscape, between social and environmental issues

Chapter 1
Landscape, an interpretative framework for a reflexive society

Marie-José FORTIN

What kind of connections exist between the concepts of landscape and environment? The question is ambitious. To begin with, we should note that this query implies that the two concepts are not equivalent. It is in fact in line with the first theoretical efforts aiming to distinguish between landscape and environment.

Such a distinction was strongly advocated by Augustin Berque (1995) and Alain Roger (1978, 1997), among others. With skill, they both showed how the landscape experience is above all a process of social and cultural mediation. Furthermore they considered that landscape would be a matter of sensitivity, fundamentally subjective in nature, while the environment would be made up of objective facts (Berque, 1991).

Roger (1997) defends this dissociation, in particular to limit the claims of an environmental science that might become too insatiable, by attempting to absorb the subject of landscape and reducing it to its physical dimension. In his opinion, the aesthetic dimension represents landscape's original contribution and basically draws its source from art and institutionalized culture. In support of this idea, Roger (1997) refers to, among other things, a proposal by Bernard Lassus, who believes one may consider a polluted area as beautiful landscape. According to such postulates, environmental considerations could be excluded from the aesthetic experience associated with landscapes. Following this line of reasoning, landscape would essentially be a sensory experience tied to formal characteristics.

Berque (1995) underlines that the dissociation between the environment (as 'fact', object of physical geography) and landscape (as a 'sensitive' relation, object of phenomenology) is more the result of a cognitive position, inspired by scientific traditions based on modern ontology, than of the experience of reality. He explains this as an historical transition phase leading to a new ontology, linking environment and landscape in a dynamic relationship. He thus puts forward the *ecoumene* theory, based on the notion of *mediance,* of which landscape would be an expression (Berque, 2000). Berque thereby opens an interesting area of inquiry[1]. Since, however, the issue is to describe the notion of landscape, we feel that his insistence on a strongly art-based conception of the aesthetic experience reduces its scope[2].

Following these theoretical efforts, we will suggest taking a broader approach to this experience, and seeing in it a wider social practice, referring to people's lives, perceptions, knowledge, relations of intersubjectivity and the materiality of their surroundings. This suggestion is the result of empirical research carried out in France and Quebec. After briefly presenting a few findings (part 1), we suggest a series of reflections to feed this theoretical debate (parts 2 and 3).

Describing industrial landscapes: interpretations by affected parties

Our research was based on a hybrid concept that challenges the ideas of landscape and environment, the concept of 'industrial landscape'[3]. One of the questions was whether a major industrial landscape, whose activities have a negative impact on the environment, could be described in a positive manner and even be considered 'beautiful'?

From the outset, the perception of landscape is conditioned by the concerned individual, social group, or institution that depicts landscape in relation to a specific geographical and historic context. For our research, this involved two communities where a major smelter for aluminium production has been built, one in France (in Dunkerque, in the Nord-Pas-de-Calais region) and the other in Quebec (at Alma, in the Saguenay–Lac-Saint-Jean region). In connection with the principles of sustainable development and environmental justice, we gave priority to the viewpoints of those who are potentially affected, i.e. residents whose homes are located near the industrial sites and the land developers[4].

One research objective was to better understand the process of interpretation of industrial landscapes and how their social meanings are built. This analysis made it possible to outline three areas, or 'chains' of relations, that appear to be especially important in this

---

[1] This, moreover coincides with a pioneering project supported by a number of other landscape researchers including G. Bertrand, H. Décamps and Y. Luginbühl.

[2] Three of the five criteria highlighted as conditions for the existence of a landscape culture refer to artistic representations (e.g. painting, gardens and literature) (Berque, 2000).

[3] This research was carried out within the framework of a doctoral thesis (Fortin, 2005) as part of a multidisciplinary research programme on social impact follow-up, in connection with the example in Quebec (www.uqac.ca/msiaa). Landscape research received funding from this programme and the Social Sciences and Humanities Research Council of Canada (SSHRC); the Fonds pour la formation de chercheurs et l'aide à la recherche (FCAR); the programme to support Quebec-France thesis co-supervision; and the Fonds d'action québécois pour le développement durable (FAQDD). The dissemination of results during the symposium in Bordeaux and within the scope of this publication was made possible through the financial support of the 'Décanat des études de cycles supérieurs et de la recherche' of the Université du Québec à Chicoutimi. We thank all these organizations for their support.

[4] Twenty or so individual and semi-directed interviews were conducted among residents of Dunkerque and four focus group interviews were conducted in Alma. These were enriched by a review of documents, direct observations on a multipartite environmental monitoring committee, individual interviews, etc.

process. They concern, first, landscape's materiality and visible forms, second, land use dynamics and third, social relations. They are described briefly below.

The materiality of visible and planned forms

According to certain theoretical approaches, the formal and plastic dimensions are decisive in the landscape experience. The research project set out to test this postulate by asking residents to give their opinion on the way the industrial site had been planned and managed, its architecture and its integration into the landscape. The material forms of the production units do indeed appear to be a first key component in their interpretation. This could hardly be otherwise, since the aluminium production complexes cover several hectares of land and are made up of buildings that measure nearly 1 km in length. Moreover, in Dunkerque, the industrial zone covers some 20 km of coast. The interviews revealed, however, that from the residents' point of view, the form could not be separated for long from other social considerations.

Trees to forget the industrial sites but not to 'hide reality'

The residents and developers we met, both in Dunkerque and Alma, generally appreciate the efforts undertaken by businesses and public authorities in major 'landscaping' projects in industrial sites and zones. Planting trees and making other landscaping improvements make it possible to reduce the cognitive presence of industry: a city with an abundance of plants and flowers helps "forget" the factories and the pollution with which people live on a daily basis. A well-planned landscape is synonymous with "tidiness". By extension, it gives the impression of no pollution. More broadly, landscape designs are perceived as a way of reinvesting a part of the profits generated by big business into the local area. They are part of the new conscience that companies should have, in connection with today's 'mentality' aiming for better integration of production sites into the local surroundings. In this sense, 'landscaping' sites is considered by local populations to be a new indicator, both manifest and visible, of companies' social responsibility.

In certain ways, landscape designs may have *concrete* effects on the quality of their environment. For example, trees located near production plants make it possible to capture part of their gaseous emissions. Similarly, for residents, plantlife is an indicator of the state of the environment. This is why they inspect their gardens, looking for any abnormalities or unusual events in their surroundings (e.g. significant loss of leaves in summer, plants suddenly turning brown, decreased fruit yields, dust, unusual odours, etc.). Plants' appearance around the edges of factories is also used in the same way, to distinguish between the factories, according to the greater or lesser degree of pollution they generate. For example, when plantlife is absent or has difficulty growing, it is considered to be a sign of the existence of pollution that prevents 'nature' from developing.

In addition, despite these various positive effects attributed to landscaping operations in industrial areas and cities, respondents anticipate the potential downside of an approach of landscape that might be too cosmetic and reductive. A number of respondents thus interpreted these operations as strategies promoted by productive businesses to protect their image, or influence the way they are perceived locally. Such a strategy may be risky because, as they warn, landscape planning and planting gardens must not be used as ways to "hide reality". And yet, in the case of industrialization, particularly in these regions that have historically been home to heavy industry, the issue of the environment is a fundamental dimension of this reality that must be taken into account by the responsible authorities. Pollution may be a problem, even in a "flower garden", as they pointed out: "The quality of life, you know, if you are living with serious pollution and you are in a flower garden (...), well, that doesn't change anything! What I mean to say is that the problem here is pollution"[5]. Such statements underline the fact that concerned residents do not confuse landscape design with a decrease in pollution.

In sum, while they appreciate the efforts made by the companies, the residents we met who live near major industrial sites do not limit the idea of landscape to that of 'landscaping', anymore than they limit it to its visible forms alone.

A visual presence revealing the symbolic conflict with one's habitat

This first observation is strengthened when one considers the divergent evaluations gathered in both cases, dealing with different levels of architectural investment. In Dunkerque, the company Pechiney worked a great deal on the architecture of the façade of the complex, which can be seen from the national highway, as well as on the landscape design (Figure 1.1). The language they chose is particularly elaborate and refined, which is rather rare in the case of heavy industrial production units. Despite this, the residents we spoke with remained rather harsh in their judgement. Although some of them appreciated the quality of the architecture, the majority thought that the industrial complex is not well-integrated into the landscape. Several complained about the lack of greenery around the complex, which makes it especially visible and imposing against the local landscape. Some of the words used to describe it include "massive", "flashy" and "mastodon".

By contrast, Alcan in Quebec did not go to such lengths to create an elaborate architecture for its factories. The façades are covered with sheets of white and green steel, including the anode factory, which is 75 m high. Lastly, plans to plant vegetation to diminish the visual impact were cut back by 50%. The huge complex indeed stands out in the Alma landscape; it can be seen from a number of places (e.g. roads, apartment buildings, recreation areas, navigable river, etc.) and it is visible for kilometres around (Figure 1.2). While they all noted this high degree of visibility, most .of the residents and developers we met generally considered the industrial complex to be well integrated in the landscape. Some of them even spoke of the Alcan complex, as an "attractive factory".

It must be noted, however, that the interpretation is quite different for certain individuals[6] and in certain situations. This is the case when the complex is visible from residential areas. The visual proximity between home and industry is unanimously and fiercely criticized: "I wish I didn't have to see this in my yard. (...) It's terrible!"[7]. In fact, the visual impact confers on to the Alcan complex a strong cognitive meaning for the affected residents. The industry seems closer to them. Spatial proximity between home

---

[5] Source: individual interview number 6, lines 588–596, Dunkerque, 4 June 2001.

[6] Four profiles of respondents were more critical of the Alcan complex's "new" landscape: residents who live close by and witnessed a technological accident during start-up, those interviewed that can see the complex from their homes, some of those who live closest to the site and, lastly, those who are strong advocates of environmental 'values'.

[7] Source: group interview, Alma, 12 March 2003.

Figure 1.1. View of the landscape design of the Pechiney complex in Dunkerque, visible from an glass walkway accessible to visitors (Photo: M.J. Fortin).

Figure 1.2. View of the Alcan complex in Alma's landscape (Photo: M.J. Fortin).

and a heavy industry is seen *a priori* as a difficult combination to live with in the eyes of all the respondents, whether or not they were personally affected. Such proximity is deemed unacceptable in contemporary society: "Those days are gone; today people are not going to accept having their homes attached to the smelter. Today's mentality won't accept that anymore"[8]. In its visible and material dimensions, landscape amplifies a symbolic conflict between habitat and heavy industry. The conflict takes on such importance, moreover, that certain people no longer see it as a landscape: "Well, I think it's sad to see such an imposing thing near the house. That's not a landscape".[9]

Overall, the remarks about landscape are put in perspective in comparison to a more vast land use dynamic, to social relations and, above all, to the specific issues that most concern those involved.

The visible landscape tested against land use dynamics and social relations

Within the contexts under study, living close to heavy industry implies particular realities. The opinions expressed by the residents encountered, both on and around the theme of landscape, made it possible to highlight various concerns, in particular related to the environment, health, nature-related territorial practices, economics, quality of life and risks. These preoccupations are joined together in a synergetic way within a very dense system, linking facts and perceptions, the components of which seem difficult to separate or delineate clearly. From the viewpoint of social actors, they are part of a major issue, namely sustainable development.

Sustainable development is not a matter of rhetoric in this case. It represents above all the ideal of a local society that people would like to maintain for the future, distant or not. When this ideal comes up against land use dynamics, it exacerbates certain tensions that sometimes become important issues in the case of specific events, depending on the context (e.g. historical, geographical, political). Along the Dunkerque coast, there is much talk of the connection between industrial discharge, air quality and people's health. Nearly two-thirds of respondents consider that the quality of the environment has declined over the past 10 or more years. The same proportion of people explain that either they or family members have experienced health problems that they feel were caused by the poor quality of the environment, in particular the air. In Alma, it is the opposite: only a minority of respondents brought up the subject of health.

The density of the industrial fabric may partially explain these contrasting observations in France and Quebec. The situation is in fact much more exacerbated in Dunkerque, with a high concentration of some 15 companies that have been classified as high-risk (Seveso Law). Yet overlying this first territorial 'fact' is the impression of having lost control of development and, consequently, feeling that the quality of one's surroundings is threatened by the ever-increasing number of industries. Thus another decisive factor is added to the residents' interpretations of the landscape: the governing capacity of small communities and their decision-makers when they are dealing with powerful multinationals.

---

[8] Source: group interview, Alma, 5 March 2003.
[9] Source: group interview, Alma, 5 March 2003.

Case studies offer similarities when it comes to the issue of social relations. In these two industrial regions, citizens have complex, changing relationships with the large firms present there, in a sort of love/hate mixture. On the one hand, people welcome a new major complex that creates jobs in a region hard hit by unemployment. For some people, the industrial project becomes the panacea for all their ills, "the light at the end of the tunnel."[10]. On the other hand, people complain about the damage to the environment and risks to health. Between these two extremes is a reasonable attitude about the "price to pay" for economic development and the fact that "you can't have everything". Shared rationality is also developed to deal with the presence of risks on a daily basis.

The concertation structures set up to encourage dialogue with major producers – despite their number and originality in Dunkerque – are not sufficient to restore a feeling of power to these citizens. Economic dependence, both structural and perceived, comes into play in the social relations revealed in discourse about industrial landscapes.

Landscape, a sensitive and reflexive interpretation

What should be gleaned from the observations of empirical research into industrial landscapes compared to the above-mentioned theoretical debate? First, as a social practice, one cannot easily limit landscape to one or more preconceived theoretical dimensions. This is because it is necessarily *contextualized*, connected to people, geographies, temporalities and social relations. Let us nonetheless attempt to resituate these observations in four steps, starting with the role of the form and material dimensions of landscape in its aesthetic appreciation.

The non-autonomy of form in aesthetic judgement

Is it possible for the landscape of a major industry whose activities have a negative impact on the environment to be described in a positive way, in aesthetic terms, as is suggested by Roger (1997). Yes, this is possible. The foundations for this positive description, however, are not those given by Roger. As is clear from the discourse of the concerned parties, the 'beauty' of a landscape is not only a function of its plastic characteristics. The two case studies certainly indicate that the landscape design of industrial sites, as well as the efforts invested in the architecture, have beneficial and tangible effects for local populations. They are appreciated and even considered as a new expression of businesses' social responsibility. This type of investment in the 'forms' of the production complex (e.g. architecture, landscape design, visual impact, etc.) are not sufficient by themselves, however, to integrate the industry into the local landscape, if the industry is considered threatening. The inhabitants who were questioned explicitly expressed their refusal to accept landscapes that are merely "decor" or an image empty of substance and disconnected from the actual experience of living in their surroundings.

From this perspective, those inhabitants do not express an aesthetic judgement on the formal, plastic dimensions of landscape in an autonomous way. Instead, these dimensions are connected to other social considerations. When they describe the landscape,

---

[10] In Alma, people have high hopes, such as slowing the mass departure of young people.

the residents and developers give their judgement on the whole, rather than its parts. Spontaneously expressed in a global, aesthetic appreciation (e.g. beautiful, ugly, sad, marvellous), at times a moral judgement (good/bad), the landscape is a concentration of different factors that convey meaning in connection with their values, preoccupations, experiences, knowledge, etc. This explains the potentially divergent descriptions given by a resident, an ecologist, an entrepreneur, an artist or an elected official.

The environment: fundamental in the landscape experience

From the viewpoint of citizens who live near these major industrial sites, the environmental issue is especially important. This is because, among other reasons, it touches upon the quality of the land and people's health. Ultimately, *the environment as it is lived, perceived and understood, appears to be fundamental in the contemporary social experience of landscapes*. The environmental issue may nonetheless be *latent* in this experience. It is rather empirical contexts that exacerbate certain tensions, so that this issue takes on more importance. This is the case of the context under study, concerning the experience of living close to large factories. Consider also the context of change created when a new industrial facility is built or, even more, situations causing uncertainty such as technological accidents and the introduction of new information. These observations highlight the fact that this environmental issue cannot be expressed in purely factual terms. From the viewpoint of social stakeholders (e.g. local population, elected officials, the developer's experts and scientific experts), their surroundings are *made of both objective data and perceptions*, in connection with their awareness and knowledge, either technical or land-related[11]. This introduces the cognitive dimension of landscape.

Sensitivity and reflexivity: the importance of knowledge in reflexive society

Thirdly, the landscape is the result of a *sensitive and reflexive experience*. In other words, the landscape experience is fundamentally a situation where people gain 'awareness' of the land. It implies an ontological distance between the first (the individual, the social group) in relation to the second (object, surroundings) along two fronts simultaneously, physiological and cognitive. The material forms first stimulate the senses, i.e. sight, touch (the experience of space and the sites of huge complexes), hearing and even smell. The sensory landscape mediatizes the object (large factories in this case) and is revealing for those concerned. Although materiality represents the first contact, it nevertheless remains an incomplete source. Indeed, as is clear in the respondents' views about industrial landscapes, the visible and tactile dimensions are quickly challenged by other sources of knowledge and known 'facts'. Knowledge resulting from formal sources (i.e.

---

[11] The *quality* of the environment is assessed on the basis of 'factual expertise', which, is itself, built on knowledge and methods that are part of a given historical context and which have undergone rapid progression in recent decades. Several sociologists (Buttle, Guay Vaillancourt) have shown how the conceptualization of these environmental issues evolves as knowledge and social 'sensitivity' progress. Let us mention only the importance granted to the problems stemming from demographic growth and deforestation in the 1960s, which today have taken a back seat to issues concerning the greenhouse effect and global warming.

experts' and scientists' learning) and informal (i.e. practices and observations of the surroundings, empirical experiences), offer a second, complementary path, to reveal the object/surroundings to the individual's conscience[12]. Forms and knowledge, two major sources for understanding landscape, are thus used by individuals and social groups to create an interpretative framework and give meaning to industrial landscapes. Such a framework, moreover, is not 'frozen' and can be adjusted.

Reflexivity in fact creates tension between action and knowledge. As Giddens (1994: 45, personal translation) explains, as a characteristic of our modern advanced societies, reflexivity stimulates "the constant examination and revision of social practices in light of new information about these same practices". For this reason, the issue of knowledge turns out to be of particular importance in the interpretation of landscape. It contributes to building shared meanings and representations among the actors, which are the basis for building social compromises concerning development. Since it is subject to change, however, knowledge may become a source of interpretation conflicts, changes to practices and social tension. Combining different types of knowledge may or may not favour the elaboration of stable representations of the object. On the one hand, convergence of the sources of knowledge makes it possible to stabilize a framework for interpretation, to build a positive or negative representation of the object-factory. This may be shared by all those involved who recognize the legitimacy of this same knowledge. Conversely, if there is divergence between the sources of knowledge (e.g. visible forms/information; types of scientific/empirical knowledge), those concerned are not able to establish a global meaning, positive or negative, for a given situation, and seek more information. This divergence may be caused by information that is contradictory, vague or unsatisfactory from the actor's viewpoint (in terms of quality, quantity or credibility)[13]. In a nutshell, reflexivity implies that the interpretation of landscape results from a dynamic balance.

'Beautiful' landscapes, or when there is no break

Fourth and finally, the depiction of the landscape is part of a mindset of integration and anticipation: social, territorial, environmental, cultural and political. This is why the factory remains an object on the landscape, i.e. it is not integrated into the landscape as long as it pollutes, threatens the quality of the environment, meaningful social practices (leisure or community's identity) or the health of inhabitants! To the contrary, in a region

---

[12] In particular, this is what explains why, in our research on industrial landscapes, the citizens we met were so eager for information. In both Alma and Dunkerque, they wanted to know about neighbouring complexes, types of discharge, and, above all, the impact on the environment, the quality of the local surroundings and health. They use various more or less formal strategies to obtain this information, ranging from attendance at information meetings organized by the developer to asking questions of friends and family members who work at the industrial complex and observing changes to their surroundings (e.g. suspicious dust in the garden, plant growth, odours and smoke coming from the factories).

[13] To illustrate this point, we can mention the case of a technological accident that occurred during the start-up phase at the Alma factory. The accidental release of white particles that were visible on the residents' lawns contradicted the information that the residents had received; according to this information, the emissions from the new complex were "fewer" than with the previous complex. The representation of a "safe" factory that had the "best technologies in the world" was undermined by this accident, at least for the residents who were eyewitnesses (Fortin, 2007).

in a situation of vulnerability, the "beautiful factory" is the one that makes it possible to revitalize the local economy without compromising individual and community lifestyles, health, well-being[14], etc.

A beautiful landscape would thus be when no break occurs between existing social and territorial dynamics and those that are desired for the future. Through material forms, it makes an *implicit* ideal become *manifest*, and it becomes a tangible reference for the senses, reason and social discourse. From this perspective, landscape aesthetics use the material framework of form to enter into the meaning of the relationship experienced with one's surroundings as well as with one and others. Artistic representations (e.g. pictorial, iconographic, literary, etc.) in some ways celebrate such landscapes, which are symbols of harmony between humans and nature, and also between humans (social relations), and a source of physical and cognitive pleasure[15].

In sum, the contemporary aesthetics of landscape as a sensitive and reflexive expression of the relationship to one's surroundings and nature, is based on a *combination of sensory and cognitive experiences*. It relies in part on the environment, in its material dimensions, both perceived and known. Our proposal does not claim to resolve the entire theoretical problem regarding the connections to be clarified between landscape and environment. At most, it attempts to better account for empirical observations to thereby stimulate new reflections.

Banking on landscape's 'duplicity' in territorial governance

To conclude, the complex and polysemic nature of landscapes has often been seen as an insurmountable conceptual difficulty. Consequently, for many experts and scientists, it is easier to take refuge behind reassuring concepts by suggesting useful divisions (objective environmental facts/subjective landscape experience) and by limiting landscape practices to their formal and visible dimensions. The consequence of taking this position is that interventions carried out in the name of landscape contribute in a minor way to the debates and problems experienced by our contemporary societies. To render such interventions useful, it becomes necessary to remove the concept of landscape from categories that are too simplistic and to work instead *with* its 'duplicity' (Daniels, cited in Matless, 2003). In the words of David Matless, "*One could instead view the doubleness of the term as a virtue, as something which is both analytically productive and makes landscape so important a matter beyond the academy*". This is the approach we recommend.

We believe that the importance of the concepts of *landscape* and *environment* lies in their complementary nature and *their interaction in action*, in particular in the exercise of social interpretations that force this interaction. The landscape offers a framework for a global interpretation that allows the subject, whether individual or collective, to combine a series of significant considerations exacerbated by a given context. The evaluation of

---

[14] On this point, see the reflection put forward by Yves Luginbühl during a Council of Europe workshop.

[15] This is perhaps even a condition for it to merit representation; only 'beautiful' landscapes deserve to be celebrated and passed on. This would explain why certain societies have few or no artistic representations of landscape, when for them it is more a synonym of oppression than well-being. This is at least our hypothesis.

the surroundings through the framework of the landscape enables actors to describe and impart meaning on a given situation. The environment is part of such considerations, integrating as much the thing-in-itself (fact), as the building of perceptions, experiences and knowledge. These ever-changing connections, lived out in the experience of landscapes and expressed through its aesthetic description, perhaps approaches the new ontology that is much desired – by Berque (2000) in particular.

For these reasons, landscape could become a framework for federating governance in our reflexive societies, as is implied moreover by the European Landscape Convention adopted recently by the Council of Europe. In the exercise of democratic debate, which is not always simple, it offers two clear advantages. First, it is conceived as a positive theme, fostering participation and mobilization. Second, it constitutes a framework for evaluation and reflection creating tension between the dynamics of territoriality-related issues and temporalities (what was, what is and what is desirable). But landscape will not truly become a "mainstream political concern", as stipulated in the Convention (point 23), until there is open public debate accompanied by concrete measures that attack the basic social issues, including that of fairness. Otherwise, "beautiful landscapes" will remain the realm of the privileged, while the others must be content with "landscapes of risk"[16]. Let us trust that we will be sufficiently creative and have the political will to make landscape a genuine 'tool' of citizen governance and sustainable development.

---

[16] Expression borrowed from Andrew Blowers (1999).

Chapter 2

Should the effects of landscape changes on biodiversity be taken seriously?

Jacques LEPART, Pascal MARTY and Mario KLESCZEWSKI

Reference is often made to the role of ideology in the defining of problems concerning landscape (e.g. Barthélémy and Jacqué, 2002). Knowledge of social representations as well as documented striking examples of changes in positions on landscape demonstrate that, over time, landscape conditions perceived today as being highly desirable were in the past considered problematic by certain social groups and vice versa (Lepart *et al.*, 2000; Trivelly, 2004).

On the basis of this observation, we may adopt two different positions. The first – a 'constructivist' approach – emphasizes the high degree of relativity of environmental issues that depend on intellectual constructions, often of short lasting, linked to debates of highly variable duration (e.g. acid rains discussed in Rudolf, 1998). In this way, identification of environmental problems linked to landscape would arise from positions on landscape that would use 'scientific facts' as rhetorical elements at the service of devices aiming at imposing one's convictions in a debate. In other words, there is no problem with the landscape but only with landscape stakeholders.

The second position – a 'realistic' approach – hypothesizes that reality is independent of all human interpretations and positions and that environmental problems identified at landscape level truly exist. It is then possible to analyse and eventually to refute the objective reality of the alleged facts, if necessary, within the framework of a scientific field, (landscape ecology, for example). Thus, dynamics at landscape scale would not be dependent on the force of conviction of an environmentalist lobby but would instead deserve to be analysed objectively.

In this text, based on an analysis of the impact of changes in formerly open and non-forested landscapes, we attempt to show that modifications of the morphology of a landscape, regardless of the diversity of viewpoints, opinions and perceptions of social groups, have consequences on their capacity to provide habitats for wild species and, therefore, on biodiversity.

Transformations of open landscapes generally fall into one of two main categories: closure of heath and semi-natural grassland landscapes following changes in land use practices, and modification or disappearance of permanent grasslands as a result of the intensification of agriculture. These landscape transformations by human users raise an issue for the conservation of other species, the non-human users of the landscapes (birds, plants, etc.).

In this text, we analyse the contribution of open landscapes to biodiversity by focusing on rare plants and habitats. After demonstrating that the erosion of biodiversity is inevitable in the current situation of open habitats, we then discuss its impact on debates, decisions and local government agencies that deal with this issue.

Defining open landscapes, measuring their dynamics

Open landscapes are transformed as a result of two processes: landscape closure and the disappearance of permanent grasslands and meadows.

What we refer to as 'landscape closure' is one of the modalities of landscape dynamics. It is preceded by older dynamics: changes in land use practices, gradual transformation of plant cover, woody plant recruitment, etc. Since the end of the 1970s (Lepart *et al.*, 2000; Le Floch, 2003), concerns with the closure of landscapes or open environments have been expressed more and more often, among scientists, land managers and others in the agricultural sphere (Lepart *et al.*, 2000).

The expression 'landscape closure' has a strong metaphoric connotation, invoking the visual, tactile and physical, a reference to impenetrability (Friedberg *et al.*, 2000). At the subjective level, it refers to the elimination of landscapes inherited from the past. As often in the case of environmental issues (Micoud, 1999, 2002), it does not distinguish between the many problems concerning rural space, i.e. loss of biodiversity, loss of aesthetic value, degradation of the living environment, threat to regional identity, disappearance of farms, rural exodus and rural countryside abandonment.

Opening and closure applied to landscape

Landscape ecology provides elements to define notions of opening and closure applied to the landscape. Within the field of landscape ecology (Forman and Godron, 1986; Forman, 1995), the landscape is considered a dynamic organization of space. It has an appearance: a geometrical arrangement that varies to some degree from one year to the next, areas planted with crops, meadows and woods linked by corridors such as field margins, hedges, slopes or banks, or degrees of fragmentation. Moreover, there are also more subtle realities: human intervention, utilization by living beings of some landscape components during part of the year, interactions between species, heat transfers linked to the shift of air masses, water flows, nutritional elements and pesticides in the ground and water table.

In ecology, the opening of plant formations is a function of the amount of light that reaches the lower strata of the vegetation: open landscapes are made up of crops, grasslands, meadows, peat bogs, steppes, tundra, etc. Closed landscapes are dominated by forest. Although hedgerow or *dehesas* landscapes are closed from the outside, they are not closed to light and are, therefore, considered to be open landscapes. *Garrigue*-type shrub formations are somewhere in between the two. Other ecological criteria could be used to define an open landscape: water flow or mineral elements, the potential for species to move within the landscape, although we tend to use the terms integration or connectivity. Finally, land use specialists emphasize the visual aspect of an open landscape.

In Europe, open landscapes were generally developed and maintained by societies that favoured the first stages of plant succession by planting crops, grazing, pasture cutting, fires, etc. In more rare cases, they are linked either to natural disturbances (floods, fires, windfall, wild herbivores, etc.), or to climatic or edaphic constraints limiting the establishment or survival of trees: alpine meadows, ombrotrophic bogs, serpentine heaths, coastal halophytic vegetation and marshes, etc.

In most cases, open formations correspond to crops (*ager*) and to the grazing or rangelands (*saltus*) of the agro-sylvo-pastoral system of Georges Kuhnholtz-Lordat (1945). These two components differ in the way they work because the reproduction of crops is entirely under human control in the '*ager*', whereas rangelands leave more room for spontaneous dynamics.

Biodiversity and landscape changes

Several types of methods exist to measure landscape changes. If we define the landscape as a portion of space seen from a point located on the surface of the earth or nearby, we can compare photos taken over a period of time (Lepart *et al.*, 1996; Debussche *et al.*, 1999). Other techniques compare and analyse cartographic-type documents (e.g. cadastral surveys and old maps) and are better adapted to an analysis of the landscape within the framework of landscape ecology, i.e. a set of somewhat interconnected vegetation units used by living beings (Lepart and Marty, 2004). Finally, even if they are unclear from a spatial point of view, land use statistics gathered through surveys make it possible to rapidly quantify changes in land use. In France, the General Agricultural Census provides data on land use categories in areas managed by farms (www.agreste.agriculture-gouv. fr), for the years 1970, 1979, 1988 and 2000. Analysing this data will allow us to assess landscape changes. Changes since 1970 in areas under permanent grass (PGA), plots with natural meadows and grasslands planted over the past 5 years are indicative of the dynamics of open landscapes.

Regarding biodiversity in France, given the lack of accurate data on the population and distribution of all taxa, we can approach the effects of landscape changes by using lists of protected species, for example, the protected plant list (Danton and Baffray, 1995), '*Livre Rouge de la Flore Menacée de France*' (*Red Book for Endangered Flora in France*) (Olivier *et al.*, 1995), and the lists of plant taxa covered by the Natura 2000 network (Collectif, 2002). We classified taxa on the basis of their sensitivity to the loss of open landscapes in order to evaluate the contribution these landscapes make to the maintenance of biodiversity in France. We completed these lists by analysing the number of plant communities linked to the major environment categories. To do this, we used the list of alliances for France proposed by Bardat *et al.* (2004), and we related these alliances with the major environment categories in which they are found. The phyto-sociological classification is based on floristic affinities and effectively reflects variations in biodiversity; it is based on a relative consensus at the alliance level. We also used the list of priority habitats recognized by Natura 2000 (Romão, 1997).

Modifications of open landscapes: processes and realities

Contraction of open spaces: towards a two-components landscape

The transformation of open landscapes takes into account the considerable changes in land use resulting from different processes, depending on the regions and the *'terroir'*. Using data from the agricultural census (Table 1.1), we can get an insight into these transformations in France: from 1970 to 2000, the portion of the utilized agricultural area (UAA) of permanent grassland area (PGA: natural meadows and grasslands planted over the past 5 years) decreased by 28% (-40,000 km^2), attesting to trends underscored by IFEN (1996). The situation is very contrasted depending on the region:
– an increase in the PGA/UAA ratio in the Mediterranean region resulting either from the reappropriation of land (related to agri-environmental measures), or to a decrease of the UAA; the dominant dynamics of land use are a continuous trend towards reforestation;
– relative stability in mountain regions, closely linked to the difficulties of growing crops on many of the 'terroirs';
– a steep decrease in the ratio, generally in favour of crops, and an increase of fallow land elsewhere, particularly in intensively farmed areas; in these regions, the PGA is barely more than 10% of the UAA.

There are considerable variations at the infraregional level. Thus, in Burgundy, grassland has increased by as much as 18% in some cantons and decreased by as much as 30% in others.

Beyond the closure aspect, landscape is affected by a binarization (Lepart and Debussche, 1992). In 2000, forests represented more than 150,000 km^2 (1992–2000: +4.3%; IFEN, 1996) as opposed to 90,000 km^2 at the beginning of the 19th century (IFN, 2000); crops occupied approximately 165,000 km^2 in 2000. Another characteristic of this binarization is the homogenization of each of the components: forest areas whose density increases as a function of the time elapsed since agricultural activities were abandoned and elsewhere the advent of more intensive and larger crop units (e.g. use of pesticides, eutrophication, etc.). On the other hand, relatively unartificialised open spaces, grassland, heaths, pastureland, alpine meadows, hedgerows and corridors decrease. Only spaces taken up by rocks and water remain the same.

New relationships between agriculture and ecological systems

Saltus, or rangeland, occasionally planted with crops, has lost its usefulness over the last two centuries through the development of forage crops, the shift away from mixed farming, the intensification of animal breeding conditions, and the concentration of farms and their refocus on more productive land within new technological frameworks.

Table 1.1. Evolution of permanent grassland area (PGA) between 1970 and 2000.

	Utilized Agricultural Area (UAA) under PGA (%)					Utilized Agricultural Area		
	1970	1979	1988	2000	Evolution 1970-2000 (%)	1970 (ha)	2000 (ha)	Evolution 1970-2000 (%)
Île-de-France	8	5	3	3	-67	642,047	583,246	-9.16
Poitou-Charentes	29	26	21	12	-60	1,833,206	1,761,867	-3.89
Bretagne	27	19	18	11	-60	1,932,026	1,701,566	-11.93
Pays-de-la-Loire	49	41	34	23	-54	2,483,456	2,169,981	-12.62
Picardie	24	19	15	12	-50	1,377,901	1,341,461	-2.64
Centre	18	16	15	10	-45	2,543,784	2,365,694	-7.00
Haute-Normandie	49	43	35	27	-45	850,063	794,026	-6.59
Alsace	40	34	28	24	-41	335,906	336,229	0.10
Basse-Normandie	80	71	63	49	-39	1,400,441	1,264,133	-9.73
Champagne-Ardenne	31	27	22	19	-38	1,532,361	1,560,325	1.82
Nord-Pas-de-Calais	32	30	25	21	-35	924,619	838,168	-9.35
Aquitaine	36	35	31	24	-34	1,627,176	1,473,396	-9.45
Lorraine	58	55	48	41	-30	1,143,602	1,132,531	-0.97
Metropolitan France	41	39	36	30	-28	29,904,200	27,856,313	-6.85
Franche-Comté	72	70	67	54	-24	706,776	667,674	-5.53
Midi-Pyrénées	36	35	32	28	-20	2,518,377	2,361,914	-6.21
Rhône-Alpes	55	55	53	46	-17	1,754,345	1,526,724	-12.97
Bourgogne	47	46	45	40	-14	1,823,765	1,775,182	-2.66
Auvergne	65	68	70	63	-2	1,568,790	1,510,577	-3.71
Limousin	58	70	70	61	5	930,753	861,021	-7.49
Languedoc-Roussillon	35	36	37	40	11	1,121,601	981,459	-12.49
Corse	66	69	76	83	25	134,100	155,888	16.25
Provence-Alpes-Côte-d'Azur	35	35	39	46	32	719,105	693,252	-3.60

Source: Agreste, General Agricultural Survey. (www.agreste.agriculture-gouv.fr)

Rangelands are particularly maintained today in mountain zones with high added-value production. They have been ploughed and cultivated in other areas or gradually replaced by forest in the Mediterranean region and some mountain zones (IFN, 2000: 5).

The decrease in wetlands, neglected since the 1950s and estimated at -50% in just a half century by the '*Commissariat Général au Plan*' (1994), has also contributed to the decreasing importance of the *saltus* without always leading to a closure of the environment, since it is generally the result of the intensification of agricultural production, in addition to industrialization, urbanization of the coast, canalization and flood control.

At the same time, large-scale farming areas have become increasingly less adapted to life forms other than cultivated species (Robinson and Sutherland, 2002). Open cultivated environments thus can less and less be comparable to natural environments and substitution habitats for other species. Species specific to certain crops disappear with them.

Consequences for biodiversity

It is difficult to define reliable indicators of changes in biodiversity. Assessing diversity in open landscapes by the sum total of the species present in the area, other than being difficult to achieve, is not entirely satisfactory since it assigns identical values to species whose conservation statuses are very different (i.e. rare species vs. ubiquitous species). Other factors must be taken into account, such as the determination of a 'heritage' value by evaluating the ratio between local abundance of a species and the distribution of this species over a reference area, the risk of extinction or the importance given by society to one species or another. We chose to use habitats and plants included on lists of rare or threatened species as indicators.

Diversity of habitats in open and closed landscapes

Open landscape vegetation is generally more diversified than forest stands. This can be seen by the fact that barely more than 10% of phyto-sociological alliances are of the forest type; all the others are linked to open environments (e.g. ledges, heaths, peat bogs, fallow land, etc.). Approximately 55% of the alliances are present in old *saltus* or *ager* and are, therefore, affected by the reforestation of these areas or by their eventual clearing for cultivation. The dynamic stages between the colonization of bare land by annual species and the establishment of the forest are multiple; each dominant species leaves its mark on the community. Turnover (species replacement rate from one year to the next) tends to decrease as cleared land is transformed into forest, mainly as a result of the life span of the species in question (Lepart and Escarré, 1983). Ecological factors such as topographic, edaphic and climatic conditions have more impact in a herbaceous environment where the buffer effect of vegetation is lower than in a forest environment. Specific elements linked to water (e.g. wet patches, temporary pools, etc.), the presence of old trees (e.g. hedges, isolated trees, etc.), considerably increase the diversity of open landscapes.

The positive effects of pastureland on biodiversity have been known since Darwin (1859): it avoids the monopolization of space and resources by just a few species. Land use by societies has contributed to maintaining the diversity of open landscapes for many years. In herbaceous environments, herbivorous species, grazing seasons, animal stocking rate and temporary cultivation, all play a major role in the composition of the vegetation and probably the fauna.

Rare and threatened species, remarkable habitats

Lists of rare, threatened or protected plants are social constructions. The authority of the lists derives from a consensus of recognized groups of experts and, therefore, they deserve to be taken into consideration in order to define the main challenges facing biodiversity.

In France, among the 439 taxa included on the protected plant list (approximately 10% of flora), only 26 are species of forest habitat (Table 1.2). All the others are associated with open environments. When these habitats are stable, these taxa (193 taxa) are, as a rule, not affected by the closure of the vegetation (Table 2a); when these open habitats depend on land use, they can be negatively affected (246 taxa) by the closure of vegetation (Table 2b).

Table 1.2. Plants protected at national level.

a. Plants not negatively affected by vegetation closure

Habitat groups	Number of taxa	% of total
Forest	26	5.9
Aquatic environments	16	3.6
Temporary standing water	21	4.8
Screes	25	5.7
Boulders, cliffs, ledges	60	13.7
Coastal gravels and sands	34	7.7
Active ombrotrophic bogs	11	2.5
Total	193	44

b. Plants affected by landscape closure

Habitat groups	Number of taxa	% of total
Heaths and shrub plant communities	23	5.2
Sclerophyllous shrub plant communities	3	0.7
Chalk grasslands and borders	65	14.8
Silicicolous grasslands and borders	36	8.2
Alpine and sub-alpine meadows	15	3.4
Wet grasslands and megaforbs	24	5.5
Mesophilic grasslands	6	1.4
Waterside vegetation	29	6.6
Low coastal marshes, springs, etc.	21	4.8
Cultivated land	19	4.3
Fallow land	5	1.1
Total	246	56

Source: Danton and Baffray (1995)

Among the 553 taxa listed in the '*Livre Rouge de la Flore Menacée de France*' (Table 1.3), 35 are linked to the forest, 258 are potentially threatened by the closure of the vegetation and the others belong to climactic open critical environments. Finally, among the 59 plant taxa included in the annexes of Natura 2000 (Table 1.4), 33 are linked to open environments that can be affected by the closure of the vegetation and 3 are linked to forest environments. Thus, these three lists built on different criteria all point to the fact that over 90% of rare and threatened species are linked to open environments. Among these species, more than half live in environments that can be affected by the closure of the vegetation or changes in land use (intensification).

Another aspect that should be looked at in more depth than is possible in this study is that these species' habitats generally do not present intermediate environmental conditions (i.e. mesophilic grasslands) but, on the contrary, are subject to somewhat considerable environmental constraints (e.g. rocky environments, marked by an excess or lack of water, excessive salt, etc.). The more marginal these environments are on farmland, the more they are threatened by the landscape closure (or eventual changes in land use).

The list of Natura 2000's priority habitats (Table 1.5) provides a slightly different perspective. Priority habitats threatened by the closure of vegetation only represent 33% of habitats (whereas the figure is 55% for alliances); this is undoubtedly linked to an explicit intention to favour those habitats referred to as 'critical'.

Table 1.3. Taxa included in the '*Livre Rouge de la flore menacée de France*' Tome 1, Espèces prioritaires.

Habitat groups	Number of taxa	% of total
Thickets and grasslands	178	32.2
Peat bogs and wetlands	34	6.1
Farmland and man-made environments	46	8.3
Total	258	46.6
Other habitats	295	63.4

NB: Species linked to wooded environments, coastal environments, aquatic environments, active ombrotrophic bogs and rocky environments are not included in this table.
Source: Olivier *et al.* (1995: 14)

Table 1.4. Taxa of community interest.

Habitat groups	Number of taxa	% of total
Forest environments	3	5%
Aquatic environments	4	6.7%
Rocky environments	15	25%
Coastal sands	3	5%
Temporary standing water	1	1.7%
Environments concerned by closure	33	56%

Source: Habitats Directive Habitat, Annex II (Collectif, 2002)

Table 1.5. Habitats of Community interest (Source: Romão, 1997).

Habitat groups	Number of habitats	Priority	% of priority habitats
Habitats not affected by the "closure of environments"			
Coastal habitats and halophytic vegetation	19	4	21%
Sea and land dunes	17	5	29%
Freshwater habitats	15	1	7%
Ombrotrophic bogs	5	2	40%
Rock and rock shelter habitats	13	2	15%
Forests	26	6	23%
Total	95	20 (21%)	
Habitats affected by the "closure of environments"			
Sclerophyllous thickets (mattorals)	7	0	0%
Natural and semi-natural grassland	17	5	29%
Temperate thickets and heaths	7	3	43%
Fens	4	3	75%
Total	35	11 (31%)	

NB: Only habitats existing in France were taken into consideration.
Source: Romão (1997)

Discussion

Landscape changes affect biodiversity

The modification of plant covers that corresponds to the expression 'landscape closure' is most often observed in the mountains and in the Mediterranean region (Delcros, 1994); elsewhere, the erosion of the semi-natural component of the landscape can be observed: intensification of agricultural practices, clearing and seeding with increasingly less use of spontaneous species. Landscape changes are characterized by the development of binary landscapes where crop zones alternate with forest in proportions ranging from entire culti-vated landscape to entire forest landscape. Intermediate types of vegetation (grasslands) remain at the same level in certain regions whereas meadows and, even more so, heaths, shrub plant communities and thickets decrease almost everywhere else. Rangelands (*saltus*) decrease and become homogenized. The rate of disappearance of meadows and permanent grasslands is comparable to that of tropical forests: in the same period the average annual rate of disappearance of tropical forests is 0.9 % (FAO, 1991).

Measures taken to limit farm over-production (e.g. rotational fallow), to improve product quality (e.g. specifications of labelled products, development of organic farming), or to avoid the conversion of grasslands (e.g. agro-environmental 'grass premium') all contribute to limiting the effects of these transformations but remain fairly marginal.

Diversity is considerably affected by these rural landscape transformations. On the one hand, the habitats are decreasing rapidly and the taxon populations that live in them are rapidly diminishing. On the other hand, the intensification of agriculture homogenizes habitats and reduces diversity by reducing environmental constraints to the advantage of a limited number of species (Benton *et al.*, 2003). The situation in France is not exceptional and research carried out in northern Europe clearly shows the effects of these two constraints on diversity (Robinson and Sutherland, 2002).

Diversity in open landscapes: a heritage

The problem with the disappearance of the *saltus* is that it is the result of a long and gradual history that began before the arrival of the first human populations. These habitats are not a simple human creation. Vera (2000) clearly showed that in the prehistoric forest, all stages of plant succession coexisted because of the influence of herbivorous populations. The regeneration of trees under the pressure of grazing required the establishment of thorny thickets that served as shelter. Tacitus spoke concisely about the importance of these thorny hedges in the forests of ancient Germany. The landscape was, therefore, not open but included large open formations. Beside herbivores, other factors such as fire or windfall played a role. Semi-natural open vegetation enabled the coexistence of species adapted to open environments that then remained in the *saltus*.

With the establishment of the first societies of breeders and farmers (Diamond, 2000), grasslands, meadows, heaths and wetlands were maintained and modified by man. Their location, often on the periphery of village communities, depended on the interests of the populations and their production systems. Land development by societies did not create diversity (at a specific scale) except for several metallophytes and cereal weeds (with debatable taxonomic status); it simply created new habitats and enabled considerable expansion (e.g. meadow vegetation, plain birds, etc.). The word 'forest', designating that which is outside the space managed by humans (*foris*) – the wild world (*sylva*) – thus took on another meaning. As of the 18th century, it came to signify 'an extensive land area planted with trees', or even 'a group of these trees'. The forest then became a highly homogenized and biologically poor environment, just at the time when its area was the most reduced: sciophilous species were able to develop whereas all of the others decreased. Within this context, the *saltus,* which continued to subsist only as a result of grazing, took on considerable importance in maintaining diversity.

By maintaining the diversity of landscapes and the diverse uses of the *saltus*, societies in the past made it possible to maintain part of the biodiversity. However, even if there has been a 'long cohabitation' between European societies and nature, the idea that diversity problems only occur in tropical regions is not valid. The toned-down version of government environmental services and environmental associations according to which agriculture enabled the survival of a diversified flora and fauna is more valid but does not necessarily imply that this is still possible. The biodiversity crisis in European landscapes occurred at exactly the same time as that of the agricultural model: the exclusive focus on marketable products led to neglect the impact of cropping practices on commensal species. The recent history of cohabitation between human and non-human users of the landscape was possible until the advent of industrial farming.

Landscape closure: Issues? Decisions? Concerned groups?

New representatives of issues facing rural areas have appeared over the last decades: associations, government services and public agencies that aim to protect the environment.

They call attention to issues linked to the conservation of species that have evolved over a long period of time in habitats maintained by wild herbivores or natural disturbances, and that then adapted to landscapes created or maintained by agricultural societies. Their point of view, based on landscape transformations and the loss of biological diversity, casts doubts on the ecological sustainability of the dominant agricultural model. This point of view does not have the legitimacy of the arguments advanced by the agricultural sector and its different support structures, i.e. exercising their prerogatives of ownership or holders of production rights, improving their income, feeding the planet, ensuring food safety, supporting scientific progress, decreasing production costs by integrating agriculture and industry, etc. This point of view was dominant for many years and its representatives were able to do as they wished. Today they meet with resistance. Groups for the protection of the environment and wild species now share common ground with other communities such as those interested in alternative forms of agriculture, consumers concerned about the quality of their food or animal well-being, protectors of cultural landscapes, farmers, manufacturers concerned with the marketing of their products, etc. Discussions began at the time when the AES (agri-environmental schemes) were established and continue in relation to fluctuations in agricultural policy. According to Le Floch (2003: 20-21), the emergence of the landscape closure theme can be linked to 'questioning strategies', i.e. to the aim of creating a debate focused on a problem as well as on possible solutions to the problem. The initiation of a debate can also act as a vehicle for sectors or groups to legitimize a role, to look for visibility or to achieve recognition by other groups (Boltanski and Thévenot, 1991).

To consider this issue only from the point of view of representations or strategies of conservation or empowerment would neglect the link between the erosion of biodiversity and landscape changes. This link, even if not indisputable, is now based on sufficiently convincing evidence.

Once this link has been clearly illustrated, we must know exactly what are its implications. It seems that France's commitment at the European level (implementation of the Natura 2000 network), as well as the international level, integrated into the national biodiversity strategy, is explicit enough to emphasize the need to address the risks of decreasing biodiversity in relation to the transformations of open landscapes. If 'landscape closure' allows a "renegotiation of social relationships to space and to nature" (Le Floch, 2003: 22), it is undoubtedly necessary to redefine the framework of the group of stakeholders concerned by the stakes at hand. Rather than the empowerment of new social categories in rural areas, it should be taken as the end of a certain type of land management with consideration of issues related to sustainability (Benson and Roe, 2000) and multifunctionality (Allaire and Dupeuble, 2003).

Conclusion

Changes that affect the properties of landscapes not only concern the closure of the landscape. They also concern the homogenization and the eutrophication of the areas that make it up. The effect of these changes on biodiversity is considerable. In addition to the characteristics that can be perceived from the outside, landscapes deal with ecological processes determined by human users and that affect non-humans as well (Latour, 1999).

The integration of these processes into the analysis of relationships between humans and landscapes makes us question our relationship to nature, that is, to other living beings. By taking the impact of landscape changes on biodiversity into consideration, man is forced to question his relationship to nature and to take the actual space around him into account.

Chapter 3

'Closing of the landscape': beyond aesthetics, challenges facing open rural spaces

Sophie LE FLOCH and Anne-Sophie DEVANNE

Since the 1970s in France, the expression 'closing of the landscape' is often heard in reference to the social debate on the future of rural areas. Used by scientists, professionals and public actors involved in the development and management of rural areas, it implicitly designates the feeling of discomfort felt by all – farmers, inhabitants, tourists – in view of the extension of wooded areas on to farmland – whether it be fallow land, spontaneous or voluntary plantations. It implies the unproved assumption that a closed landscape is ugly, that an open landscape is beautiful, and advocates the land development paradigm of 'maintaining open landscapes'.

Blowing the lid off 'landscape closing'

At the request of the Ministry of Ecology and Sustainable Development (Landscape Division), we became involved in work, the aim of which is to shed light on the expression *'landscape closing'*. In the first phase, we attempted to show the emergence and utilization conditions of this expression in the institutional context, where it is used by those concerned with legitimizing their role in rural land development (e.g. development landscape architects, scientists involved in applied multidisciplinary programmes, etc.) (Le Floch *et al.*, 2005). In a second phase, from case studies[1], we collected and analysed the testimonies of those who are often called upon but rarely heard, i.e. ordinary users of rural territories characterized by the development of wooded areas. It is this second

---

[1] The case studies reported here also received funding from the French Pyrenees National Park and the Ministry of Agriculture, Fishing and Rural Affairs.

aspect that we deal with here. Our aim is not to assess the statistic representativeness of standardized opinions ('positive' opinions/'negative' opinions about the progression of wooded areas) in a given population. Instead, it is to truly understand how these users live as well as participate in such changes, from the point of view of their genuine concerns and practices.

A huge number of studies have been carried out on how people and social groups interpret and evaluate the geographic space in which their activities take place. Only a few studies focus on the way in which spatial changes are experienced. Rare are the studies that deal with the question of the importance of the actual notions of opening/closing and the role of their reciprocal dynamics on this evaluation[2].

Some studies have obviously identified different representations (values, assumed causes and consequences, etc.) that social groups associate with the physical landscape components that could possibly contribute to closing it, for example, fallow land, forest plantations, grasslands, etc. (Brun *et al.*, 2002; Friedberg *et al.*, 2000). Even if some people put these representations in perspective with actors' strategies (Debroux, 1995; O'Rourke, 1999), they would generally not go to the extent of analysing in what way the renegotiation of the meanings to attribute to physical changes in the landscape – more than the renegotiation of the changes themselves – is also a renegotiation of the definitions that the different groups give about themselves (Greider and Garkovich, 1994). Australian research (Barlow and Cocklin, 2003) has shown that plantations challenge the ideas that existing populations have of the appropriate uses of farmland, the definition of the rural world, as well as the appearance of the rural landscape. Visual openness is not maintained and the inhabitants feel enclosed and isolated.

We hypothesize that the mobilisation and content of opening/closing ideas by users who talk about their relationships to a given space are diversified. These notions, far from being limited to aesthetic considerations, are also linked to the relationships of people to space and the relationships of people to each other. They are particularly linked to the question of closing/opening of geographic space to 'others'. Different aspects of these relationships would exist in different ways, depending on the social groups in question: for tourists or visitors, these notions would be basically reduced to aesthetic terms, whereas the inhabitants would develop multiple aspects of their spatial and social relationships.

Collecting user testimonies

Our conclusions are based on the analysis of data collected from two types of predetermined categories of users: inhabitants and visitors. Our sample was as diverse as possible[3] (age, profession, socio-geographic origin, etc.), from two rural territories

---

[2] Historical approaches attest to the existence in our culture of landscape evaluation models favouring opening, such as the 'panorama' (Briffaud, 1994).

[3] In order to increase the representativeness of the diverse ways of perceiving the extension of wooded areas, we constantly adapted the sample throughout the surveys. We encountered people identified by surveys as being 'representative' of what most people think as well as those with 'marginal' positions. We also interviewed people at random, in different types of housings (city centres, new housing estates, etc.). In every case, we were careful to take a sample that represented different ages, professions and geographical origins.

characterized by the extension of forest areas. We interviewed 28 people in the community of Villelonge (in the central Pyrenees), where spontaneous afforestation has been developing since the Second World War, and 31 people from the Plateau de Millevaches (Massif Central), where conifer forests have been planted since the 1960s. Twenty-three visitors to the central Pyrenees were also interviewed. They were chosen because they were hikers, an activity that implies – we assumed – an affinity with the physical environment.

We chose a method based on semi-structured interviews. Indeed, this approach was relevant to our research objective, i.e. to understand ways of approaching a given geographic space. Interviewees were asked to talk about their experience as either inhabitant or visitor, about the physical environment in which they live or move and the changes that they perceive, in their own words and in relation to their own activities and concerns. Each interview, lasting an average of one hour and 50 minutes, was recorded and transcribed in its entirety[4].

An empirical content analysis was made on this information, based on the following questions. Are there interviewees that speak of something that is implied in the expression of landscape closing and/or in the expression that we assume to be symmetrical to the first, that of landscape opening? Do they actually use the words 'closing/opening of the landscape', or similar terms ('obstructed landscapes', etc./'far-off landscapes', etc.)? Which physical objects or portions of space do the terms opening/closing apply to? What are the nature (aesthetic, social, ecological, etc.) and content of the message that they are trying to deliver?

In the first part, we will see how the idea of landscape closing can be used as a criterion to evaluate the aesthetic aspects of a landscape. We will take the example of inhabitants and visitors interviewed in the Pyrenees. We will then show how this idea can be used at the social level by the inhabitants on the basis of the data gathered from the two field surveys. Before concluding, we will discuss traditional categorical oppositions (e.g. farmers/non-farmers, etc.) used in studies on the human relationship with space or nature.

"Landscape closing": a challenge in terms of aesthetic contemplation

In our material, reflections linked to the idea of closing/opening of the landscape can be broken down into aesthetic categories. Mainly expressed by visitors, they are also expressed by the inhabitants under certain interview conditions, such as when they are asked to speak of leisure walks they take around their place of residence. For visitors, as well as residents taking leisure locally, these aesthetic considerations are basically the same. We discuss them on the basis of the two categories of users interviewed about spaces in the Pyrenees.

---

[4] Each visitor to the Pyrenees was interviewed three times: first, before a visit to the Pyrenees, at the person's main residence; second, after a hike in the Pyrenees, at the site; third, at the person's residence once again, several months later.

The beauty of 'empty' wide-open spaces

Hikers mainly look for the contemplation of open spaces: the bigger the spaces, the more likely they are to contribute to a particularly satisfying aesthetic experience of the mountains. Higher up and further away, the discovery of open expanses, the expectation of a panorama, are often presented as a reward. Many hikers emphasize their pleasure in arriving at a pass, on a crest, at a peak – simply to gaze into the distance. It is almost always in aesthetic (or spiritual) terms that the opening of the landscape is expressed ("a feeling of freedom", "recharging one's batteries", "proving oneself", etc.).

The open landscape that visitors like to contemplate is basically a landscape empty of others (Figure 1.3). It is not a window to others. The visual opening is precisely appreciated because there are no 'others'. Some people stress that there is nothing like arriving alone at the peak. Others stress that if they cannot be alone, they have to find less travelled paths or, in all cases, to avoid 'must sees', such as Gavarnie or Pont d'Espagne. At most, they are prepared to tolerate a few people, as long as they are 'like them': hikers that reassure them when they think they are lost, or with whom they can exchange information and hiking stories in the evening at the refuge, etc. Rare are the hikers who look for other people or signs of human life in the mountains. Whereas some people see renovated hamlets of barns as animated spaces, products of a culture and a mountain society, most see them as being picturesque ("pretty barns").

Figure 1.3. Visitors to the Pyrenees welcome the opportunity to contemplate an open landscape, empty of others (Photo: A.S. Devanne).

Through recounting the aesthetic experience of an open landscape, we often see a distinguishing process at work. When hikers speak of the pleasure of arriving "at the top", they are distinguishing themselves from others, physically and 'mentally', because they must walk and climb for a long time. The idea of testing one's limits through physical effort and, sometimes, solitude, is significant in what they say. From a social point of view, then, interviewees recognize themselves as part of a small group of 'hikers' (doing something that "not everyone can do"). The aesthetic and spiritual rewards symbolized by the contemplation of the panorama are reserved for a select few, those who have left the "crowds" of "tourists" below. Even though visitors supposedly like the mountain for its open spaces and free access to all, they actually try to show that there is a selection at work.

The ultimate in deception: not seeing anything!

Whereas opening is enhanced in the mountain experience that people look for through hiking, everything that opposes it is negatively assessed.

Fog, for example, is described as the ultimate insult: it totally destroys every mountain experience because one cannot see anything. The forest is also qualified in negative terms. Obviously, hikers can accept it as a departure point for hikes in that it gives them the time to warm up in the shade, which is particularly pleasant in summer. It can also be appreciated for its fungi, or at the aesthetic level, for its "beautiful trees", "lovely paths", "flowers", etc. (a similar vocabulary as used to refer to gardens). However, from a general point of view, it constitutes what is least appreciated during a hike in the mountains: it is dark and sad and, above all, it "obstructs the view". The forest is a "closed space", "cramped", from which we cannot "discover" the landscape. It stands in the way of an aesthetic experience of the mountains that inevitably comes down to a contemplation of open landscapes.

Moreover, when speaking about hiking, interviewees never mention a dynamic of landscape closing through the extension of woody vegetation. When they express themselves as visitors, they do not see changes in the physical landscape over time. This is also true in the case of the inhabitants. We should point out, however, that since the inhabitants are familiar with the local environment and its history, they can speak knowledgeably of forest development at other times during their interview. Then, however, the aesthetic issue is minimal, as we will see later on.

When closing puts opening in a positive light

Open spaces, therefore, are appreciated for themselves; closed spaces are not. However, analysis of the interviews shows that the latter have a greater importance than could be imagined from the way they are qualified (generally negatively) by interviewees. Appreciation of the open view is often linked to – as opposed to – the appreciation of the closed view. The way closed and open spaces alternate seems to be essential to the aesthetic experience of hikers for two reasons. First, it introduces the diversity that they look for and, secondly, because of the contrast with closed spaces, the open spaces are highlighted.

Diversity arises from the alternation between forest and open spaces, but it can also be present in the forest itself when interruptions in the plant cover occur. The forest is never more appreciated than when it provides "windows" or "gaps". Visitors that speak of appreciating forest views often contrast the "cover" of the trees to the "opening" of the gaps.

The forest is above all a transition space, an often necessary passage preceding an open environment (e.g. grassland, crests or peak). But what a pleasure to be able to "arrive" somewhere! It is as if the forest and its "blind spots" actually act as a springboard for the open view, for the "discovery of a landscape". We may also wonder if, paradoxically, woods at lower altitudes that one must "pass by" in order to "surpass oneself" are not a "foil" for an experience that takes place elsewhere (farther, higher, there where it is open) (see Figure 1.4).

Finally, people who attest to an aesthetic contemplation of the landscape express themselves not in terms of landscape closing/opening and the reciprocal evolution over a period of time, but in terms of closed/open landscapes and the spatial alternation between the two. However, aesthetic considerations are neither the most frequent nor the most developed aspects mentioned in the interviews.

"Landscape closing": a challenge in terms of open space

When interviewees speak of their relationship to space as inhabitants, the dominant aspect is social. A distinction can be made between those who speak in terms of landscape closing and those who do not speak in these terms.

Figure 1.4. The forest, foil for open spaces for hikers in the mountains (Photo: A.S. Devanne).

The demand for open (to others) space

Many of our interviewees explicitly speak of the closing and opening of the landscape. These people are inhabitants of the Plateau de Millevaches, and are extremely different. They have different professions, they come from other regions including urban areas, or have lived elsewhere before returning, or have never left the region.

Expressions that interviewees use, such as "closed", "obstructed", "disfigured", "demolished" or "blocked" landscapes, or words like "stifling", "claustrophobic", etc., allow them to express several ideas at the same time. At the aesthetic level, like at the level of meaning and symbols, they express a negative impression of the forest, not in itself, but for what it deprives them. The "fir stand" is not considered ugly in itself, but it obscures the view of far-off horizons. Being able to contemplate an open landscape is an integral part of the experience of living on the plateau. The open landscape is the one in which one can see – and even hear – others and marks of their activities (particularly agricultural ones). It is the symbol of the existence of links between people and heather is its emblem. Forest plantations that "fracture" the open landscape symbolize depopulation and isolation. They are considered "creepy", in the figurative as well as the literal sense, evoking the death of a region or, in other words, of a local society.

These people defend a vision of the plateau as an open geographic space, in the sense of being open to other people, and to other ways of seeing and doing. An idea that keeps recurring is that the mix of different people is good: people from different social and geographic backgrounds bring a particular dynamism with them, vector of strong social ties. According to them, as an economic resource, the plateau should be divided up among even more farmers. As a living environment, it should be divided up among even more inhabitants (host region). As a leisure space, everyone, including visitors, should be able to move about freely. Interviewees stress the diversity of farm products and on the effectiveness of reciprocal aid and exchange networks (farm products, for example). They give the example of Ambiance Bois, a limited liability company with worker participation based on principles of sharing (work and income). They emphasize the dynamism of associative life, the Télé Millevaches experience, etc. Their vision of the plateau is symbolized by a landscape representation characterized by opening and diversity: alternation of grasslands and fields, barrenland and heath, peat bogs, forests and lakes, in a landscape where crest lines and areas around villages and roads are free of vegetation (Figures 1.5 and 1.6).

The commitment to controlling landscape opening (to others)

Among the interviewees encountered, all do not speak in terms of 'landscape closing'.

First, one group of interviewees of the Plateau de Millevaches, hold views diametrically opposed to those of interviewees that feel uncomfortable with landscape closing. This group is formed entirely of people originally from the plateau but who generally have lived their professional lives in the large cities of France and who possess relatively large pieces of real estate, either partially or totally planted. For them, the plateau is a

Figures 1.5 and 1.6. The inhabitants of the Plateau de Millevaches who speak in terms of 'landscape closing' demand an open and diversified landscape as their image of an ideal local society (Photo: C. Chardon).

space that should remain relatively closed, i.e. privately owned by a small number of people, and to which access is a privilege and not a right. They are very attached to the principles of property rights, and denounce farmers who lack respect for their work tool, mushroom "thief" intruders in 4x4, "dropouts" that chose to live on the plateau and do not want to work but nevertheless contribute to the real estate war, etc. They do not hesitate to bring lawsuits to close public paths and to fence in parts of their property.

Secondly, none of the inhabitants of the Pyrenees speak explicitly of 'landscape closing'. When they describe the spontaneous development of vegetation in the area above the village, the inhabitants of Villelongue adopt a decidedly fatalistic tone: "it is a return to the natural order of things". The vocabulary of closing is not really used. Unlike in the case of the first interviewees of Millevaches mentioned above, it is particularly not applicable to the landscape in general, i.e. to the physical manifestation of human activities, associated with a certain ideal of local society. It is applicable to actual elements in the physical space, i.e. roads and meadows. The closing of a meadow or a road signifies that it is no longer part of the human environment in visual, practical and symbolic terms. Basically, interviewees tell us that "it is closed because we no longer needed it".

An aesthetic judgement is rarely voiced on the disappearance of views. Generally speaking, visual landmarks are rarely mentioned – never "from one barn to another", or "from a barn to the village or the valley", etc. It is also rare to observe a feeling of oppression caused by the advance of the forest. Generally, and despite a certain degree of nostalgia – that may simply be the nostalgia of youth – no regret is ever mentioned. On the contrary, interviewees are proud to recall that the local population made the choice, at a given moment in time, to reject the conditions of peasant life, difficult and miserable, and to replace them with social "progress" (a term often repeated in interviews) in relation to society as a whole. This was made possible by local industries. Since then, estrangement from the mountain space is both the condition and the product of this march towards progress. Clearings and woods symbolize this dual-edged movement.

The inhabitants of Villelongue, however, have not completely turned their backs on the mountain. The way they describe certain objects in mountain space is highly significant in this respect, explicitly qualified as "*open*" and endowed with a new qualification. Therefore, the road used by motor vehicles created in the 1980s, restored barns and cabins in the middle of "open clearings" (also commonly described as "lawns"), and, to a lesser degree, hiking paths, all symbolize the new life in the mountains; life that is more closely linked to leisure than farming activities.

If the inhabitants are attached to these forms of opening, they also clearly want them to remain moderate and controllable. Opening more would mean opening to others, to "outsiders", to people who do not live in the village and whose families are not from there. Interviewees do not want the farmers from Villelongue to have to share their summer pastures with farmers from other villages. They do not want their local roads to become a "motorway". They do not want to see increased traffic along their roads and trails caused by "sightseers" and "tourists". Without going so far as to say that these woods suit their needs, these open spaces made up of barns and their clearings distributed in an area gradually being taken over by woods can be seen as a metaphor for wanting to remain by themselves (among inhabitants of Villelongue), with everyone in his/her own place, in his/her "garden clearing", as an annex of their main residence located in the village.

The difference in vocabulary between those who express themselves in terms of landscape closing and those who do not express themselves in these terms is finally significant of the difference in the way of experiencing a space and its changes.

Differences between social groups that resist traditional oppositions

Most studies focusing on social relationships to space and nature assume a certain categorization. They designate ways of seeing and/or doing (seeing or doing the landscape, for example) specific to different large sets of individuals linked together by the same characteristic, generally linked to nature and the intensity of its practice in the space in question. These recurring categories are built around contrasting pairs, e.g. farmers/tourists, inhabitants/visitors, etc. Even where the authors are well aware that this type of categorization cannot totally express the complexity of the actual situation, studies that question it are rare (Le Floch *et al.*, 2004).

The empirical studies presented here show that users of rural areas characterized by radical changes in land use experience these changes in different ways, as much as their contributions to these spaces differ. Nevertheless, the distinctions between groups that we were able to observe do not strictly follow traditional categorical pairings.

Farmers/non-farmers

In scientific literature on representations and/or spatial practices, the distinction between farmers and those who are not farmers is clear (Bergues, 1995; Cadiou and Luginbühl, 1995; Laurent, 1994). It is assumed that the former, because they derive their income from the land, are essentially concerned with practical and economic aspects related to space and nature, and that the second are essentially concerned with aesthetic considerations. However, we must look beyond the production/consumption distinction implied by the farmer/non-farmer opposition. Thiebaut (2002) recommends not analysing the rural landscape alone as an external effect of agricultural practices. He reminds us that if the beautification of areas around farms is to everyone's advantage, then the farmer is the first to benefit from it. In the case of architectural urban spaces, Llewellyn (2003) encourages us to recognize that ordinary consumers are creators as well, in that they actively shape the meaning of the goods that they consume in different places.

In our specific case, the farmer/non-farmer distinction does not apply. The inhabitants of Villelongue speak with one voice when they express the basis of their relationship to space. Obviously, if we wanted to take a closer look, we could surely find nuances in their individual ways of perceiving space, but these nuances would not be based on the distinction between farmers/'others'. They would be the expression of a group of people with a particular sensitivity to an environment, both developed by, and resulting in, a commitment to certain activities in relation to nature (hunting, in particular, but also photography, mountain biking, etc.). In the case of Plateau de Millevaches, the two major opposition groups do not follow the farmer/non-farmer pairing, but correspond to different types of relationships to space. In one group, the individual relationship to private property is essential, whereas the others demand a certain vision of local society based more on the idea of sharing and exchange.

Old inhabitants/new inhabitants

Another frequently used criterion in the literature to build categorical oppositions is related to the seniority of inhabitant status. O'Rourke (1999) shows how the native population does not approve of the behaviour of the 'neo-rural' population that, by interfering in the land retirement process, becomes involved in something that is "not their problem". Barlow and Cocklin (2003) conclude that native inhabitants' social representations are a strong factor in resisting change – the development of eucalyptus plantations, for example – leading to a confrontation with the new arrivals.

It is perhaps this opposition between the old/new inhabitants that seemed particularly relevant to us in relation to the information that we analysed. Admittedly, our analysis does not cast direct light on this issue. In Villelongue, there are new inhabitants that the 'native' inhabitants seem to welcome rather favourably (they bring "change", "dynamism"), as long as they stay in their housing estates. However, everything points to the fact that the inhabitants are actually afraid that the people who have come from the outside will demand some control of the space extending above the village, for example, farmers from other valleys wanting summer pasture for sheep, tourists on the trails, etc.

The distinction between the 'pure-bred' inhabitants and those from the outside comes up again and again in interviews carried out on the Plateau de Millevaches. More than a mere distinction between natives/new, it is a distinction between those that have a 'migratory' history (immigrants or emigrants returning to the region) and those who do not (those that have never left the region). The idea that it is the first group who play a major role is constantly repeated, and it is true that they often participate in the most innovative activities (Ambiance Bois collective, tourist infrastructures, etc.). Even if the groups that we distinguish on the basis of the landscape closing issue do not reproduce the division between migrants/non-migrants, it is probably because migration has been part of the demography of the plateau for decades. A reciprocal acculturation between migrants and non-migrants would have had the time to water down the divisions.

Inhabitants/visitors

The contrasting pair of inhabitants/tourists is perhaps the most common (Bergues, 1995; Cadiou and Luginbühl, 1995; Donadieu, 1995). However, some authors show that the aesthetic point of view can be present in the same way in both categories: inhabitants and 'the public' appreciate the same landscape manifestations, whether they be hedges (Oreszczyn, 2000), forest (Hunziker, 1995), or the landscape in general (Kaur *et al.*, 2004). Other research works focus on the distinction between inhabitants and tourists (Aramberri, 2001; Sherlock, 2001). Sherlock (2001), for example, establishes a parallel between migrants and tourists in the same city. In this case, they would be attracted and repulsed by the same things, often for aesthetic reasons.

We also find it necessary to qualify the opposition in our research in the Pyrenees: the inhabitants, like the visitors, testify to an aesthetic contemplation of mountain space; they appreciate the view points, the wide-open spaces, going beyond the forest. The inhabitants can become "visitors on their own turf", with a distant gaze. On the Plateau de Millevaches, some of the interviewees emphasized that they made the transition from tourist to resident, following an infatuation with the beauty of the area. Finally, the status of inhabitant and visitor are not definitive. Inhabitants can be visitors elsewhere and their opinion is formed by what they see. Tourists are inhabitants somewhere and their opinion is also formed by their domestic environment.

Different concerns about the same issue: open rural space

As in the social debate on the future of rural areas, the advance of woody vegetation can lead to the formulation of a problem in terms of landscape closing in the testimonies of ordinary users. From the point of view of action, interest in the work reported here would be twofold. On the one hand, it shows that expression categories vary among users and those that dominate are not necessarily the same as those used in institutional discourses. On the other hand, it clearly demonstrates that divisions that can be made between users – farmers/non-farmers, old inhabitants/new inhabitants, inhabitants/visitors – should be approached with caution.

Among ordinary users, we cannot observe, or only to a small degree, naturalist-oriented issues that would indicate the importance of the role of clearings or plantations on the ecological wealth of the environment[5]. On the contrary, we do find a question of aesthetic contemplation of the landscape: the contemplation of open landscapes is essential to the aesthetic experience, at least in the case of high and middle-range mountain spaces where our interviews took place. However, if the aesthetic issue is important for non-resident visitors, it cannot be neglected for the inhabitants, particularly as some of them place themselves in the position of "visitors" within their own environment.

In any case, the most frequent category in the testimonies gathered is of an another kind. When people speak as inhabitants, familiar with the sites and their history, it is the social issue that is apparent. The 'closing/opening of the landscape' refers to a certain idea of society, as well as sociability, at the local level, and the idea that the physical configuration of the landscape is the product of this society. On the one hand, those that feel uncomfortable with the idea of landscape closing demand a geographic space open to others, for different activities, etc. They do not exclude a place for the forest either, but this place is not necessarily the one that land developers would attribute to it. The "postage stamp plots" established "anarchically" that public actors denounce can be appreciated by the inhabitants of Plateau de Millevaches, precisely because of the variety and even disorder that they generate. But, these inhabitants cannot deal with the idea of extensive forested areas. On the other hand, a large number of inhabitants do not feel discomfort at the idea of landscape closing, and this can even be true for the majority of inhabitants of a region, as in our example in the Pyrenees. Undoubtedly, in many cases, awareness of the meanings that the inhabitants attribute to spatial changes would make it possible for elected officials or public agents, to avoid projecting on to the inhabitants fears that do not exist, in other words, to avoid intervening when it is not necessary.

---

[5] Those conclusions fit with those of Brun *et al.* (2002).

To us, the question of landscape closing/opening refers above all to the idea of *open space*, as defined by certain authors (Le Floch and Devanne, 2004): a space open to others, with its differences and diversities, a collective resource that juxtaposes many social practices (Goheen, 2003); a space of sociability in which the distinction between traditional social categories does not really work or functions on the basis of the diversity and the interaction of these categories; places where we are in contact with natural processes, in which disorder has its place as much as order (Ward Thompson, 2002).

Chapter 4

Health, identity and sense of place: the importance of local landscapes

Simon BELL and Catharine WARD THOMPSON

Over the last 10 to 15 years in the UK, there have been two complementary developments in relation to the provision of more and better accessible 'natural' landscapes closer to where people live. One is the movement to create forests and woodlands close to towns, villages and other communities, often in post-industrial or degraded landscapes, and the other has been the desire to include the population in planning and decision-making. This movement is widely known as the 'community forest' movement and it incorporates a range from very urban to rural locations (National Community Forest, 2003; Central Scotland Forest, 2003). Common to all areas is the desire to involve local people in local places in order to generate benefits that can range from health and well-being to the creation of environmental and social capital. Various government agencies, such as the Forestry Commission, Countryside Agency and English Nature, are the prime movers in these initiatives, as well as local authorities and local communities all over the country. The programmes were established in the early 1990s and, from small beginnings have grown into substantial projects. Towards the end of the decade, however, it was recognized by government agencies that there was no real knowledge of the true nature of the benefits generated by these programmes and whether they were fulfilling the needs of the populations who were supposed to benefit.

In the UK, there are great differences in the availability of access to the countryside, natural areas and woodlands as a result of location, distances, social patterns and structures, economic and legal factors, demographics and the ethnic composition of communities. At one level, there is the difference between the 80–90% of people who live in urban areas, or who could be said to be living in urbanized conditions, and the remaining 10–20% who live in rural areas. Yet this is a simplistic division that masks more complex patterns such as rural and urban poverty, both caused by low-income jobs, but where the transport problems and lack of services in rural areas contrast with the degraded environments in urban areas. The ageing population is common to all locations but ethnic diversity is a feature of urban areas more than rural ones.

Against this background, research has been undertaken to try to understand better the relationship people have with their environment – their activities in, and their perceptions of nature, woodland, the landscape or the countryside, whatever term happens to be used (the term 'landscape' will be used from here onwards in this paper). Two major research projects in this area were carried out between 1999 and 2004 by members of the OPENspace Research Centre, devoted to research on inclusive access to the outdoor environment, based at Edinburgh College of Art and Heriot-Watt University. In addition, an important scoping study was prepared on the issue of the unrepresentative nature of participation in countryside recreation in England. These projects were funded by the Forestry Commission, English Nature and the Countryside Agency, all government agencies whose policies were moving into the social realm, away from an emphasis on resource or landscape management (timber production, nature conservation, landscape protection), although all three have been heavily involved in promoting access and recreation for many years.

These projects, while different in emphasis, each explored how people interact with their local landscape as opposed to the landscape in general or that of special or designated areas of importance. This is partly because, in order to solve problems of providing inclusive access, people must be able to visit places close to where they live, but also because it is with the places where we live that we develop the closest relationships. It is also the case that community participation and involvement can only operate at a local level. Thus, it is evidently most important to understand more about local landscapes. This does not mean that the results of the research need only apply to the localities where the research was undertaken. In both projects described here, a number of different locations and their local populations were sampled so that wider application of the results is possible.

Research methodology and theoretical basis

A common feature of the projects described in this paper is the theoretical basis, structure and methodology adopted for them. It has become a strong feature of the research carried out by OPENspace to base the methodology in Personal Construct Theory (PCT) (Kelly, 1955) and Canter's Theory of Place (Canter, 1977), and to use Facet Theory (Shye *et al.*, 1994) as a means of structuring the research itself. PCT has shown that people bring previous experiences, expectations and personal objectives in a place to any evaluation they make of the place. According to Canter (1977), the qualities of place are composed of three elements: the physical attributes of the environment, the activities in which people engage and the perceptions they have. When exploring the contribution of the local landscape to people's lives, it is necessary to consider all three elements and the interaction between them. The main advantage of using the Facet Approach in relation to this is that it facilitates the explicit structuring of the central issues in the research and their relationships to one another. While this is often considered to be inherent in scientific investigation, it is easy to miss key issues and their interrelationships unless they are explicitly expressed (Borg and Shye, 1995).

After an initial scoping of issues and the literature, in order to formulate the detailed research questions, the research was always user-led. The issues to be explored in relation to people's engagement with place were identified by local communities, either through the use of focus groups or a number of individual, semi-structured interviews. The qualitative data thus gathered could be categorized under the headings of physical landscape, people's activities and their perceptions, and were then used as the basis of a questionnaire for gathering a more representative sample of data capable of quantitative analysis. The questionnaire in each case was developed using Facet Theory.

Each research project is described briefly below, so that the common themes and central issues can be illustrated and drawn out more explicitly.

Open space and social inclusion: local woodland use in central Scotland (Ward Thompson *et al.*, 2004)

The Central Belt of Scotland lies between the cities of Edinburgh and Glasgow, roughly bounded to the north by the Ochil and Campsie Hills and to the south by the Pentland Hills. The area is characterized by a rolling, low-lying landform, a number of small to medium-sized settlements (villages and towns) and major transport corridors. The major development in a number of settlements dates from the 18th and 19th centuries, when heavy industry and coal mining dominated the economy. The area includes some 20th century New Towns and, in places, more recent growth of 'sunrise' industries have brought prosperity. However, there remains in many communities, a demographic structure with high levels of unemployment and relatively low incomes, poorer living environments and services, and low level of access to personal transport.

The area has been the subject of long-term attempts at environmental improvement and a number of forest plantations were established as commercial enterprises. The movement to create and manage community woodlands has been a more recent activity. The resulting combination of publicly and privately owned woodland and forests, which many settlements have within a relatively easy travel distance of people's homes, represent an important recreational resource potential for day-to-day use by the local population.

Research aims and objectives

The research project (Ward Thompson *et al.*, 2004) sought to answer four basic questions:
– how important is forest use to local people? What proportion of the population and which segments of the population use forests?
– which forests and woodlands do people choose to use or abuse?
– how do people use forests and what counts as use or abuse?
– what are the design and management implications for forest managers?

Case study communities

After the initial scoping stage, five communities were chosen for study to represent a range of conditions and locations typical of the area. The focus groups and questionnaire surveys were based on these, so that comparisons could be made and generalizations drawn between the different communities. Focus groups were carried out in the local communities and the questionnaire survey was undertaken in public places such as high streets, to get a representative sample of the population.

Results

Focus groups indicated that the choice of woodlands and forests for recreational purposes is mainly driven by proximity to where the users live. Woodland type, e.g. coniferous or broad-leaved, appears to have much less influence on choice of where to go. Facilities (e.g. for cycling or children's play), determine use and, although many people prefer open woodland, some users, including teenagers, like woodlands as places in which to hide or to get away from others. Woodlands have evidently been the location for many vivid and positive experiences of engagement with the landscape, often associated with childhood memories. Despite this, many people consider the opportunities for woodland enjoyment are more constrained than they were a decade or more ago. The two factors most often mentioned as preventing fuller use of woodlands are safety and forest abuse. The dumping of rubbish and general littering of many woodlands does not always completely deter people, but encounters with others who may be vandalizing or behaving anti-socially, will. Fears for safety, whether because of other people or from injury (particularly among elderly users) were mentioned as deterrents to people using woodlands alone.

The questionnaire results ($n=339$) revealed apparently distinct categories of visitors:
– those who visit daily, often for walking the dog; this is the only group who generally visit woodlands alone;
– those who visit weekly, regular visitors who have the most positive views about feeling safe, at home and free from anxiety in woodlands;
– those who visit monthly, who share some of the perception characteristics of the weekly visitors, but to a lesser degree, and who are most interested in woodlands free from rubbish with information boards and signage;
– those who visit once a year, who are least likely to visit woodlands alone, most positive about signage, and who share some characteristics with those who never visit woodlands at all in feeling less at home there.

The qualities of woodlands which attracted the strongest responses overall related to liking woodlands that are free from rubbish and feeling 'at peace' in a woodland (Figures 1.7, 1.8 and 1.9). In general, people disagreed strongly with suggestions that they consider woodlands as scary and that they fear having an accident there. This suggested that the focus group responses about fears for safety in woodlands were not typical of the population as a whole but reflected subgroups' perceptions. These were explored further. Compared with men, women had stronger preferences for signs leading through the woods and for group and family activities, such as picnics, and were less likely to walk alone. Men were generally positive about walking alone and felt much less vulnerable in woodlands than women. Crucially, the frequency of people's woodland visits as children is related to the frequency of adult visits and to how comfortable adults feel walking in woodlands alone. Usage was also affected by age: those aged over 45 years were less likely to visit for specialist outdoor activities, although the 55–64-year-old age group was an exception to this.

The strength of the relationship between childhood visits to woodlands and adult patterns of visiting was the key message to arise from the statistical analysis.

Figure 1.7. A well-designed and welcoming entrance with a view into the woodland which makes people feel comfortable about visiting.

Figure 1.8. Three elderly retired miners who enjoy spending their time in a wood planted on the site of the coal mine in which they once worked. An interesting example of environmental and social renewal.

Figure 1.9. A semi-natural oak woodland that now lies on the edge of a town and is valuable for both nature conservation and recreation, being a local nature reserve.

Nature for people: the importance of green spaces to communities in the East Midlands of England

The East Midlands of England is a region of mixed land use and landscape. It comprises the counties of Nottinghamshire, Derbyshire, Leicestershire, Lincolnshire and Northamptonshire. Lincolnshire is largely an agricultural county with a coastline and Derbyshire includes the Peak District National Park. All the counties apart from Lincolnshire have areas of significant extractive industry, especially deep coal mining and open cast iron ore working. These have left large areas of disturbed land, which have often been restored to woodland. Several large cities or towns can be found in the region, with significant proportions of ethnic minorities among their populations. The green spaces available to these populations range from waste land used informally, formal city parks, many dating from the Victorian era, country parks, often established on former industrial land, remnant woodlands now lying within the boundaries of urban areas, nature reserves of woodland, wetland, heathland and coastland, and open upland moorland (in the Peak District).

Research aims and objectives

The aim of the project as specified by English Nature, was "to specify the contribution that 'nature' in green spaces makes to people's social well-being by examining the use people make of, and the feelings that they have towards, a selected number of artificial and more natural green space sites distributed throughout the East Midlands".

Case study sites

As this was a regional study, a number of sites were selected to fall more or less equally in each of the region's counties and to represent a range of urban, suburban and rural conditions. Focus groups were carried out in the local communities and questionnaires were carried out in the different green space sites themselves.

Results

Key points raised by participants in the focus group discussions are summarized below.
Nature and green space:
– The terms 'nature' and 'green space' are hard to define.
– Definitions are influenced by cultural perceptions of the natural environment.
– Nature cannot be considered in isolation from the world of human activity.
– Green space can be land over which residents feel they have little or no control.
– Green space can be a small pocket of land in an urban area that is badly maintained and unsafe to use.
– Green spaces can also be very precious.
Social benefits:
– The key forms of anti-social behaviour are fly-tipping, litter, vandalism (Figure 1.10), dogs (faeces and running loose) and intimidation from large groups of young people.

– Anti-social behaviour can prevent the implementation of green initiatives.
– Management must be visible whilst at the same time being sensitive to the location (Figure 1.11 and 1.12).
– There is currently an imbalance between preservation and access to sites of special interest.
– Children are not encouraged to explore and take an interest in nature.
– Parental attitudes towards, and ability to undertake, nature education have changed significantly over the last 50 years.
– The educational system must take responsibility for nature education.
– There is a lack of effective interpretation.
– Green initiatives instil a sense of ownership and encourage responsible behaviour.

Figure 1.10. Burned out cars in a neglected conifer woodland are symptomatic of the problems of abuse taking place in urban fringe woodlands.

Figure 1.11. A managed mixed woodland on the edge of a small town provides a substantial recreational asset and enables people easily to gain an experience of nature.

Figure 1.12. A more developed country park on the edge of a larger town where interpretative facilities help urban people understand more about and gain a greater benefit from contact with nature.

The importance of having green spaces nearby:
– There are many social, mental and physical benefits that can be derived from access to nature and green spaces.
– All the participants felt that access to nature was important, although in some cases the knowledge of nearby nature and green spaces was enough to instil a sense of well-being.
– Members of minority ethnic groups are rarely approached to take part in green initiatives and are unsure of where to obtain information.
– Sign posting and information given at sites is often inadequate and not very informative.
– All attempts to provide inclusive access should be sensitive to the location.

The questionnaires (n=459) developed on the basis of the focus group findings revealed that many people, regardless of age or sex, visit all type of green or natural sites. However, the sample revealed disproportionately low numbers of people from black and ethnic minorities and people with disabilities as visitors. While many people visit on their own, couples and families make up the majority of visitors, the latter especially at country parks and other sites with special facilities and wildlife displays.

Women visitors were under-represented in comparison with the general population. Comparatively low numbers of unemployed people were recorded as visitors but many retired people appear to visit. The findings about women seem to confirm previous studies showing that women tend to be significantly less frequent visitors than men to woodland or countryside sites (Burgess, 1995; Ward Thompson *et al.*, 2004). It may reflect concerns over safety expressed by women in the focus groups, and women's responses in the attitudinal section of the questionnaire, where feelings of vulnerability were also rated strongly.

Teenaged children are also infrequent visitors compared with younger children. One of the possible causes of this is that what urban teenagers frequently consider 'outdoor' places to visit are in fact indoor spaces such as arcades and malls (Travlou, 2003). It may be a particular phenomenon of this age group. Læssøe and Iversen (2003), in an in-depth qualitative study of the importance of nature in everyday life, found that youth generates a discontinuity with the nature relationships of childhood because a lot of energy is put into social relations during this phase. The findings also bear out other research into the relationships teenagers and children have with outdoor places such as woodlands (Bell *et al.*, 2003).

Few people from black and ethnic minorities visit any of the green spaces. This seems to follow a common pattern in the UK, as there is a range of evidence from the literature that black and minority ethnic communities in the UK do not participate in visiting the countryside and other natural open spaces and related activities, proportionate to their numbers in society (Countryside Agency, 2004).

The main reasons people visit green spaces are to walk the dog, to exercise, and for the pleasure of being in a park or close to nature. Dog walking is most popular at local sites and in woodlands and country parks, but less frequent at nature reserves. Reducing stress and relaxing are significant reasons for visiting green spaces and represent one of the main social values. The importance of dog walking in relation to green spaces has been corroborated by other studies (Ward Thompson *et al.*, 2004; OPENspace Research Centre, 2003), and cannot be under-estimated. In this study, focus groups identified dog fouling as being a key form of anti-social behaviour, so the tensions found elsewhere between dog-owners and other green space users seemed to surface here too (Ward Thompson *et al.*, 2004). This, however, is not the only problem associated with dogs; a study by Madge (1997) showed that the fear of coming into contact with animals, and in particular dangerous dogs, was much higher for African-Caribbean and Asian groups than white groups.

A key message from the project was that people think of nature in quite a broad way. Nature includes physical characteristics, wildlife and also perceptions and emotions, especially peacefulness and other terms associated with the calming or de-stressing value of nature. When talking about 'social values' people tended to focus on 'anti-social uses'. There is much evidence that sites need to be well managed (but not over managed), welcoming, provide information and have a natural appearance if people are to obtain the best value from them. Sites close to home are preferred, especially by those who used to visit frequently when children.

Summary and conclusions

The above projects may be set in a broader context by the results of a scoping study for the English Countryside Agency carried out by OPENspace Research Centre, to survey barriers to access and enjoyment of the countryside (Countryside Agency, 2004). This study found evidence that many groups in the UK – young adults, low-income groups, black and other ethnic minority communities, people with disabilities, older people, and women – do not participate in countryside and related activities proportionate to their numbers in society (Ageyman, 1990; Breakell, 2002; Countryside Agency, 1998, 2002; DTZ Pieda, 2001; Slee *et al.*, 2002). However, exclusion cannot automatically be inferred from under-representation; a group that is under-represented may not feel excluded if it has full access but still declines to participate in countryside activities. Social, physical and psychological barriers significantly influence the way that people perceive the countryside and how they make choices over whether or not to use it. Many barriers to access and participation were identified in the literature but it also suggests that participation in countryside activities can offer a range of benefits, including enhanced physical health and general well-being, the development of social and personal skills, enhanced community development and cohesion and improved quality of life. The research projects outlined above add to the evidence which supports and amplifies some of these issues.

The research projects briefly described in this paper demonstrate that the relationship that people have with the physical environment or landscape is a personal one, constructed through the medium of social and economic activities and perceptions. Canter's three elements of place, the physical attributes of the environment, the activities people engage in and the perceptions they have, are all important in exploring the contribution of the local landscape to people's lives. The results show that, in exploring people's attitudes to, and perceptions of their local area, qualities of the social environment may appear to override in importance many qualities of the physical environment. This has been shown in previous studies of place (Donald, 1994a, 1994b; Scott, 1998) and it is accepted that, while the physical environment has a significant role to play in everyday life, this role may not always be explicit unless a feature of the physical environment obstructs, prevents or otherwise interferes with a person's objective. This is not to diminish the value of the physical landscape, whose natural qualities were clearly appreciated by people in both projects, but it suggests that the social context must also be taken into account by planners, designers and managers.

Evidence of the benefits to health and well-being provided by green areas is increasing. Feeling at peace and getting away from stress was associated in the projects with relaxation and nature – seeing it, being in natural places and learning about it. This supports findings from other studies where it has been shown that leisure activities in natural settings or exposure to natural features have important stress reduction or restorative effects (Kaplan, 1995; Sheets and Manzer, 1991; Ulrich, 1981, 1984; Ulrich *et al.*, 1991). However, it is important to provide the right kind of green areas in the right places. Locally accessible green areas are particularly important, especially for children.

There are significant associations between the type and degree of use of green spaces by people now and how frequently they visited such sites when children. This suggests that if children are not being allowed or encouraged to visit natural areas or other parks by themselves, they are less likely to develop a habit that will continue into adulthood. Using green areas and becoming comfortable with them as a child has significant implications for continuing use as an adult and the related potential for maintaining or improving health later in life. However, there are many social groups whose level of use, even of accessible, local green areas is low and efforts need to be continued to explore why that is and to remove barriers to use. Green areas need to be local, safe and inclusive, free from the signs of litter and abusive behaviour so as to become true places, part of the social as well as the physical landscape.

Chapter 5
Where does grandmother live? An experience through the landscape of Veneto's 'città diffusa'

Benedetta CASTIGLIONI and Viviana FERRARIO

In Veneto's 'città dffusa'[1]

As has been well-noted, in the last 10 years of the 20th century, Veneto, within the more general context of the north-east of Italy, underwent fast and uncontrolled development, characterized by the rapid transformation of the rural economy into an industrial economy based on small and medium-sized enterprises. At the same time, this rapid change created upheaval both within a social and a territorial context. Both of these have played fundamental roles in the process of development of this area. The traditional distribution of settlements across the Veneto countryside is now being replaced by small family-run businesses, which constitute the dominant productive force of the entire area. Residential buildings and factories are constantly on the increase, spreading and growing ever larger at the expense of agricultural land. Pre-existing smallholdings are becoming increasingly bigger and a great number of new residential and business building lots are springing up in between existing smallholdings, especially alongside the busy roads. In a manner approaching anarchy, the density of settlements is increasing with no clear logic. This has been caused largely by urban planning that has all too often followed, rather than led, these developments. The widespread construction of disparate, yet highly urban elements on to a predominantly rural social fabric has dramatically transformed the Veneto landscape. This has made any attempt at interpreting such changes difficult: the traditional categories of town and country seem to have lost their meaning and been

---

[1] This study was the subject of a presentation at the International Conference 'From Knowledge of Landscape to Landscape Action', held in Bordeaux, 2–4 December 2004, and is the result of a shared reflection on the previous work conducted by the authors. In particular, V. Ferrario was responsible for the first two paragraphs, and B. Castiglioni the last two.

replaced by a kind of 'diffused city', or 'nebulous settlement', in this variously defined phenomenon[2].

But what relationship is there between the population and the Veneto landscape as it appears today? How is this process of change being perceived?

The European Landscape Convention considers the population and the way it perceives the landscape of great importance, even for 'ordinary landscapes', whose meaning and value are already clearly recognized. To speak about the landscape in the diffused city does not, therefore, seem quite as absurd as it might do at first sight. But is there an awareness of this by those who live there? Does the landscape really condition the lives of its inhabitants? After such a rapid change and uncontrolled urban development do we see, reflected in the local population, a corresponding disorientation and loss of reference points? Drawing on the issues raised by the European Landscape Convention, these questions formed the basis of our analysis, the initial results of which are presented here.

The study compares a number of local situations in different parts of the "metropolis of central Veneto"[3]. The first case study used to explore answers to these questions was the small village of Vigorovea, in the municipality of Sant'Angelo di Piove di Sacco, about 15 km south-east of Padova[4].

By comparing current and historical maps, it was possible to observe the changes, the development of the villages, the 'urban expansion' linked to primary and secondary roads. The origins of place names were analysed as a means of linking inhabitants to their surroundings, while photographic documentation enabled other interesting evidence to be gathered[5]. All these data were compared with the results of repeated discussions with the population. Some inhabitants (of different ages and origins, residents and non-residents of Vigorovea) were involved individually in informal conversations, structured around a number of points their own homes and those of their immediate family, their

---

[2] The extraordinary economic and urban development of the north-east of Italy and the 'diffused city' which covers the central Veneto plain (and in particular, the area of land which rests between the cities of Padova, Mestre, Treviso and Castelfranco) have been the focus of study of many different academic disciplines in the last few years. With specific reference to territory, research and publication of findings have been ongoing since the beginning of the 1990s, starting with the analyses of Francesco Indovina and the studies of Bernardo Secchi. Cited here by way of example of differing approaches are Munarin and Tosi (2001) and Vallerani and Varotto (2005). A comprehensive bibliography collected for the fourteenth course on management of the landscape by the Benetton Foundation for Study and Research, in 2003, can be consulted online (http://www.fbsr. it/file/corso03_189bibliografia.pdf).

[3] As recently defined by Indovina *et al.* (2005), where the "Veneto metropolis" is compared with other conurbations in southern Europe.

[4] The studies of Vigorovea were carried out during spring and summer of 2004. During the course of 2005, a second case for study was taken into consideration: the village of Biadene in the municipality of Montebelluna (Castiglioni and Ferrario, 2005), situated in the strip of foothills along the northern margin of this town. The third case currently being studied moves the investigation to a more consolidated area of the extended Veneto city.

[5] The changes at Vigorovea were observed not only from a plan perspective achieved by studying maps and land photographs, but an attempt was also made to follow its more commonplace landscapes "searching for clues of the relationship which link the inhabitants to their place of residence" (Boeri, 1997); the camera was an invaluable tool for capturing these clues, a number of which are visible in the accompanying photographs. Images, place names, words, sounds and opinions were collected in an attempt to verify the possibility of constructing a small 'eclectic atlas' (Boeri, 1997.) of Vigorovea.

daily movements, the reference points in the population centres and movements to nearby towns; their perception and evaluation of the local landscape and its recent changes, their perception and evaluation of the relationship between the city and the country within the context of the diffused city, their personal feelings associated with their place of residence; their thoughts in general on the concept of landscape.

Avoiding those topics (or even references) which could have led the interviews down the 'blind alleys' of local arguments about land management or other current controversial topics, the main approach was to bring out first and foremost strictly personal points of view, linked to experience. This approach often aroused a certain initial surprise in the people being interviewed, who were not used to addressing themselves to these topics and expressing opinions on them. The local residents were not used to making connections between the reality of their daily lives, opinions and personal experiences, real or imagined perceptions and the problems associated with land management: all these ideas were generally considered to be distinctly separate issues.

Despite the emergence of a significant lack of understanding about the close interrelationship between questions concerning the landscape, the interviews shed a great deal of light on the numerous complex problems which distinguish the relationship between the population and their surroundings.

Recent changes to the landscape

The regional affairs of Vigorovea

The village and area surrounding Vigorovea present a number of significant characteristics for our study.

– The presence of a main road (the SR516) represents a form of 'urban magnet' along which the signs of the transformation of the landscape are particularly evident (Figure 1.13).

– The absence of particular recognizable landscape 'values' make Vigorovea a classic example of a typical 'everyday landscape', and, as such, highly representative of many other areas of the Veneto plain.

– Its peripheral geographic position with respect to a strict definition of the Veneto diffused city makes Vigorovea an ideal subject for studying the embryonic phase of urban sprawl, and thus offers an interesting case for comparison with more consolidated areas of 'sprawl'[6].

Vigo de Rovea (literally 'village of oak trees') was for centuries a small medieval settlement along the road that runs from Padova across territories originally under the power of the Bishop; a resting place, maybe with an inn for pilgrims, and a handful of farm houses spread across the countryside.

Administratively speaking, Vigorovea became part of the municipality of Sant'Angelo di Piove di Sacco only in the 19th century. It is situated next to the main road, where the church and *osteria* are situated.

---

[6] A geographically oriented bibliography is presented in Munarin and Tosi (2001: 218).

Figure 1.13. The main road passing through Vigorovea.

The tramway from Padova to Piove di Sacco was built in the 1930s along the main road, consolidating its importance as perceived by the local population. The stops along the tramway became new reference points: their names were documented on official maps, where they were perpetuated long after the tramway itself was dismantled.

After the Second World War, the focal points, traditionally centred around the church and along the main road, started to become less and less compatible with the rapidly increasing heavy traffic.

The idea that Vigorovea should be developed away from the main road has been a feature of local planning from as far back as the first regulatory plans of the 1950s, with varying results. During the 1960s, the economic boom years, an industrial estate was built just to the south-east, close to the main road, which was important for the whole area of Piove di Sacco[7]. The local population preferred factory work to rural labour, which became only a secondary activity for many people.

This stage marked the start of the expansion of Vigorovea, principally directed towards Sant'Angelo: the first residential development, the 'Raggio di Sole' ('Ray of Sun') district, is a prime example of this trend. The most consistent part of the development took place, however, during the 1970s with an increase in the population density of pre-existing residential areas, which expanded radially from central points as well as in a

---

[7] The so-called 'Piovese industrial estate', which serves the entire area surrounding Piove di Sacco (Saccisica), the regional capital with a total population of around 15,000.

linear fashion (Figures 1.14 and 1.15). The sprawl was absorbed painlessly by the loose-knit residential area, where houses were originally fairly widely distributed.

During recent years, however, the increase in the number of buildings has been much more rapid: two new residential areas were built to the north-east of the main road, which finally fulfils the provisions of the original urban plan. The two areas have recently merged into a kind of 'new Vigorovea', in many ways detached and isolated, like a piece of autonomous city, which proclaims itself the new centre by virtue of having a building with a red-orange pediment, and a rather dull brick-paved main 'square' (Figure 1.16).

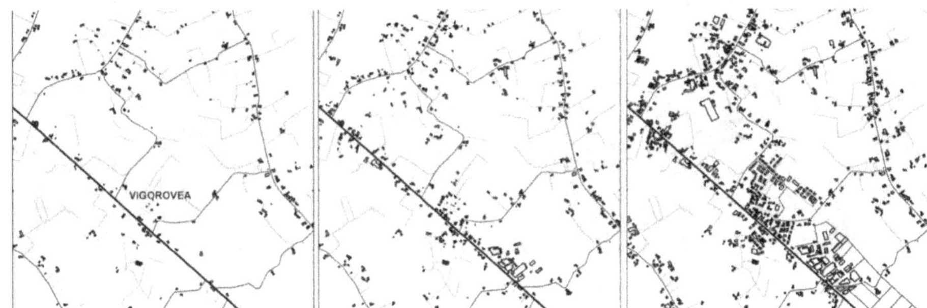

Figure 1.14. From diffused settlements to a 'diffused city': the development of Vigorovea, reconstructed from official maps (1910–2004).

Figure 1.15. An old farm building, a reminder of the traditional rural landscape.

Figure 1.16. Mother Teresa of Calcutta Square; the shrine is in the foreground.

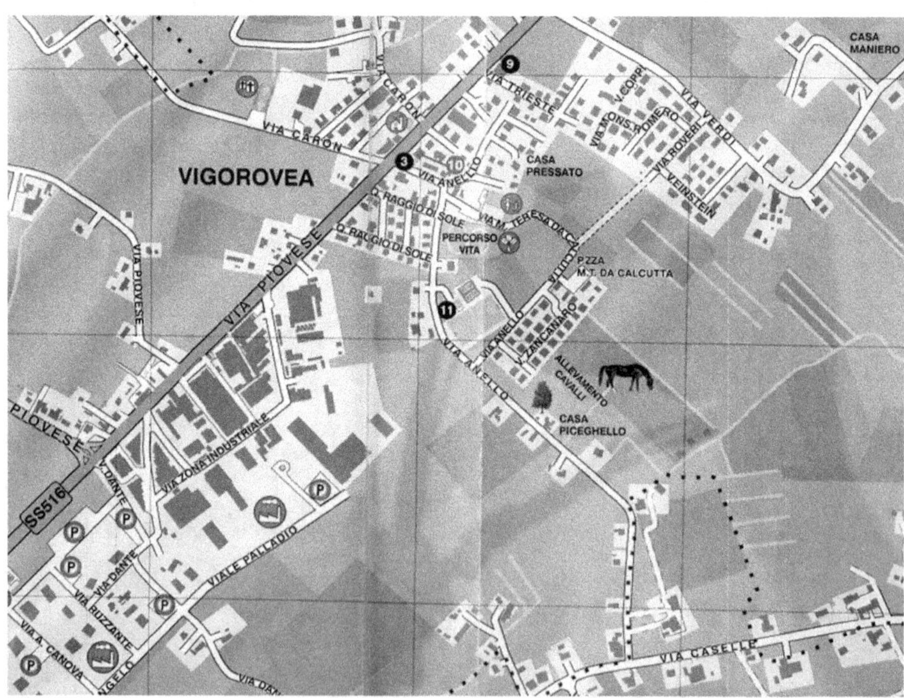

Figure 1.17. Details of the council map of the area. Two different types of place names can be seen: family names associated with rural farmhouses and the more recent names of residential areas. The built-up areas encroach ever more on the green of the agrarian landscape.

Figure 1.18. Road signs crowd the new areas under construction.

Vigorovea is now well on the way to a process of development whose outcome is very obvious to the local population. The council map, in its graphic ingenuity, describes the process underway quite clearly. The clear surfaces, which are a background to the buildings, mean asphalt, pavements, road markings, flower beds, fences, private gardens, urban furniture: all the standard, dreary ingredients of low-quality street furniture, which encroaches on to the green countryside of the agrarian landscape (Figures 1.16 and 1.18).

Different ways of naming places

The persistence of place naming convention encloses ancient meanings in a protective shell. It seems to confirm for us, however, that the value of studying place names does not stop only with an 'archaeological' meaning, shedding light on past activities, reference points or past values, but that they can also be indicators of changes in progress and may also be a window on the future. Observing how places come to be named is a useful way of understanding ongoing processes of change, especially where new names are given to 'new places'.

In Vigorovea, there exists an older layer of place names, which describes the first ancient shapes of the humanised natural landscape[8], crystallized by a century and a half of official map-making. On to this, one can find another system of place names based on

---

[8] Arzerini (small embankments), Ardoneghe (property of dominus), Celeseo (cherry wood) and Vigorovea (village of oaks) are just some examples.

land ownership, also adopted from military maps from the Istituto Geografico Militare, which correspond to a system of references still partially being used, which are linked more to people than to places. This system is based on the presence of farmhouses and their inhabitants, using the surnames of the resident families, for example, Casa (Home) Maniero, Casa Menegotti, etc. Such place names enable us to develop a picture of how this countryside must have looked for many years in the past, when farmhouses dotted a rural landscape continually travelled across by a well-established population using a dense network of footpaths which criss-crossed the fields. Compared with this, the current rural spaces are in fact much less 'inhabited'[9].

This reference system holds up in a rather static world, where one is born, lives and dies in the same place. Family names, on the other hand, change on the sequence of official maps: from the end of the Second World War to the present day, the continual updating of the names of, houses, on maps is a reflection of a changing world and an indication of the difficulties this naming system faced when trying to adapt to the increasing complexities of residential development and new lifestyles.

Today the development of the diffused city produces a new layer of place names, superimposed on the previous ones, which gives an official name to new roads, new lots of land and new residential areas. It is this very 'administrative' naming of places that seems to be an interesting indicator of some aspects of what it means to live within the diffused city. The naming of village roads after the lives and achievements of famous people from the past is a custom which dates back to the 19th century: the name of a road takes on the status of a 'monument', an encouragement to remember. Sometimes this is based on official national history, or on some aspect of local history whose importance has been recognized more recently.

Three roads in one of the new districts of Vigorovea (the so-called 'di Via Verdi' district) have been named after three very different personalities: Albert Einstein, monsignor Oscar Romero and Fausto Coppi (a champion Italian cyclist of the early 1950s). The village square created off the main road, now the functional centre of Vigorovea, is named after Mother Teresa of Calcutta; a short distance away one finds a street named Tono Zancanaro, a noted contemporary engraver from Padova linked to the communist party. What emerges then is a heterogeneous 'fruit salad' system for naming places, which appears to be removed from local contexts and – at least for the time being – unable to produce a new form of coherent naming system.

Some places with a special meaning for meeting and socialization, which have never needed naming – their name linked to their functional value sufficing – have been renamed to sustain some ideological position. The church courtyard, always known simply as 'square', has recently become 'the Square of the Family' (not the Holy Family!), official support for this basic structure of traditional society.

Some common names attributed to places, which until a short time ago, were fields have their origins in urban language (e.g. district, park, industrial estate, etc.). Even this exporting of terminology can represent a kind of cultural colonization, or even self-colonization, with urban origins: an urban mark on something that is no longer countryside but is certainly not (yet?) a city.

---

[9] Oral testaments confirm "Once upon a time if you called out, your voice could be heard from one house to another".

Vigorovea, for example, has a public garden, a small plot in front of the school. The pupils call it 'the little gardens', but a sign at the entrance gives its name as 'Rainbow Park'. So the name used tends to identify it as a place of enjoyment rather akin to a nightclub. Its 'urban' character (both in shape and name) clashes with the neighbouring field of maize.

Next to the Fausto Coppi road emerges the faint memory of a lost landscape, suggested by the name 'via dei roveri' ('oak tree road'). This is possibly the effect of the long attempt to heighten awareness of the value of traditions, which for at least 20 years, has tried to rectify the assumed loss of identity that would accompany the progressive loss of the traditional rural landscape. But what do a hero of Italian cycle racing and the rural landscape have in common? Maybe as they both belong to the past, they are both worthy of a place in the memory of local residents – and, indeed, in the place name. Indeed, the naming operation of 'via dei roveri' (a new/old name) has all the elements of an old style reconstruction.

Our observations at Vigorovea confirm that the naming process (even the modern one, or rather the comparison and continual interchange between official and spontaneous names) is evidence rich with meaning (Cassi and Marcaccini, 1998: 17). Its stratification and contradictions are indicators of the complex relationship between the population and its surroundings.

The landscape perceived

Living and moving through the landscape

Most people that were interviewed had always lived in the village or come from nearby villages, within a radius of not more than 15 km. The physical nearness to parents and relatives is extremely widespread and very often the two generations live in close proximity in the same house, in keeping with tradition[10].

With short trips being the norm, mostly by car, they can maintain close links with the family network (often associated with specific needs and essential to compensate for the lack of efficient public services for children or the elderly), and also to carry out most other daily activities, such as going to work, taking children to school or going shopping. The network of roads which can be travelled by car is quite dense, often comprising pre-existing country byways which have now been asphalted[11].

---

[10] The habit of living near one's immediate family is a typical phenomenon of the Veneto (but is also typical of all of Italy and the Iberian peninsula), often observed by researchers. In the north-east of Italy, 67% of couples married between 1993 and 1997 live less than 1 km from at least one member of their immediate family, and 40% live in the same building (Barbagli *et al.*, 2003: 173). Thus a strong mutual help network is created based on family relation, which constitutes an important support, substituting for the lack of efficient, readily available public services for families.

[11] The network of secondary roads is useful for getting around by car and avoiding the often heavy traffic encountered on the main roads. The inhabitants of Vigorovea did not highlight traffic as one of the main problems. By contrast, in other parts of Veneto where the urban sprawl is denser and in a more advanced degree of growth, local residents have to deal with traffic congestion which is so bad that it makes moving around virtually impossible. Residents in these areas have started to become much more critical of the growth and shape of the diffused city.

Observation of this phenomenon lead to an important consideration: the population perceived the area as an extension of the traditional rural house inhabited by a patriarchal family, like a large, interrupted, farmyard, around which most activities are carried out and most relationships are structured. With journeys of only 10 minutes links can be maintained with all the elements from the previous rural context (work, family, shops, church, leisure time), which today covers a larger area.

Each of the places visited is well-defined; private houses are always surrounded by secure and obvious fencing, which inserts a dense network of small boundaries into the fabric of the land[12]. There was an awareness that this constitutes a change: the Vigorovea of the past was described as an open place, devoid of boundary markers, with farms and courtyards, where everyone knew everyone else.

The numerous daily trips are based on a few poor reference points, not always clearly identified or unequivocal: the church, the main road, the traffic light and the new Mother Teresa square. Despite this, residents do not seem to get lost when moving around in this complex landscape; it emerged that more importance is given to the points of departure and arrival than to the route followed. It is a world perceived and lived almost as in a kind of 'hypertext'[13], without ordered hierarchical references, structured tightly around their own well-known home and the homes of their close relatives and other familiar places, from which they can 'skip' from one point to another. It is a landscape constructed by points, connected to each other, in which, however, the connection, or rather the territorial context, is gradually losing importance.

This diffused city shows its ambiguity and its paradoxes. On the one hand, it is so widespread that it covers a large part of the Veneto plain, with ill-defined borders, giving a sense of great uniformity and vastness. On the other hand, it is totally discontinuous, made up of points interrupted by frequent mini-borders, and experienced and perceived in this total discontinuity[14].

Vigorovea, old and new

Almost all the people interviewed felt that the large changes experienced in Vigorovea in recent years were important, and linked especially to the new residential areas. These were seen as an important event for the village, more for the social implications – obviously generated by a steady increase in the number of inhabitants – than for the impact on the landscape. People who had always lived in Vigorovea (both old and young) observed these changes with a certain degree of perplexity: some people regarded the presence of new blocks of flats as a negative issue because "there is less countryside, less green", while others considered this development a positive thing and saw the new flats as a quaint addition.

---

[12] As an example of how such numerous 'private' borders are becoming a typical feature, one needs only refer to the words of the mayor: "urban development in the coming years will be made up of small enclosed areas, accessible with magnetic badges". Many researchers are studying this phenomenon, for example, Boeri (2003).

[13] Giuseppe Dematteis, *Le Nuove Forme Urbano/territoriali Europee e il Caso Veneto,* a conference held during the Fourteenth Course on the Management of the Landscape 'Nella Città Diffusa. Idee, Indagini, Proposte per la Nebulosa Insediativa Veneta', Benetton Foundation for Study and Research, Treviso, September 2003. Interpretation of the landscape as a hypertext can also be found in Cassatella (2001) and Castiglioni (2002).

[14] This ambiguity can be interpreted as the differing perceptions from inside and outside the 'intermediate' German city, as noted by Sieverts (2003).

Whatever the opinion, their reaction to the newly arrived families was not entirely positive. The new inhabitants, for their part, were aware that they are contributing to a significant change in the village, but they tend not to show that they feel they are out of place.

However, a slightly passive reaction was observed regarding these processes of transformation, as if the phenomenon was not something that had direct implications for the population. Although a widespread nostalgia for the past (and for the relative idyll of the landscape), was perceived, this did not seem to be connected with what is happening and did not appear to influence the people's opinions. The words of the mayor are illuminating: he explicitly distinguished between the highly nostalgic landscape of his childhood in the 1950s and 1960s (which he himself called 'micro-landscapes'), and the 'macro-landscapes' which as administrator he is now responsible for organizing and managing to meet the needs of the population, the trends of a developing economy, political issues, etc. It is an almost schizophrenic task, but this approach really appears as a sort of schizophrenia, accepted and absorbed with no evident conflict.

What seemed to emerge from the interviews is that the concept of 'past' is generally 'compressed' until it contains both the recent and remote past. It represents the history of the village, to be remembered and passed down to children and grandchildren[15]. Archaeological findings as well as the farming tools of great-grandparents also belong to this past. The difference between today and the so-called traditional rural world is perceived as greater than the difference between this rural world and what is known about the medieval world, or about the Roman or pre-Roman periods. This past is referred to as something from a very long time ago.

New symbols

In one corner of the vast Piazza Madre Teresa di Calcutta is located a shrine (Figure 4) with a reproduction of the Pietà di Michelangelo and a distinctive surround (white columns, a cupola in wrought iron and polycarbonate, a small altar, benches arranged in a semi-circle, flowers and candles), which makes it both a place for prayer and for meeting. It was built fairly recently, following an agreement between the parish priest and the village council.

Most of the new inhabitants of Vigorovea and many of the older residents considered this to be one of the most beautiful places in the village. It is appreciated both for its beauty and as a meeting place. Its importance is such that the new square has become recognized as the new centre of Vigorovea, parallel to the church situated on the main road, which marked the old centre of the village.

This shrine, therefore, has an important symbolic value and contributes to the meaning of the location, both for older residents as well as the new arrivals. This combines also with its obvious religious meaning, a factor still prevalent in Veneto society. The new residents find something in the shrine that links them to their roots, thus reducing potential feelings of disorientation, while for those residents who have lived there for some time, it represents a link between all the new events and the old traditions. The elderly farmer in Figure 1.19, for example, is making the sign of the cross while he passes the shrine.

---

[15] The role of recording the past is, however, considered to be more the job of institutions such as schools and public authorities rather than the family.

Figure 1.19. The contrast between old and new landscapes, and old and new lifestyles.

Somewhat absurdly, it is the shrine, an object completely typical of a traditional rural landscape, which has conferred on the Piazza Madre Teresa its value as an 'urban centre'.

City or country?

When asked to describe or classify their home town as either 'city' or 'country', those asked replied with a degree of difficulty, proposing at the same time some interesting definitions.

– "A rural urban place" is the definition suggested by the mayor.

– "Along the main road is town, when you turn off on to a minor road it becomes country".

– "Once it was country, now it still isn't city; it's more like a suburb, everything is becoming one big suburb".

– "It is 'outside' city".

– "Once Vigorovea used to be hidden, today it is like a chick which is growing" (sic!)

The residents are generally happy with the place in which they live, content with the quiet and green of the fields. Almost no-one would go and live in the centre of Padova, describing the city as "suffocating". There was a clear distinction between Padova, clearly identified as 'city' and 'the rest of the world', a landscape without limits, spread out in all directions, an omnipresent/equidistant/shapeless suburb.

It is in fact interesting to highlight that, when talking about suburbs, no reference was made to a centre (Padova, it would be logical to assume), towards which such a suburb gravitates, nor was reference made to an urban belt, but rather to something essentially difficult to define. There is, however, the need to use a term that contains references to 'towns', given the number of elements in the local landscape which are more typical of a city.

Rooted in the perception of 'country', Vigorovea is involved, as is most of the Veneto diffused city, in an ambiguous process, or challenge, between becoming a city and at the same time avoiding the idea of being 'city'.

"As far as I know there ain't no landscape 'roun' these parts!"

"As far as I know there ain't no landscape 'roun' these parts!"[16] is the highly telling reply given by a resident of Vigorovea to one of our questions regarding the definition and description of the local landscape, and the more and less beautiful aspects of it. As with most of the people we met, the question seemed unexpected and was met with surprise and a certain difficulty in formulating a response.

The 'most beautiful places' in the village were listed based more on their usefulness rather than on any aesthetic aspect: even the small industrial estate was cited as among the most beautiful places! The new flats and family houses were viewed positively by those who lived there, while generally viewed negatively by those who saw them being built and who were somewhat puzzled and a little suspicious of the changes taking place.

There was a degree of confusion between the categories of 'beautiful' and 'good-useful'; the residents did not seem to be accustomed to looking around them, often seeming virtually unaware of their surroundings. Aesthetic evaluation was given only to an individual's own house (always positive), while the landscape of Vigorovea was considered merely 'normal', with nothing beautiful or ugly from a visual point of view. Just one young person (relatively well-educated) replied that "the landscape is the features possessed by any environment" intuitively suggesting that the shape, aesthetics and perception of landscape are all important. Almost all the other people involved seemed to live in an un-aesthetic way[17] regarding their relationship with the landscape. The landscape is something distant, which exists only at the seaside or in the mountains, linked to holidays and tourism, or to important landmarks or sites of historical importance (not present in the village); it possibly still exists in the memories of traditional landscape, belonging only to the past. At most, the idea of landscape is connected to the most natural or 'rural' elements or to gardens, but certainly not to a built landscape: "it is not possible to define a residential quarter as a landscape".

---

[16] Literal translation of local Italian dialect: *"Che sapia mi, paesagio qua no ghe ne xé!"*/*"Che io sappia, di paesaggio qui non ce n'è!"*

[17] Thomas Sieverts, *Non Più Città, Non Più Campagna. Le Città "Intermedie"*, a conference held at the Fourteenth Course on the Management of the Landscape 'Nella Città Diffusa. Idee, Indagini, Proposte per la Nebulosa Insediativa Veneta', The Benetton Foundation for Study and Research, Treviso, September 2003.

This general lack of awareness regarding the landscape does not imply that there is no link between the people and the place where they live. The local inhabitants like Vigorovea; most would not live anywhere else. Above all, they love their homes, surrounded by gardens, and that is all they need to live well, resolve any problem of local identity or adapt to new locations.

As observed above, the residents were very much tied to their own microcosms, to their own micro-landscape, a place just like many others in the middle of the macro-landscape, to which much less attention is paid. As highlighted in the place names, the network of familiar micro-landscapes, which are linked more to people than to places, prevents the risk of disorientation and sense of displacement. The very local scale (behind fences and gates) and the wider scale, perceived as something not connected to themselves (Turco, 2003), have nothing to do with each other and are perceived as pertinent to limits which are quite separate from each other.

The diffused city, because of its disorder and lack of reference points, can be compared to the dark and obscure forest in the story of Little Red Riding Hood. Although the inhabitants do not feel 'lost' in this forest, the metaphor expresses the existence of a troubled and inconsistent relationship with the place in which they live and with its image.

In the light of the initial results of this research, one could possibly re-read some of the fundamental points presented in the European Landscape Convention (and in the Explanatory Report). If the document makes explicit reference to "landscapes of daily lives", which would include those of Vigorovea, sentences such as "the landscape is a key element of individual and social well-being" or "the quality of landscape has an important bearing on the success of economic and social initiatives, whether public or private", are not backed up by any immediate evidence in this particular corner of the diffused city.

Is it true that the inhabitants of Vigorovea are "no longer prepared to tolerate the alteration of their surroundings, which is the result of technical and economic decisions made without their involvement"? Do they really "wish to enjoy high quality landscapes and to play an active part in the development of landscapes"? On the contrary, it would seem that there is no particular interest in the quality of the landscape, nor that such quality is perceived as a right or a duty.

Vigorovea is only a first study case, but the questions arising from this research press for wider consideration. They cannot be left out when these topics are broached in the Venetian diffused city and, perhaps, more generally in the European ordinary landscapes.

From the European Landscape Convention, an impression filters through that there is a kind of inherent "wish for landscape", whose existence is taken for granted. However, this might be worth considering as being a goal to be achieved in itself.

In fact, if it is true that "the well-being of landscapes is closely linked to the level of public awareness", then the "poor condition" of the landscape of Vigorovea (and maybe of all the towns and villages of the diffused city) could be explained by the seemingly low level of awareness on the part of the local population.

Vigorovea, therefore, needs decisive and wide-ranging action, if this level of awareness is to be raised. This should start not from a superficial, ill-defined 'Veneto culture', but rather from appropriate courses of action designed to teach people "how to see", which would help them to understand the meaning of "the marks left by humans on the face of the Earth" (Turri, 1974: 15).

Section 2

Public spaces in the cityscape

Chapter 1
What role does plant landscape play in urban policy?

Nathalie BLANC, Marianne COHEN and Sandrine GLATRON

Can landscape become a major consideration in urban policy? This is the focus of the research that we carried out in Paris at three different sites, each with its distinct social components and morphology. This research, which involved a limited number of interviews, measurements and sites, is of an exploratory nature (Box 1). Our iterative approach does not pretend to be an analysis of a social system: instead, we have attempted, through the observation of different levels of analysis and by the comparison of these observations and different areas of knowledge[1], to propose new ways for addressing issues concerning the urban environment and how it is depicted. We used vegetation and landscape as our tools. The study of urban vegetation will make it possible to evaluate the contribution of plants to the urban landscape as well as the extent to which landscape can play a role in urban policy, whereas the analysis of environmental and urban policies aims at defining the role of landscape in legislation. This analysis is to be weighed against its role in the development of urban space. In this way, vegetation constitutes a communication hub between different dimensions of the urban space, from the social to the morphological and from the economic to the ecological, as well as between the different actors, from the private to the public sectors, and at different levels, from the micro to the macro. This research, which was undertaken with the aim of evaluating the place of the plant and pollution in urban landscape representations, was also a means for analysing ordinary practices that contribute to the development of this landscape. This research made it possible for us to look at previously unexplored approaches to ecological requalification and recomposition of a particular urban landscape, the city of Paris. These observations are part of an attempt to evaluate the impact of the city on ecosystems (Sauvez, 2001), as well as on the future quality of urban life , for an ever-growing population.

---

[1] This team brought together human geographers, biogeographers, physicists and architects studying the place and the role of the urban landscape in joint perceptions as well as in policies.

Box 1. Key criteria used in the study.

Study sites

Site	Location	Urban morphology	Social context	Biophysical characteristics
Lagrange	5th arrondissement	Haussmannian-style avenues in a medieval neighbourhood	Limited except for the 'Lagrange Pure Air' Association	Banks of the Seine; little vegetation; major car traffic within the site
Peupliers	13th arrondissement	Former working-class homes converted into 'townhouses'	Considerable: Peupliers Association; community choir	Small individual gardens; car traffic on the periphery of the site
Pinel	13th arrondissement	Group housing from the 1970s	Long-standing renters; role of the square as a meeting place	Community garden; car traffic on the periphery of the site

Interviews with the inhabitants
In the social survey, we inquired about the role of landscape, pollution and vegetation in the different types of housing, through open interviews. On the basis of these questions, our survey focused on the relationship to urban materiality and sensory means – sights, smells, sounds, etc. – of perceiving the city. Our questions were directed, thus, on the opposition between vegetation and pollution and the question of the landscape. Interviews carried out represented sometimes a considerable proportion of the areas studied (approximately 100 interviews).
Moreover, another type of survey was carried out on a dozen people, identified at the time of the botanical survey as being particularly involved in plant management: custodians and caretakers, inhabitants working in a community garden, and city gardeners. These surveys are considered to be of a dual nature since they cover several domains and tie the botanical survey to questions about practices and representations concerning these plants.

Collection and treatment of floristic data in the city
The floristic survey was carried out on the cadastral map of the three sites: 475 taxa (gymnosperms, angiosperms, pteridophytes, bryophytes) were found, barely 30 of which are spontaneous indigenous species, a minute part of the 769 spontaneous species of the same groups observed in Paris (Moret, 2004). The number of units of each taxon was counted per 'project space unit' (PSU), in other words, areas ranging from one to several square metres (e.g. the balcony of an apartment building, a street planter, a cluster in a square, a total of 296 PSU for the three sites). In public spaces, vegetation was identified by variety (coll., A. Girard), whereas in private spaces, it was identified by the species or genus because of the observation distance.

These data were subjected to two treatments.
*The calculation of Shannon's index (Blondel, 1995) was carried out on the PSU where plants were observed. The index varies from 0 (one species found) to 5 (many species found in balanced frequency).

The Correspondence Analysis was applied to samples with fewer than three species and to species encountered fewer than three times. This selection made it possible to reduce the dispersion of information. Treatments were carried out separately for public (36 samples and 50 taxa) and private (78 samples and 41 taxa) spaces.

The plant as a central element of urban practice

The plant is clearly at the crossroads of many different issues within the greater Paris area. It is a tool for professionals in development and urbanism. It brings 'nature' to inhabitants lacking it in their local environment. It is a component of the urban landscape. Finally, it is a resource for biodiversity at the ecological level. This complex issue in Paris, and undoubtedly in other French and European population centres, makes it a major research focus and valuable tool for the democratization of intervention in the Paris area in which different stakeholders with divergent interests can each play a role. They are, therefore, invited to make a real contribution to urban space. It is with this in mind that we attempted to develop a general typology of the Parisian landscape, with the plant as the starting point. This typology takes into account the distribution of vegetation, its floristic diversity and, above all, the contribution of stakeholders, such as gardeners, landscape architects and ordinary citizens.

The introduction of plants, at the initiative of institutions (municipalities or landowners) is considered to be within the framework of urban development (Stefulesco, 1993) and part of urban public space. On the other hand, spontaneous vegetation enhances the structured aspect of other types of vegetation. The recent recognition of this phenomenon reveals the reversal of a number of preconceived ideas, leaving room for 'natural' processes to participate in the construction of public space (Lizet *et al.*, 1997). If we refer to the first type of vegetation as being 'institutional', the second is clearly wild. The natural gardens that have become popular in some municipalities today are hybrids; they are the result of a semi-spontaneous plant dynamic and an institutional project, motivated by different considerations (e.g. ecological correctness, concern for economic management, institutionalization and social control of leisure areas for inhabitants). In addition to these two categories, we can add the vegetation introduced by the inhabitants on their balconies or windowsills, visible from the public space and referred to as 'proximity vegetation'.

The interpretation of this landscape makes the city appear as a mosaic of social and biophysical environments – produced by, and producers of – a set of institutions (i.e. work, leisure and business), as well as many individual actions. We can distinguish areas within the urban space, depending on how they have integrated nature: those devoted to social interaction and, therefore, group production; 'exhibition' sites, in the 'Modern Movement' tradition in architecture; and, 'discreet' spaces resulting from the desire of individuals to become part of the public space and to make a real contribution to the city. Once again, the latter can be interpreted as the desire to make one's space public, the desire for recognition.

Institutions responsible for plant management in Paris

In Paris, the importance of the 'Haussmannian' heritage in terms of plants and planting directives is a major obstacle to any new development. Even today, all city planning must respect this tradition. This is the case for the landscaping of the Avenue d'Italie which aims at re-establishing its 'Haussmanian character' by planting trees along the boulevard. The city gardeners adhere to pre-existing models and designs, linked to standards prevalent in garden competitions, for example, gardens laid out by colour theme, or a rock garden designed in the 19th century tradition. The Square Viviani, at the intersection of several major roadways and across from the Cathedral of Notre Dame, is a good example: it is always in flower and bordered by rare species, within an historical perspective that must be maintained, for its historical as well as its touristy value.

Gardeners choose plants on the basis of their ornamental value from a catalogue that does not take their ecological requirements into consideration, or by replacing plants with the same species (e.g. after the storm of 1999). The biophysical factors of the environment are overlooked, for example, dryness of the soil, effects of shading and masking by buildings, sensitivity to pollution, etc. Occasionally, renewal takes place within an urban development operation, as was the case of the Square Mesureur created on the site of the former Cité Dorée, in the 13th arrondissement. The time of planting and the historical significance can be determined by the floral composition of plant clusters. Trends play an important role in the choice of plants: the Mexican orange tree (*Choisya ternata*) has been a popular choice since the 1990s. As for public housing, we can observe that the major group housing operations of the 1970s resulted in the green spaces of the modern movement. Closely mown lawns and the *Cotoneaster-Lonicera-Pyracantha* trio form the signature plantscapes of this period of public housing. This choice is not innocuous pollution accumulates on the leaves of the evergreens, the seasons, markers of biophysical status, are indistinguishable, and flowering and fruiting are often absent, limiting interactions with the avifauna.

This first typology is verified by the factorial analysis of floristic surveys (carried out on clusters in green spaces, planters in public spaces and gardens in communal areas at our three study sites (Box 2.1). The factorial analysis of the vegetation of public spaces (clusters in a square, planters in the street, groups and rows of trees) gives acceptable results, given the wide dispersion of floristic data (percentage of variance described by the first four axes: 33%; canonical correlation of axis 1: 0.93). The most characteristic clusters are those where just a few species are found in abundance, in fashion of a formal garden.

Axis 1 thus includes two types of clusters from formal French-type gardens, one with its hundreds of seasonal shade-intolerant flowers and rare species, and its equivalent with shade-tolerant plants. Axis 2 describes the green garden of the 1970s with its repetitions of evergreen shrubs.

On the other hand, garden clusters that do not fit this criterion are less well described. This is the case for clusters enhanced by the 'unorthodox practices' of city gardeners and inhabitants (closer to origin on the positive pole of axis 2, Figure 2.1), including: the purchase of plants by the inhabitants who then plant them in an immediate public space, either independently or with the help of the building custodian or city gardener. These acts, appropriation factors, are pretexts for the for developing new social contacts and provide the impetus for group interactions. Other practices include: the management of biological stock by gardeners, instead of ordering new plants and discarding the old ones, they recover and exchange plants in the different gardens for which they are responsible, the exchange of city plants/country plants, and the tolerance (mandatory) of gardeners for weeds[2].

This is also the case for interior gardens in private courtyards in the 5th arrondissement (located at the origin of the factorial map, Figure 2.1). Here we find a profusion of plants grown because of the importance placed on the living environment by the inhabitants. Abandoned or discarded plants, were recovered and planted in a courtyard garden by the building custodian, increasing the floristic diversity. Also found was the propagation by an inhabitant of species introduced by a landscape gardener (Camille Muller).

In the same vein as these 'unorthodox practices', the Paris-Nature Service, which includes scientists trained in the latest techniques of biological pest control or sensitivity to nature in the city, is preparing what could be a new concept of nature in the city through vegetation surveys and the development of natural and wild gardens. Thus, the 'natural garden' near the Père-Lachaise cemetery or the "'wild garden' of the 18th arrondissement of Paris transports visitors to the countryside, at least for the length of their visit. For these gardeners, it is a question of controlling spontaneous dynamics (which, could result in invasion by plants such as ivy from the 'wild garden' which must be cut back every year), or in the creation, from an educational point of view, of a diversity of environments (e.g. the natural garden, the pond in the wild garden). The species diversity of these gardens, although largely induced by man, is extraordinary with regard to their area: wild garden 156 species; natural garden 428 (City of Paris, 2004).

'Gardener' inhabitants

Urban morphology interferes with the way in which inhabitants decide whether or not to include plants in the space to which they have access. Thus, within the Lagrange site (Box 2.1), across from the Cathedral of Notre Dame, a few flowered balconies and courtyards are the exception in a space that ordinarily has little vegetation; the inhabitants explain the rare occurrence of plants to the effects of the wind or pigeon droppings (only about 20% of the buildings' storeys have plants). The central location of the neighbourhood also seems to exclude 'nature' for the inhabitants who feel that nature is confined to the 'countryside'. The high pollution level is an additional argument for not growing plants and for limiting contacts with the outdoors. However, the comparison of floristic surveys from 2000 and 2002 show a trend towards introducing vegetation.

Unlike Lagrange, gardening is generalized and practised by many in the Square des Peupliers (Table 2.1), (all the inhabitants have at least one outdoor plant), and may even form the basis for a social space, like the Place Pinel (nearly half of the buildings' storeys have plants).

At our three study sites, the vegetation introduced by the inhabitants can be broken down into three categories. Multivariate analyses of floristic surveys made on window-sills, balconies and pavements confirm this fact (Figure 2.2, Box 2.1). However, the

---

[2] The use of pesticides is reserved for specialized services, as is other heavy-duty work such as the trimming of large trees.

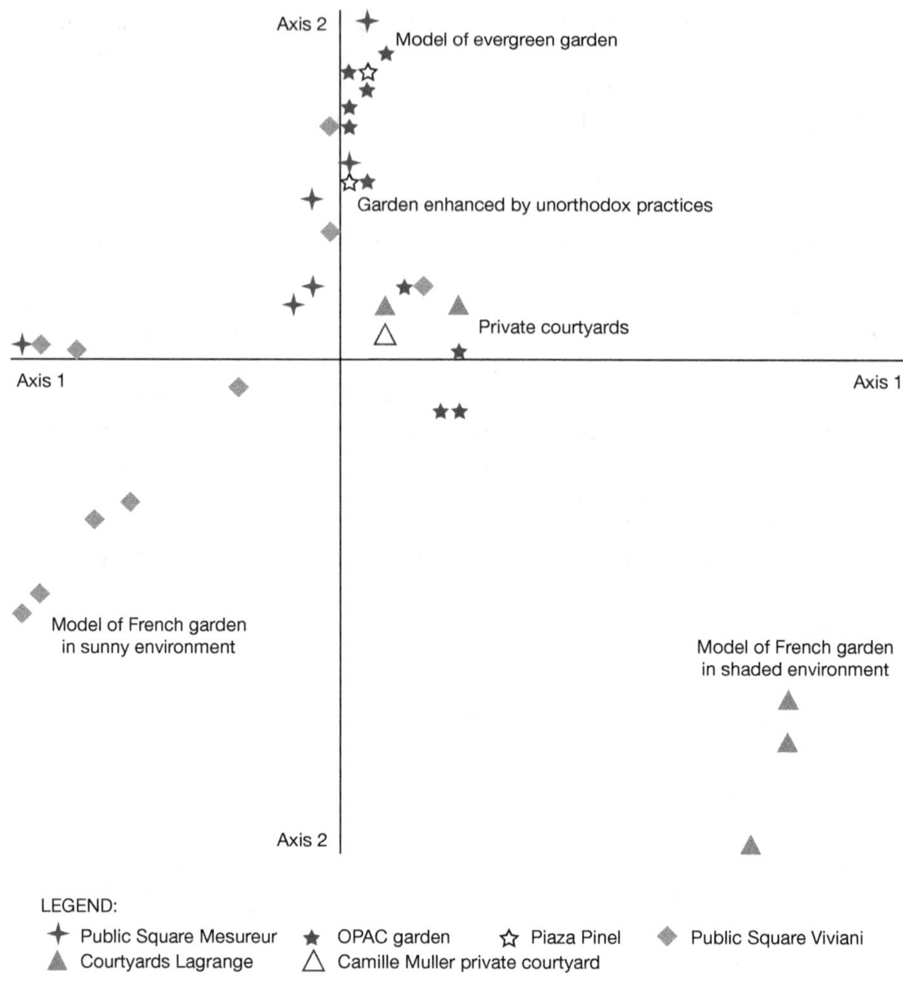

Figure 2.1. Factorial analysis (axes 1–2) flora of collective spaces.

statistical results are less satisfactory than the ones described above because the dispersion of information is greater (percentage of variance described by the first four axes: 21%; canonical correlation of axis 1: 0.80). As in the previous case, the first axis of the correspondence analysis describes types of vegetation with repetitions of species, where the inhabitants reproduce on their local scale, the cultural model of a formal garden.

The first model corresponds to the vegetation installed by shopkeepers who, in the tourist section of the Rue Lagrange, mark the entrances to their shops, sometimes even creating terraces, with Mediterranean evergreen shrubs (e.g. laurustinus or Japanese pittosporum) or the traditional thuya, forming an 'unchangeable' decor (negative pole of axis 1, Figure 2.2). The inhabitants of this central neighbourhood choose other plants, which define the two other models: the flowered balcony with repetitions of seasonal species (e.g. geraniums, petunias and snapdragons; positive pole of axis 1, Figure 2.2)

Table 2.1. Differences between representations based on knowledge and ordinary representations.

	Original criterion	Representation	Formulation	Practices
Scientific vegetation	Plant composition (horticultural and spontaneous species) Floristic diversity Landscape diversity	Social practices of gardening and cultural models Vegetation dynamics (spontaneous)	Statistical or cartographic treatments	Clarification Models Maps
Ordinary vegetation	Plants Colour	Nature Garden of Eden Disorder, jungle	Positive ("I love that!") or negative (weeds, invasion)	To garden or not to garden

and the urban terrace of Mediterranean shrubs, topiary plants or popular exotic plants (e.g. oleander, olive tree and *Dipladenia*; positive pole of axis 2, Figure 2.2). Around the Place Pinel, the balconies of the public housing project of the OPAC (Public Agency for Development and Construction) are not really put to good use for plants. In this case, the choice of plants corresponds to the conventional model of the 'flowered balcony' (Figure 2.2). The vegetation here acts as decor, in the same way as that of the OPAC garden.

In contrast, the Square des Peupliers, with its little houses and small front yards, gives an impression of the 'countryside in Paris'. The inhabitants that have been there for a long time oversee (sometimes with difficulty) the vegetation that is already there. Vegetation is both a common property and a protection against the promiscuity of neighbours (and tourists), linked to the community aspect of living of the square[3]. On the other hand, recent inhabitants of these very expensive 'townhouses' are actively involved in gardening. They do this, not because of community feeling, but plant for their own sense of well-being, as well as for the decorative display. These dynamics can be seen in the multivariate analyses. Traditional yards are a luxuriant mixture of conventional horticultural species, for example, shrubs (Japanese kerria, forsythia, pyracantha, roses, etc.), climbers (wisteria, Virginia creeper, broad-leafed ivy, etc.), and spontaneous species. Located at the centre of the factorial map 1–2 (Figure 2.2), they are described by axis 3 of the analysis (not shown). Other yards with the addition of popular Mediterranean plants are linked to the 'urban terrace' model (positive pole of axis 2, Figure 2.2).

From the point of view of the floristic diversity expressed by Shannon's index (Box 2.1), we can observe an analogy between public space and private space: biodiversity is generally low in the private space of the Lagrange neighbourhood and the ILM (low-rent public apartment buildings managed by OPAC), which is true of the group space as well (Figure 2.3). However, we can observe several elaborately flowered courtyards and balconies in the private space of the Lagrange neighbourhood (expensive model of the 'urban terrace'), whose population is more affluent than that of the Pinel neighbourhood. In the Square des Peupliers, morphology leads to a floristic diversification but the role of the inhabitants, who indulge in varying degrees of gardening and tolerate weeds to varying extents, explains the highly contrasted situations observed.

---

[3] Including a shared concern for children, celebrations in the square, the choir as well as the Peupliers Association (which extends outside the square).

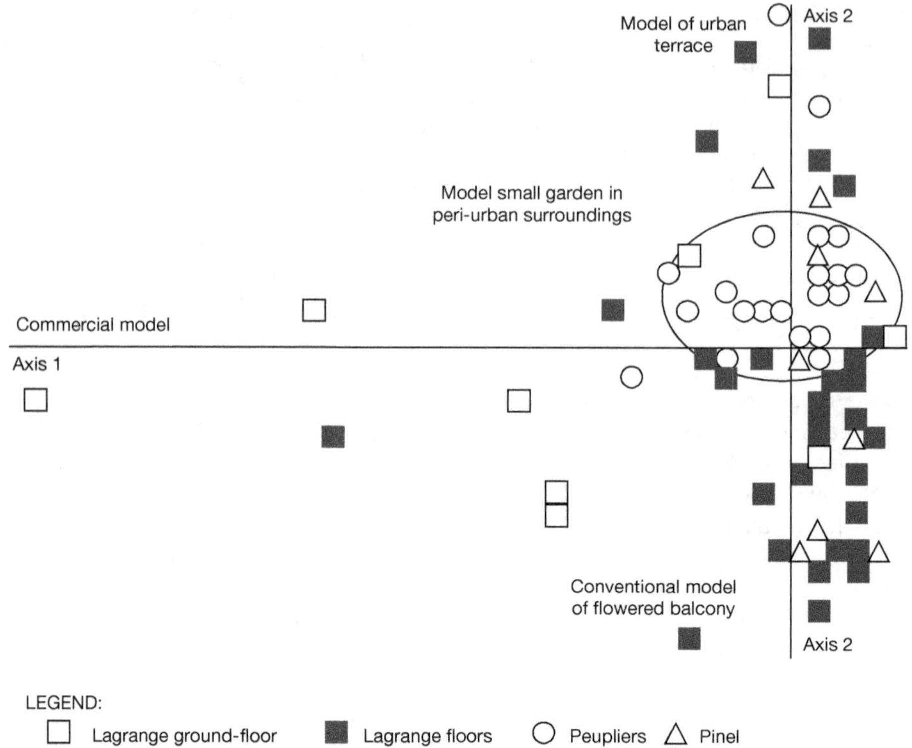

Figure 2.2. Factorial analysis (axes 1-2) flora of private spaces.

This index expresses the diversity of mainly introduced species and, therefore, the role of the inhabitants, gardeners and managers of these spaces in relation, on the one hand, to their means (technical, economic, etc.) and, on the other, to the idea that they have of the city and the garden (e.g. preference for a formal garden as opposed to a Mediterranean garden).

However, the functional role and historical interest of these introduced species is poorly understood. Spontaneous species represent less than 10% of the species surveyed. Only a few introduced species (less than 10) spontaneously multiply at the three case-study sites. As a result, the significance of the results at the ecological level is limited, even though the combination of composite species of different heights is more conducive to attracting wild fauna than the monospecific plantscapes of seasonal species. In this case, these "new ecological communities" (Kinzig and Grove, 2001) are specific to the city, dense and largely created by man. This approach is obviously exploratory because it is limited to a single scale that of small-size space units (in the order of a square metre). It can only be understood from an extended sample, a regional scale and the verification of the ecological functionalities of plant species.

Several conclusions can be drawn from these multidisciplinary observations. The plant that participates in the urban landscape is the product of many important stakeholders

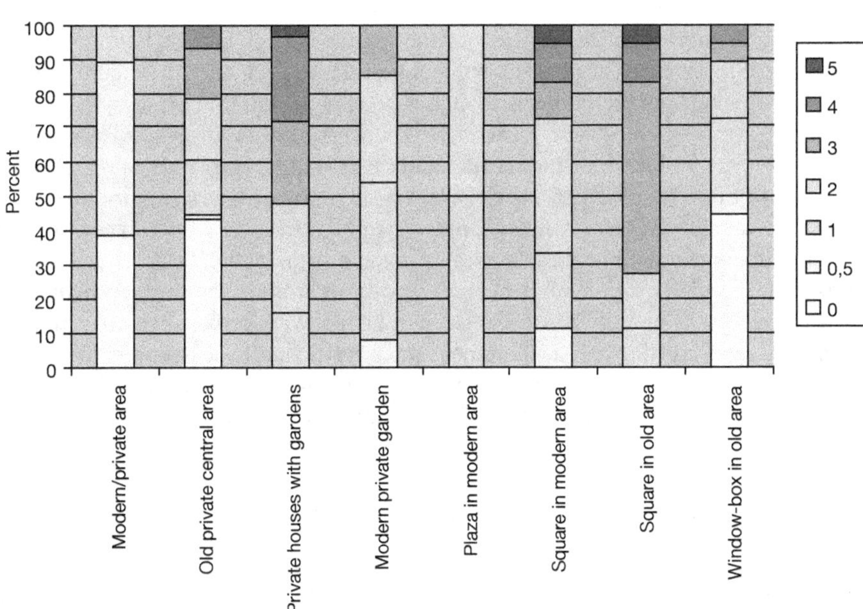

Figure 2.3. Biodiversity index frequency in private and public spaces.

including professionals. It can, therefore, become a consideration in the negotiated and shared urban landscape.

With this in mind, how do we see the future of the urban landscape? How is it considered by the legislator as well as by the inhabitants? What is the importance of a plant or a building? How can we give the landscape the respect that it is due, in other words, the possibility of it becoming a common reference frame, a tool for dialogue and the negotiated production of the city? The question of landscape allows us to take urban reality into consideration in its full complexity, e.g. social, aesthetic, etc. However, as is obvious in the legislation, landscape is often reduced to its morphological, nostalgic and outdated aspects in relation to the city; it is slow to assimilate today's challenges. Urban development today is synonymous with urban spread that creates hybrid territories, neither urban nor rural. Moreover, both on the periphery as well as within the city, private environments proliferate at the expense of free public spaces (Mangin, 2005). Within this context, we feel that it is important to make the plant and collective construction of the urban landscape one of the underlying principles for the development of a new public space, aesthetic and ecological at the same time, bringing together the quality of urban life, the inhabitants and stakeholders from the private sector, as well as professionals and local elected officials, in the cause of shared values and common interest. The retrospective analysis of urban landscape policies that follows shows that this objective is at odds with the tradition of intervention of French legislation.

Landscape policy: circumventing the city

Our research shows that the city is not the main focus of French landscape policy. Generally speaking, experience has shown that that the term 'landscape' mainly refers to the rural environment, to the point that when we speak of 'urban' landscape with regard to the city, it is considered to be a secondary concern. In the absence of a real urban landscape policy, we looked for the terms 'landscape' and 'urban landscape' in the legal corpus[4], where these terms could possibly be found, in other words, environmental law and urbanism considered to be indicative of public policy. We were thus able to show that urban landscape has a double and sometimes, even a triple context. Urban landscape is first linked to the idea of protection of rural territories from urban development associated with destructive modernity; it is also associated with 'urban beautification' and the necessary renewal of urban development today; and finally, it is linked to questions of urban identity. However, in each of these cases, urban space is only an exception in landscape policies, which mainly focus on the rural environment where they find their visibility and their rationale.

The landscape as an aspect of environmental policy

Within the framework of environmental law, the 'urban landscape' (or the idea of landscape applied to the city) is mainly the extension of a category that originally designated spaces to be protected, on the one hand, and rural territories, on the other, and still does today.

In the category of territories to be protected, landscape finds it roots at the beginning of the 20th century with site protection measures based on those devoted to monuments and sites (the law of 1906 on picturesque sites, the law of 1930 on natural monuments and sites, and the law of 1943 on the protection of the surroundings of historical monuments).

However, the development of landscape as such in public policy is fairly recent and subsequent to the 1970s. This issue emerged at the same time as a growing awareness among the population and public authorities of 'environment problems'. The focus of the law of 10 July 1976 related to environmental protection. In 1977, the law of 3 January on architecture specifies that "the respect of natural and urban landscapes as well as of heritage are of public interest". The framework remains that for the protection of essentially rural territories sometimes considered as the environment.

After a profusion of texts at the end of the 1970s, it was not until 1993, with the law of 8 January on the landscape, that urban landscape was explicitly mentioned. This law, however, is based on an evolution centred around protection of the environment. It is the

---

[4] Out of 57 codes including the *Code Général des Impôts* (Tax Code), only five use the term 'landscape' and only three on a regular basis: the *Code de l'Urbanisme* (Urbanism Code), linked to a set of practices, refers to urbanism 46 times (but we observed that references to the landscape are generally related to corrections and additions to the original texts rather than actual and distinct contributions), the *Code Rural* (Rural Code), linked to a type of territory, and the *Code de l'Environnement* (Environment Code), includes 40 references. From there, the study of the place of the landscape in national urbanism and environmental policies was extended. These codes were chosen, the first because it refers to the management of heritage and natural species, and the second because it refers to the city.

interpretation at the legislative level of the action plan in favour of landscape protection and restoration of September 1992, based on the National Environment Plan (September 1990). Moreover, it contributes to incorporating the idea of landscape as a basic element of environmental policies into different public and private activities. From the point of view of means and tools, the law proposes extending to the landscape, the possibilities of action proposed by the ZPPAU (Zone for the Protection of Architectural and Urban Heritage) that has existed since 1983 and to which the adjective 'landscape' was thus added[5], encouraging landscape to be taken into consideration in land management plans. Finally, the law institutes development and protection measures for landscapes that mainly concern "territories of particular landscape value", for which landscape elements must be defined. In order to do this, different texts used as a basis for the enforcement of the 'landscape law'[6], attempt to define them. Two major ideas become apparent: that of the actual landscape, on the one hand, and that of how it is perceived, on the other. We must, therefore, protect a territorial entity that gives the impression of unity and coherence.

The Barnier Law No 95-101, of 2 February 1995, related to the enforcement of environmental protection, reaffirms the concepts of landscape laid out above. Thus, the landscape is a category of space defined as deserving protection within the framework of the social and legal recognition of an environmental issue, in the same way as environments, natural spaces and sites. Just like the 'site', landscape category has a strong aesthetic connotation. It is linked to a historical context, as opposed to the terms 'environment' and 'natural spaces', which also express a link to a historical context, but which have a more naturalistic connotation, even if the biophysical components of the landscape are no longer visible. The two terms, 'environment' and 'natural space', also evoke the idea of natural processes and ecology, just like the concepts of 'biodiversity' and 'biological balance'. The city is always marginal and it was only with the memorandum of March 1995[7] that the city was directly faced with an official text concerning the landscape that dealt with the issues of alignments and access to the city.

Finally, this first body of legislation shows the development of defensive approaches in relation to urban development to the extent that the terms 'landscape' and 'city' are often considered paradoxical. Even in the case of landscape measures at city access roads or along roadways, the landscape is a means to hide urban ugliness – urbanization that is almost considered a disease. "Modern landscape forms" are seen to be without identity, indistinguishable or interchangeable. Among these modern forms, the major transportation and energy networks are particularly targeted. Jacqueline Morand-Deviller (1994) observes that "our heritage is threatened by the outrages that manufacturers and engineers inflict upon us". In fact, this vision of the landscape is also a conservative vision of the world. It is true that the attraction of the rural landscape has always existed in France and is part of the traditional opposition between the city and countryside. Beyond being categories that are totally opposed to each other, the city is synonymous with built-up space, whereas the countryside is a world of fields and woods; the connotation of the countryside as a healthy environment because of its proximity to nature has been around

---

[5] In 2003, 63 ZPPAUP ((Zone for the Protection of Architectural, Urban and Landscape Heritage) existed. See *infra* on ZPPAUP.

[6] Decree of 11 April 1994, Memorandum 94-283 of 21 November 1994.

[7] Memorandum 95-24 of 21 March 1995, on landscape contracts.

for a long time and comes back in vogue periodically. The city and the countryside are opposed to each other in collective representations and from a preconceived idea of economic forces: the city, the depository of capital, exploits the countryside. The rejection of the city from the aesthetic point of view, linked to the fact that it is not qualified as a landscape, corresponds to the importance of rural France and the disqualification of the city, reducing it to a mere source of leisure activities, work and socializing. Nevertheless, this opposition dims in the new vision of the city: the nature/city where buildings are closely linked to plant spaces (Chalas, 2005).

The urban landscape: beautification or urban art

Environmental law only grants a very small place to the city as landscape; instead, the city is an anti-landscape. What role does urban law reserve for it?

In the body of environmental legislation, the urban landscape is particularly linked to the question of urban aesthetics. During the 1960s, several measures concerning urban landscapes were passed. They mainly dealt with isolated elements and focused on the control of building permits[8], or had a wider scope when they concerned the neighbourhood, with the law of 4 August 1962, known as the Malraux Law that instituted the procedure for the preservation of historical neighbourhoods. Other laws with the aim of preserving historical heritage were passed during the 1980s; the law of 1983 related to the distribution of know-how between communities, departments, regions and the government "gives communities the possibility of proposing protected zones (ZPPAU)". These laws instituted the principle of protection for elements ranging from the isolated to a group of elements, integrating the idea of the urban landscape, which was already the aim of the Malraux Law, contrary to the landscape law that considers the terms 'urban' and 'landscape' separately with their own specificities.

Within the body of legislation related to urbanism, as in that concerning the environment, landscape issues critical of socio-technical forms of progress are diametrically opposed to urban sprawl and haphazard building practices. This evolution goes hand-in-hand with the emergence and, even the development of questions, revolving around the meaning of the city and even the idea of the city. This recognition of the morpho-sensitive and identity aspects of the city is in keeping with the issues concerning the beginnings of urbanism, such as those of 'urban art' or 'urban beautification'[9] (Gaudin, 1991). Urban planning thus took on a new form and was no longer strictly functionalist as it had been since the end of the Second World War.

The idea that urban growth is harmful, contrary to harmonious development and an awareness of the aesthetic value of land development, is evident in these laws. It is

---

[8] Decree 58-1467 of 31 December 1958 (article R111-21 of the Urbanism Code) [introduces] the infringement of "natural and urban landscapes" as a legal justification for refusing a construction permit.

[9] See the Cornudet Law of 31 March 1919, that requires all cities with more than 10,000 inhabitants to have an "extension, development and beautification plan". This general measure is part of a set of laws that had accumulated by the end of the war. It is an extension of a debate begun in 1909 and other laws that aim at efficiency: the law of 1917 that establishes stricter rules for the location of unhealthy establishments and, at the same time, lays the foundation for elementary zoning; the law of November 1918, that takes up the traditional arguments of Haussmanization and, therefore, establishes expropriation by zone and imposes the principle of the recovery of increased property value.

necessary to control this ex-urbanization; the landscape is a key element in the control of urban planning, linking the urban and the rural, economic development and land protection. Many measures related to urban sprawl, to the transition between the rural and the urban, to the insertion of the urban into the 'rural'[10], bear witness to this evolution. This trend is part of the extension of the 'Environment Code' where "landscape" designates a part of the land to be protected. Observation of the city from the strictly visual point of view is a means for developing an aesthetic approach to land development. Close to the idea of identity, the idea of the meaning of the city becomes more and more evident. Notwithstanding, this idea is often linked to a particular urban morphology. It focuses on a rather nostalgic vision of urban neighbourhoods.

Despite the limitations attributed to the meaning of urban landscape, as described above, and in a broader sense, despite the small place given to urban landscape policies, we can still not ignore the importance of the 'urban landscape'. First of all, it is often referred to by politicians as well as certain city managers. It is likely that the difficulties encountered by public authorities in resolving the urban crisis or just in simply keeping up with urban developments, and the trend towards a progressive but visible privatization, force them to resort to the landscape as a means of organizing space at the level of a city or an urban region. Finally, landscape architects use this expression more and more often; on the one hand, they no longer restrict themselves to the creation of garden or park spaces, regardless of how structured they are: ordinary urban space as well as rural space constitutes a new space for exercising their talents, a new professional market. On the other hand, urban environment issues are becoming increasingly important and landscape architects appear to be more and more interested in the reconciliation of city dwellers with the city and nature, symbolized here by the introduction of plants into the urban landscape (Donadieu, 1999). At least at these two levels, the development of public authority and the construction of the city as a collective ecological property, the landscape issue is a resource for urban policy.

The plantscape, a tool for ecological dialogue

If the urban landscape is not limited to its morphological acceptance purely aesthetic or simply decorative – it can be instrumental in favouring a wide range of interventions: city planners and city dwellers making the most of patrimonial or aesthetic resources, as well as inhabitants concerned with their environment and its sensorial wealth, more sensitive in the end to the quality of their immediate environment than to the composition of an urban landscape. It is in these terms and, to the extent that the plant can be a tool leading to multiple re-appropriations of urban space as well as a planning tool at the service of landscape development, that we can consider its importance for the construction of a shared urban landscape (Sirianni and Friedland, 2001). This sharing seems to be a means of rehabilitating the social bonds in the city and, more simply, neighbourhoods in relation to their morphological and social components. In any case, it appears

---

[10] For Bernard Lassus, the author of the report to the National Landscape Council (Conseil National du Paysage) in 1991, the landscape allows us to conceive of projects that integrate the rural and the urban, from the most to the least inhabited.

to us to be a means of developing forms for the appropriation of urban space with the aim of improving its attractiveness. We must focus as much on the representations that disqualify the city as on the urban living environment, its morphological and mesological components, as well as its social and democratic ones. The ecological requalification of the city must go hand-in-hand with the rehabilitation of its living environments. This must be considered a central issue if we hope to create a 'sustainable urban landscape' in the future.

More generally speaking, these observations are in keeping with the development of land management projects started in the 1970s with the rejection of technocratic ideas of spatial intervention. Nevertheless, we can observe the difficulties involved in taking account of social demand in the 1980s. Approaches based on dialogue and the establishment of shared objectives were developed within the framework of urban planning during the 1990s, leading to the creation of a new public debate. Within the radical formulations of these approaches, we credit all the participants – and not just the experts – with the know-how to define the objectives and, as a result, to establish links between knowledge of a heterogeneous nature (e.g. ordinary knowledge, expert knowledge, procedural knowledge, etc.) and action. The importance given here to the urban landscape contributes to this professional evolution in a unique way. The multidimensionality of the landscape – symbolic, shared, professional, aesthetic, ecological and economic – reveals its importance in going beyond the context of strict technical expertise. In addition to environmental issues, the urban landscape raises the question of singular and shared language, like that of the tools, to describe, understand and take action within the urban framework (Pousin, 2000). It is, therefore, a precious tool for the democratic construction of new public spaces contributing to the ecological quality of the city.

Scientific and political challenges

In this sense, one of our major objectives is to demonstrate the links between an ecological issue in terms of biodiversity and a political and individual issue: the social link and well-being. These links are that much more important because none of the stakeholders incorporate them into their own practices. Institutional stakeholders responsible for urban ecology have difficulties in taking the biodiversity introduced by the inhabitants into account (and more generally, domestic biodiversity, that of horticultural plants), whereas the latter never refer to the ecological quality of their city to justify their need for a garden. Nevertheless, this relationship revealed by an interdisciplinary analysis is fundamental in terms of urban policy and landscape. In terms of urban policy, it makes it possible to underline several approaches to what could be an urban ecology policy at the city level. It illustrates a general issue, that of biodiversity (Parisian in our case), and makes it possible to define one of the milestones: the inhabitants and their desire for a garden. The other milestone is political and favours wildlife in the city. The two are far from being incompatible: the 'gardener' inhabitants do not necessarily destroy spontaneous plant species and value some wild animal species. From this point of view, the resulting landscape does not represent a political action in conflict with the desires of the inhabitants. At the symbolic as well as at the real level, it can become a means of action within its own environment and a frame of reference for a community. In this case, the plant is a tool for landscape action at different levels: the individual and the balcony, the neighbourhood and the city in general. However, the relationship between levels, from the political point of view, in terms of land management as well as from the ecological point of view, is a central issue and made possible by landscape management. No plot of land can pretend to be closed today; all land, no matter how small it may be, is part of an open horizon.

The necessity for a dialogue between political and scientific issues is even greater today because of the number of theories that advocate a separation between scientific knowledge and lay knowledge. Our analysis shows that there are many levels of discussion on the environment and that they do not all follow the same logic. The question is not one of finding links between these different representations as some would wish us to believe, transforming each aspect into a specialized category of public intervention, but instead concerns a 'practical' recognition of spaces created by the action of the different categories of contributors (urban, in particular) to conceive a common intervention framework. This mutual recognition of the logic behind actions may make it possible to by-pass divisions due to traditional categories of specialization. Our results mainly show that the relationship of city dwellers to the landscape is strongly linked to the choice of means of locomotion, i.e. walking, using a car or taking public transportation. Even if it is mainly outstanding historical and tourist sites that are designated as landscapes, we are aware that the use of the term especially depends on the inhabitant's attitude toward the city – whether he or she considers it as just a place to live, or a resource in terms of services and sociability. The landscape is, therefore, not just an object. It is a perspective, a critical distance. It is a way of claiming the city for oneself.

A comparison of the representations of vegetation by the scientists on our team and the interviewees confirm this analysis. Ordinary representations of vegetation involve a daily practice of the city, whereas the scientific approach consists of objectifying and formalising site inventories, such as they present themselves to the observer, even if the latter has been careful to choose a set of observations compatible with the inhabitant (cadastral plan, itinerary).

A partnership between public action and private action

Finally, our results illustrate several possible experimental approaches to the public/ private partnership in the production of a city, renewed by the recognition of its natural dimension, which would contribute to the modification of the standard procedures for public intervention (Wiel, 2003). It could be a matter of developing the enthusiasm of the inhabitants for gardening in order to develop an ecological awareness and encourage a group reflection on the utilization of gardens as public spaces and as reserves of biodiversity. The implementation of the latter could be the aim of negotiations involving many different representations of the city. It could be applied in the case of gardens (family-style or otherwise) as well as in large parks or in urban wastelands that would change status from 'abandoned' to 'appropriated'. The institutions would have an orientation role to play, to advise the inhabitants on the most interesting horticultural species from the ecological point of view (e.g. those that interact particularly well with the fauna) as well as to appropriate practices (e.g. tolerance for spontaneous species, low consumption of water and pesticides, waste management, etc.). This proposal, which aims at the co-construction of garden spaces, is based on our observations on the involvement of certain inhabitants with regard to revegetation, and on its ecological effects leading to floristic diversification. However, our research also showed that not all city dwellers are receptive to this natural dimension, in other words, 'sensitive' to the city. These partnerships could also be extended to the private sector since the increasing use of plants in commercial and private business spaces demonstrates their possible contribution to the socio-ecosystem of the city.

From this point of view, the plant is a consensual material for the development of new urban spaces. At the symbolic level, it is supposed to protect against pollution and contribute to a vision of a paradisiacal garden, as a place of reconciliation between the different forms of life. At the ecological level, it constitutes an environment capable of being home to different species of animals, but does not always protect against pollution. Finally, it contributes to interpenetration between the city and countryside, another way to deal with urban sprawl. At the social and political level, it is tempting to speculate that the development of gardening activities is a consequence of greater economic wealth, as opposed to gardening to supplement diet, as observed in disadvantaged urban populations, i.e. urban vegetable gardens in Russia and Central Europe (Luginbuhl, 2003).

The participation of the population in the management of urban spaces also raises the problem of the role of land management specialists. Should their know-how, the result of professional experience, be accompanied by a negotiation role, in other words, the initiation of collective projects in which the inhabitants would participate?

Conclusion

The challenge represented today by sustainable urban development policies and those dealing with the improvement of the living environment have led us to explore different avenues. The involvement of the inhabitants in the sites and the living environment, and environmental and local development, are two of the many possibilities. We can thus focus on the issue of direct democracy; to do this, we must explore the spatial and biophysical bases. This involvement of the inhabitants as a condition for the material and symbolic redevelopment of the living environments and the city as a whole, is not the only approach to this endeavour. Urban ecological development forces us to think in terms of multiscale relationships: the role of the plant at the local level is important at the global level. The world is made up of localities. On the other hand, the local does not exist without a frame of reference. The earth is just one of them. Another point: the valorisation of biophysical interdependencies and links between living beings cannot take place unless there is a scientific representation of these established facts. The interdisciplinarity addressed in this paper attempts to contribute to this, although it is a very complex issue, since some of the categories involved, biodiversity, for example, must be reanalysed in the urban context. Thus, the consideration of biodiversity as an indicator of the effective operation of ecosystems (Levêque, 2001), determined by the richness in spontaneous species (Moret, 2004), is difficult to transpose to the city. Biodiversity in this case is a social construction, a function of the practices and the representation of the inhabitants, and it is closely linked to the presence of introduced species; whether these mixtures of spontaneous and introduced species can function in "new ecological communities" (Kunzig and Grove, 2001.) remains to be seen. Innovation and extension beyond nature/culture boundaries, in both scientific and lay terms, lead, by means of an exploratory approach and constant re-evaluation and updating, to unanticipated methodological and disciplinary challenges.

We have thus attempted to evaluate the sustainable city from the scientific, intellectual and political points of view.

Chapter 2
Green cityscapes and social inclusion in three major metropolitan areas of Switzerland

Klaus SEELAND and Nicolas BALLESTEROS

The process of urbanization is associated with an increase in distinct social groups with specific needs regarding the infrastructure of green cityscapes and socially adequate public urban green spaces. The urban population shows a decreasing tolerance towards motorized private transport and asks for close-to-nature recreation in urban environments. Thus, the design of public urban green spaces must take into account the changing demands and expectations of urban citizens. In order to offer adequate recreation areas which meet the wide range of expectations they have of their surroundings, socio-demographic studies are needed. In a cross-cultural research project, the urban green spaces of Geneva (French-speaking), Lugano (Italian-speaking), and Zurich (German-speaking) were looked at for their ability to generate recreational amenities and potentials to create a cityscape for social inclusion.

The large and still increasing proportion of younger migrants, asylum seekers and resident foreigners in large Swiss and other metropolitan areas of Europe (e.g. Fassmann *et al.*, 2002; Häussermann, 2000) has been visible in the respective urban cityscapes for several decades. In our survey, it seemed plausible to assume that the demands of foreigners regarding green space design differ from those of the Swiss population and that there are also differences in its use and perception between the various language regions within Switzerland. The research project focused on the demands of elderly Swiss people, who represent a growing section of the population, and the attitudes of Swiss youth, foreigners, unemployed people and Swiss who speak a language different from that of the region in which they live.

A general increase in leisure time, increasing unemployment and changing, as well as diversifying, urban lifestyles create new demands on public urban and peri-urban cityscapes. Crime, drug addiction, so-called 'new poverty' and deviant behaviour of marginalized social groups resulting in conflicts and vandalism place a strain on these areas. Many of these occurrences are at least partly due to problems of integration and are aggravated because of an inadequate supply of public urban space and green space in particular (Gobster, 1998; Huissoud *et al.*, 1999). A spontaneous settlement on the green embankment of the River Sihl in Zurich was established at the end of July 2005, when hundreds of youth demonstrated their ideas on the use of green space. Kuo and Sullivan (2001), for instance, showed that urban dwellers living in green surroundings are less aggressive than people living in an environment devoid of trees, parks, meadows or forests. In Chicago, prosperous areas enjoy greener surroundings than poor areas, which can also be seen in various other cities around the world (Iverson and Cook, 2000).

Zurich, the largest conurbation in Switzerland, on which this paper especially focuses, has developed an integration policy and taken measures to promote the inclusion of social groups that are causing problems. Several communities in the vicinity of Zurich (e.g. Schlieren) have integration offices which help to realize the integration objectives of the city of Zurich by giving practical assistance and counselling (e.g. Präsidialdepartement der Stadt Zürich, 2002). The mobile social service 'security – intervention – prevention' (Sozialdepartement der Stadt Zürich, 2001) actively contributes to making the public spaces of the city safer. As integration and communication between foreigners primarily takes place in public spaces, urban and peri-urban parks may thus have social integrative effects. The potential of public urban green spaces to avoid or reduce the number of conflicts and have a positive effect on the peaceful coexistence and socio-cultural inclusion of diverse social groups, was scientifically studied in two large surveys in our research project.

Method

The definition of the study areas using GIS analysis

A first step in investigating the social potential of public urban green space was a geographical information system (GIS) analysis of the three cities covered by this survey in which green space data and socio-demographic data were correlated. The social categories to be examined were determined by characteristics of social groups for which statistical data were available from the communal administrations of the three metropolitan areas, i.e. elderly people, adolescents, unemployed people, foreigners and Swiss people having a mother tongue different from that spoken in their place of residence.

The heterogeneity of urban quarters as regards the groups mentioned above was related to the availability of green spaces and thus their social integrative potential was determined. The highest social potential was attributed to areas where the most heterogeneous composition of these social groups in relation to the available green space could be found. To define the municipalities of Geneva, Lugano and Zurich as study areas would also have to include the peri-urban areas. The official boundaries of the municipalities would not be comparable because Zurich, for example, has merged with other communities in its vicinity which has not happened to the same extent in Lugano, for instance.

To make the three metropolitan areas comparable, therefore, a standard number of residents living in 1 ha area based on Swiss census data was defined. This information is available in a raster format of 100 m by 100 m edge lengths with GEOSTAT (the service giving spatial data provided by the Swiss Federal Statistical Office). Thus 50 persons per ha were chosen to define the resident density in a municipal area. The selected cells were transformed into polygons and their sidelines extended (buffered) to 500 m on each side. The polygons were merged to obtain a larger area and then buffered back again (buffer of -500 m), so that the actual city area would not be artificially enlarged.

For the five social categories studied in this research, the percentage in 1 ha was calculated for all inhabitants of the city. The quartiles of these values were then evaluated. The percentage values of the lowest quartiles were substituted by value 1, the percentage values between the lowest and the higher quartile were substituted by value 2 and the percentage values over the highest quartile were given value 3.

Thus a value for each hectare and each social group were obtained and the values of the five social categories were summarized to give one single indicator. The higher the value of this sum, the higher was the heterogeneity of the five social categories in this hectare, i.e. the assumed potential for social inclusion.

Since this research focused on green cityscapes and on the potential for social inclusion in green spaces, it was essential to locate the distribution of urban green spaces in the three metropolitan areas. This was done with the help of green space data from vector25, the digital landscape model in a vector format provided by the Swiss Federal Office of Topography, the content and geometry of which is based on the National Map of Switzerland (scale 1:25,000). This digital landscape model was divided into eight layers and only trees were considered in this survey. Green spaces were thus defined as areas within a radius of 50 m of a group of two or more trees.

In a subsequent step, the two datasets were correlated to obtain the potential for social inclusion of each green space. The study area was structured into polygons, each indicating the inhabitant's catchment area of a green space. This was a simplifying assumption in that it was supposed that persons living in one part of the city visited the nearest green space to their home for recreation.

The social potential of a green space was thus defined to be the majority value in each of the hectare plots of the green space's catchment area. Figure 2.4 shows the distribution of public urban green spaces and their potential for social inclusion per related green space area in Zurich. Germann-Chiari and Seeland (2004) describe this GIS approach in more detail.

Interviews in public urban green spaces

In the three large metropolitan areas of the different language regions of Switzerland, two public urban green spaces were chosen in each of the following types of communities: city centres, rich, suburban and peri-urban areas. Within these spatial categories, those with a highly heterogeneous composition of the five social categories were selected and comprised 5 large city parks, 7 urban forests and 15 smaller green spaces with playgrounds for children. In summer and autumn 2001, 1186 face-to-face interviews were conducted in these areas.

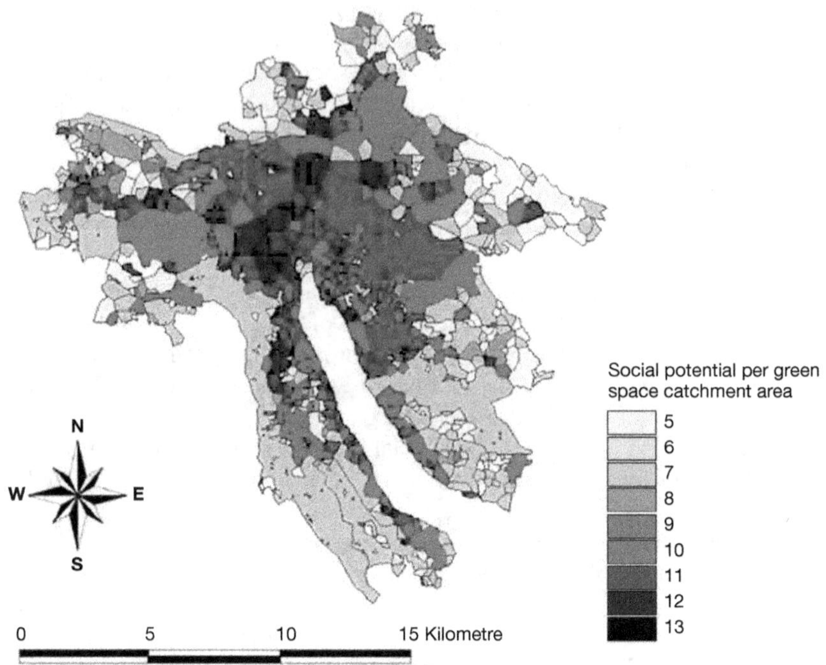

Figure 2.4. The social potential per green space for each public green space in the wider city area of Zurich. The darker the colour, the higher the potential for inclusion of the social groups considered in this survey.

Postal survey with questionnaires

After the results of the oral interviews were analysed, a postal survey with a standard questionnaire was carried out in 2002. These questionnaires were sent to residents selected by random sampling in the same communities where the interviews were conducted to give a control group of potential public green space users. About 1330 questionnaires were returned by respondents, a completion rate of 16%.

Results and discussion

In Lugano, the share of foreigners among the interviewees was equivalent to the 26% living in this canton according to official statistics. In contrast, in Geneva (mean: 19%) and Zurich (mean: 9%), the share of foreigners among the interviewees in our survey was smaller. Most of the foreigners were citizens of neighbouring or other EU countries, i.e. Italy, France, Germany and Spain. Only a few people from non-European countries were interviewed.

A considerable number of unemployed persons, pensioners, housewives and students were met in public urban green spaces, although the number in this category of respondents was smaller than their share in official statistics. In Zurich, the share of unemployed people is smaller than in the other two metropolitan areas, and particularly smaller than in Geneva.

Social contacts

In cities, contact between individuals is predominantly made in public spaces (Bahrdt, 1968; Jacobs, 1993), although sometimes this view is challenged (e.g. Thum, 1980; von Seggern and Tessin, 2002). According to Simon (1997), who surveyed a quarter in a suburb close to Paris, the behaviour in parks reflects the coexistence of people within this multiethnic quarter. In children's playgrounds, mothers are spatially separated in different ethnic groups; only the children play together sometimes. Similar behaviour could be observed in a study of Bertrand Park in the city centre of Geneva. The supervisors of the children, i.e. domestic servants from South America or English-speaking au pairs, remained separate and did not have any visible contact with the local inhabitants.

Ward Thompson (2002) pleads for a central role for urban public places in society serving as meeting places for foreigners, offering options for either remaining alone or participating in public life. Lischner (1994) found in his study that more people gather in public places where there is, for instance, an artist painting, than in places without any cultural attractions.

Making social contacts in green spaces

The most interviewees who had personal contacts with park visitors previously unknown to them were found in Geneva, followed by those in Zurich and Lugano and no significant differences could be found between age groups, foreigners or Swiss.

The results of the postal survey with questionnaires showed that respondents in Zurich and Geneva are more satisfied with the social contacts in their communities and likewise the coexistence with foreigners than those in Lugano (Figures 2.5 and 2.6).

Figures 2.5 and 2.6 indicate that the ratio of persons dissatisfied with social contacts and their coexistence with foreigners in public spaces is rather high and therefore, there is a certain potential for social inclusion with adequate measures. Similarly remarkable is the number of respondents who are undecided on whether opportunities for social contacts and better coexistence with foreigners should be improved. This uncertainty has to be seen against a background of availability of suitable public urban green spaces for social inclusion, about which the Luganese are most sceptical. If the community typology is taken into account, 77% of the interviewees in city centres had already had conversations with previously unknown green space visitors; in peri-urban areas, the figure is 55%, in suburban areas 47%, and in rich communities 63%.

In Zurich, contacts were most frequent in the city centre and decreased in rich, suburban, and peri-urban communities, where they were least frequent. In two city centre communities of Geneva, 84% of interviewees had had conversations with other visitors, the figure was about 66% for rich and suburban communities. In Lugano, about 60% of all respondents in peri-urban and suburban communities had never made any contact with other visitors to public green spaces. In the city centre of Lugano, on the other hand, about 75% of all respondents reported to having had conversations with other visitors, the figure for rich communities was about 57%.

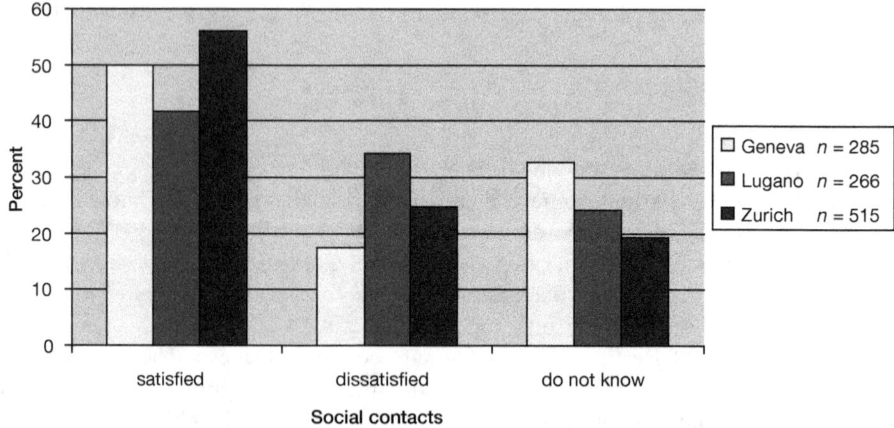

Figure 2.5. Satisfaction with social contacts; responses to the questionnaire survey (*n*=1066).

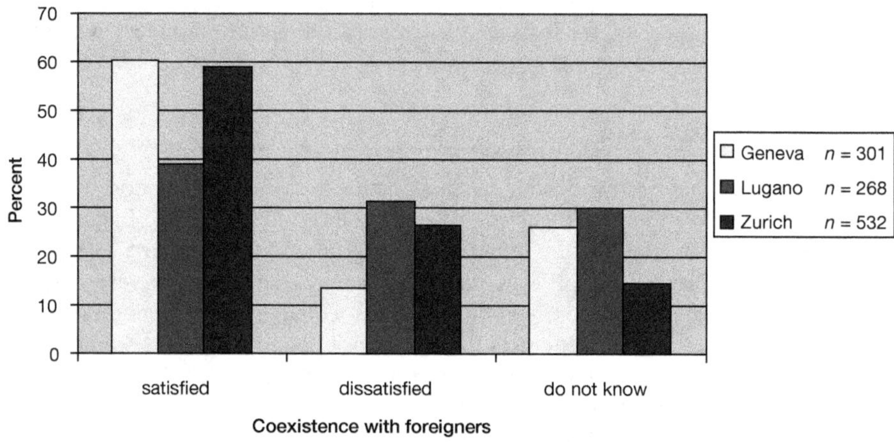

Figure 2.6. Coexistence with foreigners; responses to the questionnaire survey (*n*=1101).

Only 16% of all interviewed visitors to public parks and urban forests in Geneva, Lugano and Zurich did not wish to have any contact with social groups, such as adolescents, foreigners, drug addicts and dog owners, the most frequently mentioned groups. Vandalism and littering were exclusively ascribed to adolescents. Aggressive dogs and dog's faeces were perceived as a nuisance. Foreigners were considered to be messy and left litter, and also there were communication problems which made contact difficult. On the other hand, people with children and children in general were considered by far the most sympathetic group by other park visitors.

Other research projects covering the same themes as our survey show similar results. Public space in urban areas plays an important role for mobility, public parties and festivals, exhibitions, local identity and sociability (Bassand *et al.*, 1999.). Kweon *et al.* (1998) described the correlation between the length of stay in green spaces and the social inclusion of seniors. Their studies showed that seniors in public green spaces stayed for a longer period of time and established more relationships with neighbours. The elderly people were, therefore, better integrated into public social life. Furthermore, they suggested that elderly people would profit even more, if they could actively contribute to the design of the green space in their surroundings. Pfister (2003) emphasizes the correlation between health and a better quality of life through maintaining a relationship with nature in especially designed green spaces around elderly people's homes in the city of Zurich. The park aesthetics of intensely well kept lawns and flowerbeds were highly valued by elderly Swiss citizens. In order not to exclude them from this social life, certain pre-conditions must be fulfilled in the design, such as proximity to home, the arrangement of benches, and aspects of security (Cooper Marcus and Francis, 1998).

For the youth of two communities close to Zurich, Swiss as well as foreign, it was found that the cityscape design matters to them as does the design of public green space that facilitates social inclusion (Dübendorfer, 2001). Youth prefer open green space as a playground for sports and enjoyment, so called 'loose-to-fit' spaces, with no particular use and none or only a few permanent structures. It is through spending their leisure time in green spaces that Swiss and foreign youth come to terms with each other rather than in their classrooms or by visiting them in their home.

De la Chevallerie (1999) highlighted the role of garden plots and their social functions for the inclusion of singles and foreigners. In 'international gardens' such as in Göttingen, Germany, the integration of foreign migrants is achieved through joint gardening activities and by cultivating plants from their countries of origin (Müller, 2002). In five family gardens in Montreal, it was observed by Bouvier-Daclon and Sénécal (2001) that the owners did not necessarily have more contact with each other, but that they were easier to establish through the exchange of fruits and vegetables, seeds, recipes and horticultural advice.

Coley *et al.* (1997) described the impact of green spaces on the behaviour of the resident population of Chicago, where adolescents and mixed racial groups meet significantly more often in parks near their homes, which increases the possibilities for closer contacts between them. Floyd and Shinew (1999) reported that black and white Americans meet more frequently with members of the other group in their neighbourhood or in parks than in church or at home. Similarly, Kuo *et al.* (1998) showed that public parks near homes facilitate contacts between neighbours. Milchert (1998) analysed parks as multicultural space in which different ethnic groups and users pursue different activities and suggested that these should be more encouraged and structured with park programmes.

Cityscapes for social inclusion

The interviewees of Lugano are the most critical of the suitability of public green spaces for facilitating contacts among visitors. The majority of respondents believed that the parks in Lugano are not suitable for making contacts between people, whereas those of Zurich and Geneva considered their parks to be adequate for this purpose (Geneva 73%, Lugano 32%, Zurich 57%). However, the Luganese were principally in favour of making contacts in urban and peri-urban parks (Lugano 40%, Geneva 18%, Zurich 31%). The same conclusion was drawn by Paniga (2003) after interviewing a sample of

Lugano's inhabitants about their perception of the park and urban green space design and about their wishes for the future.

There was only a minority interested in more social contacts in public spaces (18% of respondents) in Geneva. This was confirmed by Bassand *et al.* (2001) who stated that about 60% of all the persons interviewed preferred not to have any contact with strangers in public urban green spaces. There were reservations about an increase in events and entertainment activities in public green spaces, and the level of education seemed to be an important factor in making more or fewer social contacts in public urban green spaces. About 68% of interviewees with a university education did not wish to have social contacts with strangers or to make new acquaintances in public green spaces (Bassand *et al.* (2001).

Cultural events may be a catalyst for establishing contacts between park visitors (Lischner, 1994; Tessin, 2003). The inhabitants of Geneva and Zurich, however, did not favour any such new activities in public urban green spaces (51.6% and 54.5%, respectively). The inhabitants of Lugano were more open towards them (28% in favour). The age of respondents was found to have an influence on this opinion: elderly inhabitants tended to object to a larger number of activities (61–80 years old: 56.7%) rather than younger people (21–40 years old: 38.6%). In comparison to foreigners (31.8%), Swiss people were more sceptical about activities in public urban green spaces (49.2%). This was particularly confirmed by interviewees in Geneva. In the study by Bassand *et al.* (2001), about 13% of respondents in Geneva declared that either socializing (discussing, meeting friends, making contacts) or good companionship (visiting performances, exhibitions, participating in festivities) are sufficient incentives to visit parks and public places, no special attractions are needed. As regards sociability and companionship, our study showed that the Luganese and foreigners in all Swiss language regions were more interested in these kinds of activities. Gehrke (2001) and Tessin (2003) believed that activities in public green spaces address new user groups. As most activities in public urban parks and other green spaces are practised individually, there is potential for integrating particular social groups by offering them appropriate activities.

Hanhörster and Mölder (2000) found in their study that intercultural communication happens mostly in residential areas and thus conflicts can be avoided if all cultural groups are integrated by participating in joint activities and decision-making. By establishing gardens in less attractive environments, residential areas may be upgraded and communication between different social groups can be enhanced (Hanhörster, 2001). Spitthöver (2003) in his studies on segregation tendencies described four different parks in a quarter of the German city of Kassel with a large share of foreigners. Each park had its particular range of visitor, and for each of the parks an ethnic, social or age and gender specific segregation could be observed. A mix of the different social groups or of foreigners with Germans could not be observed; visitors either use particular areas of the parks, pursue different activities, or visit at different hours of the day (Spitthöver, 2003).

There are also examples of conflicts that may lead to a negotiated coexistence, if not to social inclusion: A very popular picnic spot at the edge of the forest near Zurich was occupied by foreigners from Serbia, Macedonia and Turkey almost every weekend, which led to tensions with the Swiss residents close to this spot. When the log cabin at this picnic spot was burnt down one day, a co-ordinated mediation initiative of both the

Swiss and foreigners settled the case and negotiated a more equal use between all users (Dürrenmatt *et al.*, 2001).

In a quarter of the city of Zurich predominantly inhabited by low-income groups, both Swiss and foreigners, the use and expectations concerning the amenities of the cityscapes and green space use were found to differ widely. Two sample surveys found this park frequented by about 20–50% foreigners, approximately 60% women and almost 75% of visitors were between 26 and 40 years old. The main differences between Swiss and foreign visitors were that the Swiss were attracted by natural beauty and the design of the green space whereas the foreigners were mostly interested in having picnics and socializing. As far as their wishes for this green space were concerned, the Swiss emphasized design, maintenance and security, while the foreigners focused more on opportunities for children to play and be active. If the visitors to green spaces in the three metropolitan areas (Swiss and foreigners combined) are compared, the reasons for their attraction to these spaces vary considerably (Figure 2.7). This comparison does not indicate a clear correlation between language of origin and the reasons for attraction to green space, but does reflect cultural diversity and the multitude of preferences and expectations in multicultural societies.

The most heterogeneous distribution of the social categories examined in this research project and, therefore, the greatest potential for social integration, was found in the city centres of all the three major Swiss metropolitan areas. This is a result of the large green space areas in the inner cities on the one hand, and the settlement dynamics on the urban fringe that has an impact on the cityscape and its quality of life, on the other. There is a trend for Swiss middle and upper-class people to settle in suburban and peri-urban communities, because of the often lower income tax rates compared with rates in the city centres. Thus, the urban periphery has a more homogeneous population than the city centres where social groups are more mixed. Thus the potential for social integration

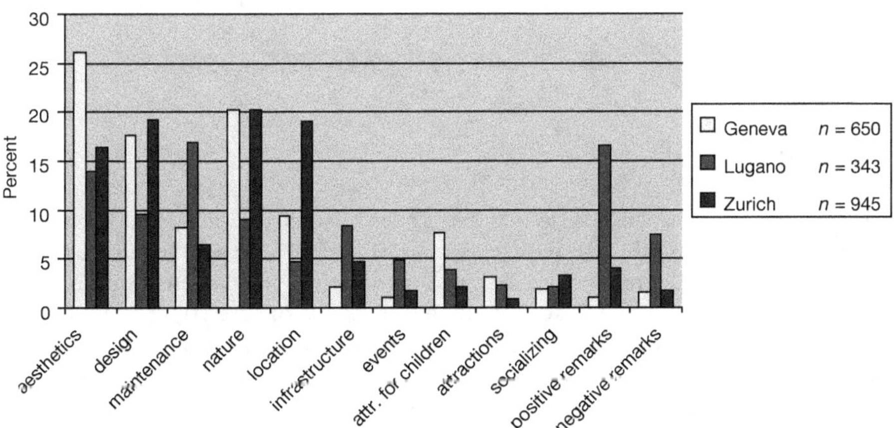

Figure 2.7. Reason for the attractiveness of public green spaces in three Swiss metropolitan areas. *n*=1330; 277 missing values; multiple responses obtained by postal survey with standardized questionnaires.

is greatest in the cityscapes of the core city centres. It is difficult to find a socially and naturally suitable environment for that purpose on the periphery. Arend (2003) stated that to achieve a better mix in resident populations, not only the origin of the inhabitants must be taken into consideration, but also the stratum, size of the household, living conditions and age of foreigners and Swiss.

Looking at the maps of the social potential per green space catchment area of every conurbation, it is obvious that the biggest social potential for inclusion is in their centres. It is unlikely that this is a statistical artefact, because only percentage values were used in the GIS analysis. This can, therefore, be taken as a structural characteristic of these metropolitan areas. If we look at the social potential of the city of Zurich, for instance, it is highest for the jobless, foreigners and Swiss from other language regions of the country. Young and old people are almost equally distributed in each of the three metropolitan areas.

There is a different situation in Geneva with more or less the same percentage of green space within reach of each social catchment area. Many foreigners in the city of Geneva are diplomats, whereas in Zurich there is a split between those living in the upper stratum of society or blue collar workers at the bottom of the service and industrial sector. In the city of Zurich, there is a significant linear correlation ($R^2=0.877$) between a high social heterogeneity of the five social categories investigated and the mean percentage value of public green spaces within the social catchment area. The same result was found in Lugano, but the correlation is non-significant ($R^2=0.332$). The potential for social inclusion in Geneva with a much more homogeneous social structure ($R^2=0.027$) is much lower.

Provided there is enough political will and a green space policy in these cities that is also largely a social policy, the "inhospitality of cities" (Mitscherlich, 1965) that partly results from the fact that the original function of towns to offer an identity to its inhabitants cannot be fulfilled adequately anymore, may be overcome. Social fragmentation, isolation and the formation of ghettos lead to a separation of urban life-worlds that can be counterbalanced by identifying a common basis for the different cultures and generations to design a public urban (green) space in the larger context of green cityscapes that meets the needs of post-industrial urban societies (Solecki and Welch, 1995).

Conclusion

To achieve social inclusion in the cityscapes of densely populated metropolitan areas and particularly in the public green spaces of the city centres, the attractiveness of these areas should be increased and designed according to the contemporary needs and demands of a multicultural and socially highly diverse structure (Bukow *et al.*, 2001). Other multicultural countries, such as, for example, the USA, that show parallels to the situation in Switzerland, have long since gone in this direction. Examples of these ventures are cited in the above mentioned references and complement our survey. In view of the continuing immigration of foreigners over the coming years and perhaps decades, particularly from developing and transition countries in Eastern Europe, Switzerland will have to adjust its planning strategies for its cityscapes accordingly. Social inclusion, and this has to be reflected more in Swiss migration policy, does not mean the complete absorption of immigrants into one's own multi(!)cultural set-up, but to let them actively participate in it and import parts of their own cultures. A multicultural nation like Switzerland with its four cultures has successfully undergone this process over the last 700 years. It is, however, a political challenge to apply these experiences to actual processes which include people coming from an international cultural spectrum. Public urban green spaces should be more strongly considered in these processes, if they are to play the role they could.

Starting points for a social policy for public urban green spaces are indicated by the results of this study. However, they need to be developed further during the course of more detailed research. In this context, experimental pilot projects will be important. They may reveal new perspectives beyond the conventional uses of public urban green spaces accepted so far. Their socio-political functions have several aspects which have to be considered. Firstly, public green spaces must be maintained as quiet green recreation areas for all in the cities. However, they must allow for new demands. In the socially heterogeneous areas of urban areas, particularly in city centres, space for activities should be designed to enhance the potential for social inclusion. The main objective is to offer activities which allow the maximum participation of all sections of an urban society and increase the social use value of public urban green spaces. A better quality of life and a more peaceful coexistence of the different sections of the multicultural society in Switzerland are goals that deserve to be pursued by a co-ordinated societal effort for a design, planning and maintenance policy for public urban green spaces.

Acknowledgement

This research project was conducted within the framework of COST (European Co-operation in the Field of Scientific and Technical Research) Action E12 and financially supported by the Swiss Federal Office for Education and Science.

The following collaborators of the Human Environment Systems Institute, Chair of Forest Policy and Economics have contributed to the data collection, compilation and analysis, report writing, and various other aspects of this research venture: Deborah Demeter, Christina Germann-Chiari, who wrote a draft on the methodology of this research, Evelyn Kamber, Ueli Mauderli, Kuno Moser, Iva Sedlak, Christian Stocker and Esther Thalmann. Many thanks to all.

Chapter 3
The public gardens in Biskra, Algeria: from elitist meeting place to no man's land

Farida NACEUR

In the early 1960s, in her work *The Death and Life of Great American Cities*, Jane Jacobs (1961) raised the problem of public parks and gardens that were starting to become unsafe areas in the USA. Since then, there has been increasing research into this issue, in particular to clarify the possible link between these areas and the phenomena of violence and insecurity. Recent work has challenged the theses about the crime-generating effect of urban parks by demonstrating that they may, to the contrary, contribute to crime reduction (Kuo and Sullivan, 2001) by easing stress and anxiety, and may even help re-create social ties in urban settings where anonymity and isolation are the rule (Davis, 1992). These findings have attracted the attention of leaders in developed countries when it comes to the role played by such areas, which are today the focus of renewed interest.

In contrast to this renewed awareness in developed countries, public gardens and squares in Algeria continue to fall into neglect and abandon. Everywhere in Algerian cities, we are witnessing a genuine decline of parks and gardens that are threatened with deterioration. Yet the inhabitants of these cities continue to lament the lack, or even total absence, of places where they can go for recreation and relaxation when they are not working. In Biskra, Landon Park, which played such a critical role in the city's history, has been abandoned by the inhabitants, who moreover complain constantly of the dull, lacklustre and deserted feeling of their Saharan city. The city's main public park, the 5 July Gardens, is in a similar state.

In spite of the strong impact of these areas on the urban structure, why are the parks ignored and deliberately avoided by the citizens for whom they were built? How have they been diverted from their original purpose to become unsafe areas or no man's lands? These are the questions we will attempt to answer in this paper.

How the city of Biskra evolved and its socio-political context during the colonial period

The city of Biskra is located in south-eastern Algeria, more precisely at the foot of the Saharan Atlas mountain range that marks the limit between northern and southern Algeria. The city has earned the name 'gateway to the desert' thanks to its geographic location, which throughout the different periods of its existence has made it a meeting point for exchanges between north and south, east and west (Léon l'Africain, 1977). Currently Biskra is the principal town in a wilaya[1] covering an area of 21,671 km².

This geographic position gives it an arid climate with very warm, dry summers (maximum average temperature of 43.5 °C) and mild winters. This dry but pleasant climate during the winter season, the luminosity of the sky, the palm trees and the camels are all natural attributes that make it attractive to foreigners. During the colonial period, Biskra saw a considerable flow of tourists who flocked to the city from the end of autumn to the beginning of spring. It became a 'thermal and climatic station' thanks to the hot sulphur springs of the Hammam-Salahine spa, whose reputation spread all the way to Europe.

In addition to its natural setting, the ancient city of Biskra enjoys a particular feature that differentiates it from other cities and oases in southern Algeria. The design of the old city is based on compact groups of homes, ingeniously arranged inside the palm grove around two features: the mosque along the main street and the watercourses that irrigate the gardens inside the palm grove. The city of Biskra is thus organized around water and plant-life.

Biskra's strong appeal during the colonial period was stimulated by ambitious urban planning projects that made it a major oasis, both picturesque and touristy, for European holidaymakers. In 1932, the city underwent its first comprehensive urban planning project, known as the Dervaux Plan (Courtillaot, 1979), which aimed to transform Biskra into 'a tourists' paradise'. This project included plans to build a group of facilities, hotels and areas for recreation and games, such as the casino that was inaugurated in 1893, as well as many new fountains and public parks. The project was, however, never completed because of difficulties concerning rights to develop the land, the cost involved and the over-ambitious scope of the operation (Agli, 1988). In addition to the hotels and recreation areas, the city also built modern transportation facilities, namely an airfield and a railway station.

The transformation of Biskra: from tourist spot to industrial centre during the post-independence period

The national political orientation favouring the development of industry and the fact that Biskra was made the wilaya's principal town in 1974 led to its transformation from a tourist and agrarian centre to a centre for industry and services. This policy caused a decline in the architectural and picturesque flavour of Biskra's old city. Indeed, starting at

---

[1] Wilaya: administrative subdivision.

that time, the rural exodus caused rapid, uncontrolled urbanization and a massive inflow of migrants from rural areas looking for jobs in the city.

This phenomenon had serious consequences, both for the palm grove, which was invaded by buildings made of armed concrete (Bencheikh, 1999), and for all of the old city. The newly arrived immigrants found ideal refuge in the old homes that had been abandoned following the 1969 floods (Saouli, 1989). Buildings sprouted up, without any regulatory control, and without first establishing drainage systems, which resulted in the uncontrolled flow of household wastewater and sewage into the waterways bordering the city (Alkama, 1995).

The accelerated urbanization of the early 1970s caused a major upheaval in the urban structure, as well as the way of life. The large-scale migratory flows created a high degree of social mixing and heterogeneity, which made newcomers slower to accept the city-dwellers' long-standing customs, values and standards. Social relations were deeply affected by the disintegration of community life and the juxtaposition of distinct populations with different lifestyles. Prior to this, the immigrants' integration into the city was a slow process that allowed them to adapt to city practices, which was no longer possible as the phenomenon accelerated. The immigrants imposed their rural lifestyle in the city (Boutefnouchet, 1985) and caused difficulties for even the most solidly anchored urban practices. Social pressure mounted against certain more modern social practices, such as taking walks in public parks, which was seen in a negative light.

Landon Park and its changing appeal
as a result of social changes

Created in 1872 by Count Landon, Landon Park is currently overseen by the city of Biskra. It is located in the Châtenier neighbourhood (south-eastern Biskra) and covers an area of 4 ha, which is entirely fenced in. When the Count settled in Biskra, he wanted to build a garden all around his main home in the style of an English garden; he wanted to fill it with exotic plants of various origins. Starting in the colonial era, the park was designed to accommodate tropical, subtropical and Mediterranean species of both ornamental and utilitarian plants. Today it ranks second only to the experimental Hama Gardens in Algiers for its floristic wealth. A study of its flora, carried out by a team at the Centre for Scientific and Technical Research in Arid Regions (CRSTRA), identified 52 plant species in the arborescent stratum of the park, which contains, in particular, numerous palm trees. The introduction of these species required constant care and, consequently, abundant and costly manpower. Improvements made to the property were the results of long years of labour, which put up fencing around the outside of the park, created pathways and lawns, conveyed potable water, created an irrigation network, built bridges, etc.

The park's main use subsequently changed and it became a place where well-to-do European families came to relax during the winter season. Several support facilities were added (e.g. a church, restaurant, bar, etc.), which made it even more attractive during the entire colonial period. Landon Park became famous across Europe thanks to the tourists who visited it.

The situation gradually shifted with the advent of industry and the decline in tourism. During the first years following Algeria's independence and the departure of Europeans, families from Biskra continued to use the park. With the massive arrival of immigrants, new standards frowned upon strolling in public gardens, and the park had fewer visitors. In addition, there were changes in how the facilities were managed and utilized, which caused them to rapidly deteriorate. As a result, not only did the families of Biskra deliberately avoid the gardens, but also foreign visitors and newcomers did not even know of their existence.

Landon Park was not alone; the same situation affected all the public gardens, including the 5 July Gardens, despite the fact that they were centrally located, in an activity and services area frequented by many people.

Impact of new uses for the park

During the colonial era, the park fulfilled various functions. A restaurant located at the main entry attracted the public and brought people into the park. In the central activity area, the church and the Count's home were the point of convergence for the different flows. From this central point, areas where one could find calm or pursue leisure activities extended into the depths of the park, such as a lake with swans and ducks, tree nursery and play area. Thanks to this layout, visiting the park was a process of continual discovery, accentuated by the curvilinear and narrow paths. The tree nursery was the culmination point in this process; it was the garden's outdoor museum and final surprise.

Following Algeria's independence, a series of transformations took place and certain parts of the park were entirely abandoned The current situation is as follows:
– the restaurant and bar have been converted into a library and theatre club;
– the church now houses a fine arts association;
– the Count's home has become the headquarters of an environmental protection association;
– the kitchen has been abandoned;
– the duck pond and the swan lake are in poor condition – the lake has dried up;
– the tree nursery has become a vast wasteland.

An analysis of the new uses being made of these areas shows that, first, the decision to set up certain cultural activities, such as the theatre club or fine arts association, was not well thought-out. The wish to impart an artistic and cultural flavour was not in keeping with the original purpose of the botanical gardens, a situation that was only made worse by closing down the tree nursery. As a result, the park lost its identity. Setting up these new activities in the park did not lead to an increase in the number and types of visitors. In Biskra, as in all of Algeria, cultural and artistic facilities suffer from a profound lack of interest. By relegating these activities to the depths of the park, they were even more distant and cut off from the city. This had an impact on the image of the park itself, which appeared to be closed-off and unappealing.

An analysis of practices also reveals that the park no longer fulfils its most basic functions, including the most ordinary ones, such as eating, drinking, disposing of rubbish, etc. The lack of necessary facilities has had an influence not only on who visits the park, but has also led to various forms of decline and numerous material traces, such as graffiti, broken objects, presence of rubbish, urine, faecal matter, bottles of alcohol, etc., can be found in the areas least exposed to the public eye.

In addition, the lack of play areas has encouraged children to carry out acts of vandalism (Naceur, 1997), which Tony Marshall (quoted by Griffiths and Shapland, 1979) likens to another kind of play. Certain children have taken to hanging from the trees, roughhousing, marking the walls, throwing balls or rocks, and various acts of damage and destruction.

An analysis of current visitors to Landon Park and their habits

A study of visitors to the park was undertaken using the behavioural observation method (Lynch, 1982). The purpose was to determine not only how many people come to the park but also to identify visitors and the way they make use of or appropriate the space, formally or informally. Following this, analysis focused on identifying the areas and cycles of activities, in order to test problems of access or the existence of exclusion zones within the park, depending on different social groups. Observations on the ground were gathered in March and April 2003, during spring when the climate in this region is pleasant. A distinction was made between two periods: the weekend days (Thursday afternoon and Friday) and two weekdays (Monday and Wednesday). Observations were made for an average of 5 h/day. During the entire 34-day period of observation, only 214 people were counted in the park, with an average frequency rate that was slightly higher during the weekend than on weekdays. One may thus estimate that, on average, six people visit the park daily, for a city population of 172,905 inhabitants (RGPH, 1998).

A significantly reduced female presence

The breakdown by gender was very unbalanced in the sample group. Females, including children, represented only 21% of all visitors, and only 27 adult women were counted, i.e. less than one woman per day. The adult female population represented only 12% of all those observed.

The rare women who were observed came with their families when they brought their children to the park, and they remained in the constant company of their husbands. Generally speaking, these were young couples with two or fewer children. This is not representative of Biskra families, which, according to the last census, average seven people per household. Sometimes men bring their children to stroll in the park while the women remain confined to the home. In addition, while men may enjoy a variety of activities (parlour games, reading, walking, resting on the lawn, conversation, sports activities), for women the only possible activity was walking on the paths.

Analysing who comes to the park is a good indicator of the way in which it is perceived. The low presence of vulnerable groups, and women in particular, is a sign of the apprehension and degree of danger inhabitants associate with this location.

A significant presence of small groups of unaccompanied young boys

Eighty boys under the age of 15 were counted during the observation period, which represents 33% of the sample group. The boys were not all accompanied by their parents;

groups of boys under the age of 14 were observed on several occasions, wandering in the park with no supervision. They committed acts of social transgression, such as smoking cigarettes, playing rowdy games, climbing the walls, chasing birds, throwing stones at passers-by, etc. One may wonder whether in the future these groups will commit more serious transgressions, or even turn to petty delinquency.

A majority presence of male adolescents

The male population observed in Landon Park included retired men but above all young men or adolescents, often in a group (Figure 2.8). Some adolescents who had stopped attending school appropriated the strategic points where the flow of visitors converged, in order to sell tea or cigarettes, setting up their merchandise on the ground. Other young people gathered in groups to listen to or play music at high volumes, which we observed on three occasions during the survey.

These ways of occupying the park bother other users, who accuse the youth of disturbing its calm atmosphere. Investigations carried out among those who live nearby or manage the park revealed that the presence of these groups created other fears. They are suspected of hiding from sight to consume drugs or alcohol and commit acts of damage or destruction, such as lighting fires, soiling the park and leaving rubbish (Figure 2.9). The young men may carry knives, and the media have reported cases of such gatherings degenerating into fights with assault and bodily injury, and even one death.

In general, urban parks can become the preferred locations for gatherings of the most marginalized or 'fringe' categories, which was observed in similar fashion near the 5 July Gardens: it is commonplace to encounter homeless persons, sitting or sleeping on the benches along the edges of the park.

Analysis of the spatial and temporal division of activities

Observation of the spatial division of activities showed a great deal of heterogeneity in the utilization of the park by different social groups. Traditional park visitors are found in those sections of the park most exposed to the public and they avoid, in particular, the

Figure 2.8. Traces of fire and alcohol consumption.
Figure 2.9. Unaccompanied children in Park Landon.

smaller paths and twisting, winding alleyways. This is consistent with similar observations made in the 5 July Gardens, where the area with the largest number of visitors is along the main (lengthwise) path.

In addition, Landon Park receives most of its visitors in the daytime, with almost none in the evening, which is also the case for the 5 July Gardens. This situation is not specific to Biskra. In numerous Algerian cities, including the capital city, people no longer go to public areas after nightfall. The security situation in the entire country has largely contributed to discouraging cities from functioning at night. In Biskra, this trend is accentuated by the lack of lighting in the two parks under study. These two factors confirm that Landon Park is perceived above all as a place of insecurity and a haunt for 'fringe people'.

Conclusion

These results show that the decline in the number of visitors to Landon Park, like other gardens in Biskra, is the result of a long process that has led to the current state of deterioration. The political choices made in the transformation of the city from a tourist centre in the past to become a centre for industry and services, underlie the many upheavals that have had a profound effect on the society there, but that have also had repercussions on the way in which public areas have been planned and managed.

The impact of rural ways of life has deeply affected customs, lifestyles and urban values (Benatia, 1976). The inhabitants themselves have largely contributed to this long process of deterioration, which has resulted in the profound malaise felt by the population of Biskra, as is true of other Algerian cities; idle youth, adolescents going astray, uncertainty concerning women's status. Such conditions do not encourage people to visit parks, particularly women (Naceur, 2004). The low number of visitors only heightens feelings of insecurity and prevents any form of informal social control (Clarke, 1999).

At the political level, from Algerian Independence to the present, the environment sector has never been considered a priority. Unlike European countries in which it has been shown that nature can constitute a point of reference for citizens in relation to time and the seasons (Sablet, 1988), here nature is not considered by planners to be a fundamental need or aspiration for city-dwellers. As a result, the community authorities neglect the upkeep of parks and gardens. In their thesis, *Broken Windows*, James Wilson and Georges Kelling (1982) showed that disregard and neglect increase the risk of deterioration. A public area that is well kept will ultimately be respected. The lack of both formal and informal controls is a factor that also increases the risk of vandalism in parks.

The convergence of all these factors has given rise to a situation in which authorities and inhabitants have been unable to appropriate this rich colonial heritage, which today has been neglected, ignored, damaged and abandoned to social transgressions.

Mechanisms leading to the transformation of open space in the metropolitan region of Vienna, Austria: is there a need for a new management paradigm?

Tanja TÖTZER and Ute GIGLER

Aside from rural, coastal and mountainous areas, urban areas constitute the fourth main type of landscape in Europe (European Commission, 1999). Urban areas are an important issue for the European Union, because Europe is one of the most urbanized continents and increasing urbanization severely influences the environment and modifies the landscape through development of transport infrastructure, construction of industries, and other built-up areas. Due to the high level of urbanization, more than 70% of Europe's population is urban (European Environment Agency, 2002). However, only one third of the population lives in major metropolises and about another third lives in small and medium-sized cities outside conurbations (European Commission, 1999). Thus, cities and urban areas represent an essential part of our living space and shape our landscape.

European cities are very heterogeneous with regard to their historical development, current policies, urban fabric and future potential. Although cities may differ, they have several characteristics in common including dynamism and growth (Barredo *et al.*, 2003). Thus, the development of metropolitan areas is a highly dynamic issue. Urban regions have to deal with continuous change. Migration, demographic shifts, economic growth and technological changes cause transitions in spatial patterns. In addition to external pressures, cities have to cope with many different interests which affect the scarce and finite resource of space within the metropolitan area. The pressure investors exert on still unsealed areas is extremely high. Real estate speculators are keen on upgrading their possessions and cities strive to attract new inhabitants and companies.

The Murbandy/Moland database[1] demonstrates that the general trend throughout Europe has been a decrease in green areas over the last four decades due to construction activities (European Environment Agency, 2002). In Europe, the current sealing rate through urban expansion and infrastructure is high – it is estimated that in the Netherlands the rate is approximately 36 ha/day, in Germany 120 ha/day, in Austria 35 ha/day and in Switzerland 10 ha/day (Montanarella, 1999). Sealing takes place at the expense of the environment and consumes open space such as farmland and natural areas. Aside from urban sprawl and contaminated land, loss of green space is identified as the most relevant and urgent issue in relation to urban land use (European Environment Agency, 2002). Therefore, special attention should be paid to open space within urban and peri-urban areas.

Open space fulfils an essential role within the urban fabric because it serves recreational, production and climatic functions. The enhancement of green areas has the potential to mitigate the adverse effects of urbanization, making cities more attractive to live in, reversing urban sprawl and reducing transport demand (see EU-project BUGS – Benefits of Urban Green Space, http://www.vito.be/bugs/). Maintaining and improving open space is thus a pre-condition for increasing resilience in urban systems.

This paper will demonstrate successes and challenges facing the implementation of sustainable urban development in the metropolitan region of Vienna. We will describe current strategies and instruments for maintaining open space in Vienna and its surrounding suburban region in Lower Austria and examine the gap between planned, perceived and actual spatial development. A new management paradigm holds promise for overcoming current barriers to preserve green space in urban areas.

The metropolitan region of Vienna

Setting

The metropolitan region of Vienna lies in the eastern part of Austria, which is a very fertile area with good climatic, topographic and soil conditions for agricultural use. Spatially and functionally, the urban and suburban regions of Vienna must be viewed as one unit because of the intense movement of people, goods and natural resources occurring between them. However, their respective administrative frameworks are completely different. The metropolitan region is divided into the core City of Vienna and the suburban region, which belongs to the province of Lower Austria (Figure 2.10). The city and Lower Austria have to be considered jointly when assessing strategies to maintain urban and peri-urban open space.

Development patterns in the City of Vienna

In the last census (2001), Vienna had 1.55 million residents. After decades of continuous decline, Vienna's population has stabilized in the last 10 years. In the 1970s

---

[1] The Murbandy (Monitoring Urban Dynamics)/Moland (Monitoring Land Use Change) database was developed by the European Commission's Joint Research Centre to assess spatial data on 25 European cities and urban areas.

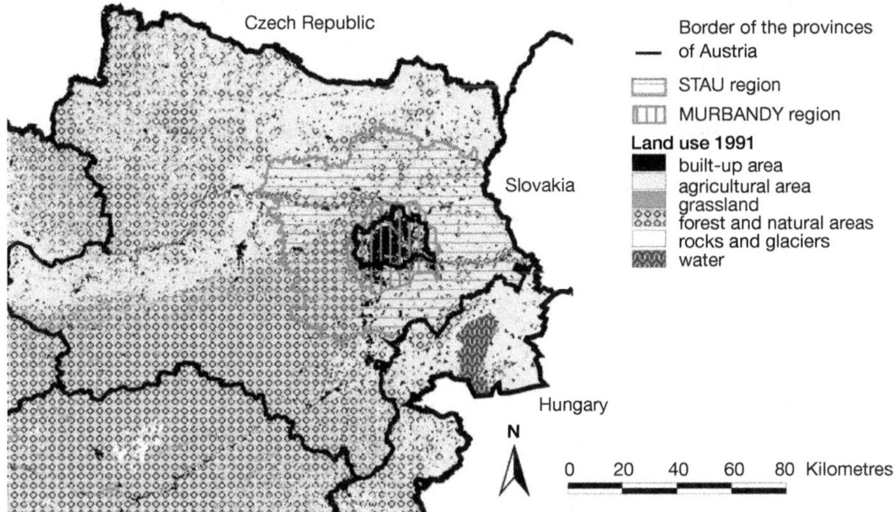

Figure 2.10. The metropolitan region of Vienna: land use and case-study areas of the STAU and MURBANDY projects.

and 1980s particularly, a noticeable migration from the core city to the surrounding municipalities took place. As a consequence of this suburbanization trend, the population of Vienna decreased. However, contrary to expectations, construction activity in Vienna did not reduce simultaneously. Figure 2.11 shows a large increase in number of apartments and residential area between 1971 and 2001.

Figure 2.12 illustrates where growth of residential areas took place between 1958 and 1997: the built-up area increased mainly in the southern and eastern parts of Vienna. The map is based on data from the MURBANDY project,[2] where development patterns of the greater Vienna region[3] were analysed based on remote sensing data of 1958, 1971, 1986, and 1997.

Parallel to the growth of residential and industrial/commercial areas, open space in the city decreased (Table 2.2). However, Vienna can still be called a 'green city'; approximately 50%[4] of the city area is open space at this point. For a better understanding of historical developments and future risks to the urban green space in Vienna, the category 'open space' needs to be split into several subcategories. Only agricultural areas and natural grassland were reduced between 1958 and 1997, while other types of open space, such as green urban areas, forest areas, sport and leisure facilities and vineyards even

---

[2] MURBANDY: <u>M</u>onitoring <u>U</u>rban <u>D</u>ynamics; a research project of DG JRC-SAI, Italy. The Vienna region was analysed by ARC Seibersdorf Research GmbH.

[3] The study region encompasses Vienna and part of the suburban region of Vienna in Lower Austria, see Figure 1.

[4] Compare: MA41 (Realnutzungskartierung 2001): 48.4%; MA22 (BiotopMonitoring 1996–2002): 51.3%; MURBANDY 1997: 47.2%.

grew (Table 2.3). An overlay of maps showing growth in built-up areas and former land use on these areas in a geographical information system (GIS) reveals that 87% of those areas converted to urban uses between 1958 and 1997 were former agricultural areas (Steinnocher *et al.*, 1999).

Why is agricultural land at a particularly high risk of being converted to urban use? What are the strategies of the City of Vienna to protect such areas?

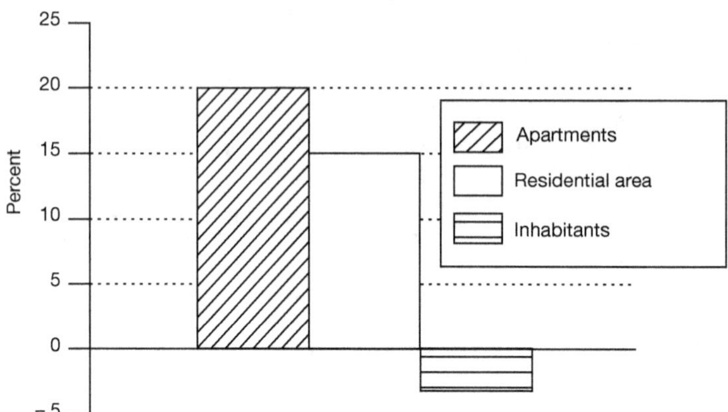

Figure 2.11. Residential development and population growth in the City of Vienna between 1971 and 2001.

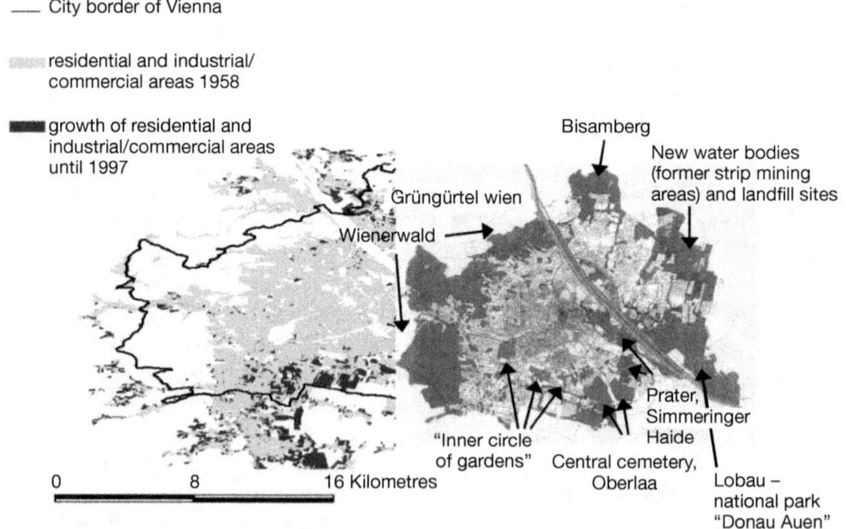

Figure 2.12. Comparison of changes in the built-up area of the Vienna region between 1958 and 1997 (left); areas of the 'green belt of Vienna' (right).

Table 2.2. Land use change between 1958 and 1997 in the city of Vienna.

Land use (%)	1958	1971	1986	1997	Trend
Open space	57.9	54.7	50.3	48.2	▼
Built-up area	35.3	37.4	39.5	40.9	▲
Industrial and commercial area	4.2	5	6.6	7.3	▲
Water bodies	2.7	2.9	3.6	3.6	▲

Source: MURBANDY
Note: Totals may not add to 100 due to rounding

Table 2.3. Changes in different categories of open space (1958-1997, City of Vienna).

Categories of open sapce in % (as proportion of total open space)	1958	1971	1986	1997	Trend
Green urban areas	5.1	6.2	7.2	8.1	▲
Forest areas	33.7	36.3	37.5	38.6	▲
Agricultural areas	41.1	37.5	34.6	32.0	▼
Natural grassland	12.5	11.0	8.6	8.4	▼
Sport and leisure facilities	2.4	2.7	5.4	7.0	▲
Vineyards	3.1	3.2	3.6	3.8	–
Other (e.g. abandoned land, dump sites etc.)	2.1	3.0	3.1	2.1	–

Source: MURBANDY
Note: Totals may not add to 100 due to rounding

Open space preservation instruments in Vienna

Politicians in Vienna are aware of the necessity of maintaining open space within city borders. Since the 1985 version of the Urban Development Plan (STEP1985[5]), aims and strategies for the protection of urban green space have been included in the plan. However, the instrument is not legally binding. Special areas such as the 'Wienerwald' are protected through the 'green belt of Vienna' instrument, which was implemented in 1995. Implementation of the green belt instrument is supported by land purchases by the city itself and land use restrictions through the zoning tool SWW[6]. Land purchases are an effective way of maintaining open space but are limited by the city's tight budget. At present, areas of the 'Wienerwald' and the Lobau are owned by the city. Parts of the green belt are also protected through a SWW-dedication, which mainly includes those areas that already have conservation status. The SWW-dedication encompasses for the most part the forest areas of the 'Wienerwald' in the northern and north-western part of the city, smaller green areas at the Wiener Berg and Laaer Berg in the south, parts of Danube Island, Danube meadows of the Lobau, green spaces in the north-eastern part of Vienna and the Bisamberg in the north (see Figure 3). The level of compliance with this instrument is high; changes in zoning require substantial reasons and development pressure can be one such reason.

---

[5] STEP=Stadtentwicklungsplan; urban development plan.
[6] SWW= Schutzgebiet Wald und Wiesengürtel: protected forest and meadow areas.

Pressures on open space in Vienna

Not all open space is protected as described above. Agricultural areas in particular are not sufficiently protected with the instruments currently available (Meyer-Cech and Seher, 2003). A number of reasons for the decline in agricultural areas become evident, both Austria-wide and within the City of Vienna borders. Since 1960, more than 40 farmers on average have discontinued farming every year, and agricultural areas have been reduced by nearly 90 ha/year in Vienna.[7] Nevertheless, approximately one third of the demand for vegetables in Vienna can be produced within city borders. More than 50% of vegetable farmers are situated in Vienna; mainly in the 11th and 22nd districts. Vegetable farming is the most productive agricultural use with a high yield per area and those areas are less prone to be converted to other uses. Land growing other crops is much more at risk of succumbing to market pressures. Those agricultural areas are still viewed as reserves for other uses in the future and farmers are more willing to give up their rights (Meindl and Jedelsky, 2003). Those parts of the city that have the highest amount of agriculturally used land (namely the 21st and 22nd districts in the east and the 10th district in the south) are, at the same time, those districts facing the strongest development pressures. Therefore, pressure on farmland is high. The increasing gap in land prices between green areas and land zoned for building makes the situation even worse. Low income for farmers, high pressure for transition, conflicts with neighbouring residents, and an insufficient transport infrastructure are all factors leading to a decline in agriculture in the City of Vienna.

Open space preservation within city borders is controversial, but highly regulated and rather effective. What is the situation in those suburban areas that are strongly linked with the core city?

Development patterns and pressures in the suburban region of Vienna

Particularly in the last 30 years, a very high settlement dynamic can be observed in the suburban region of Vienna, which can be attributed to migration from the core city. However, settlement growth is not only due to an increase in population, but also due to lifestyle or value changes in society. A boom in the construction of single family houses, smaller household sizes and an associated increase in the space required has lead to an average increase in land consumption per capita of 25% in the suburban region of Vienna (Steinnocher and Köstl, 2002). This trend toward greater consumption of land is illustrated in Figure 2.13 indicating a rise in area dedicated to residential uses and a substantial increase in the number of apartments built even though the population declined.

In the STAU-project[8], the research team examined the driving forces leading to the observed spatial development between 1968 and 1999 in the suburban region of Vienna. The analyses concentrated on the interactions between different land uses, such as residential, industrial and agricultural, and on migration and commuter patterns between the

---

[7] see: http://www.wien.gv.at/stadtentwicklung/landwirtschaft/zukunft.htm?S0=landwirtschaft#P0

[8] STAU-Vienna: City-suburban-relations and development in the Vienna region; a research project within the Austrian Landscape Research funded by the Austrian Federal Ministry for Education, Science and Culture.

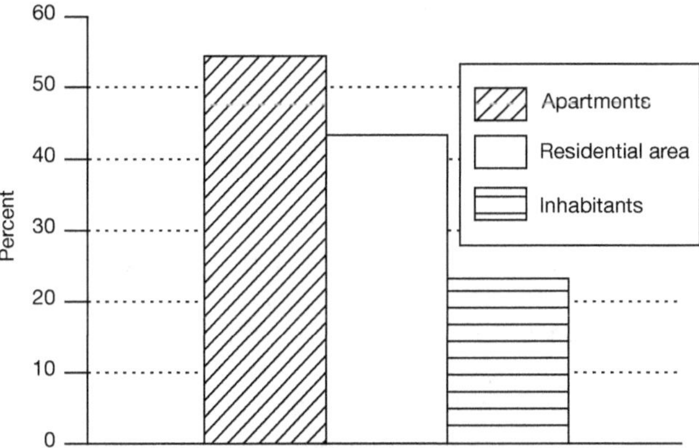

Figure 2.13. Residential development and population growth in the suburban region of Vienna between 1971 and 2001.

suburban region and the core city. Although the suburban region is still characterized by a lower density and a higher environmental quality than the core city, increasing traffic and loss of open space are the main environmental challenges (Loibl and Tötzer, 2003). The driving force behind the growth is migration from the core city, which can be explained by the attractiveness of suburban areas. Attractiveness is defined by many factors including land price, accessibility, centrally located government offices and an appealing landscape. Analyses show that areas situated on the border of existing residential areas surrounded by agricultural areas have the highest likelihood of being transformed into built-up areas. As in the core city, the largest proportion of areas transformed were former agricultural areas: in the neighbourhood of existing residential areas 77% of farmland has been converted to other uses, further away from those areas, but still within municipal boundaries, the figure is 82%.

Local politicians very much depend on tax income from businesses and residents in their municipality. Therefore, they have a strong interest in providing land to newcomers, both residents and businesses. Thus, vying for companies and residents is essential and a highly competitive issue for the city and surrounding municipalities. In addition, suburban municipalities compete amongst themselves. The municipalities are willing to provide investors and new inhabitants with sufficient land at the expense of the less lucrative agricultural land. This illustrates that the land use trends described above are very much a result of economic and political interests. Planning interests appear to be secondary.

Planning instruments applying to the suburban region of Vienna

Although the city and suburban areas of Vienna must be viewed as one unit, their respective administrative and legal frameworks are completely different. Therefore, mechanisms to protect open space differ between Lower Austria and Vienna. In contrast to forest areas or highly valuable natural areas, agricultural land is not under strict legally binding protection in Lower Austria. Farmland is only indirectly protected by spatial planning instruments, mainly through zoning, which lies in the responsibility of the mayors of each municipality. Unfortunately, the actions of local politicians are not always in compliance with strategies for implementing sustainable development on a larger scale.

To control and direct local planning, there exists another planning instrument at the provincial level, namely the Federal Regional Planning Act. The act is legally binding for municipalities and provides favoured development principles and objectives for the whole province of Lower Austria, which should be implemented locally through zoning. The law emphasizes compact settlement development, protection of agricultural land, and concentration of settlement growth on traffic axes. The document also includes urban growth boundaries for some regions in Lower Austria. These highly ambitious goals stand in stark contrast to reality. Actual development patterns (Figure 2.14) reveal that the instrument has been unable to curb urban sprawl throughout the last 30 years. Figure 5 shows changes in population density in suburban municipalities over three decades. The figure depicts number of inhabitants per hectare of residential area rather than total area. Municipalities with a tendency to dense development are depicted below the blue line (population density in 1971 is lower than in 2001) and those with a tendency to sprawl

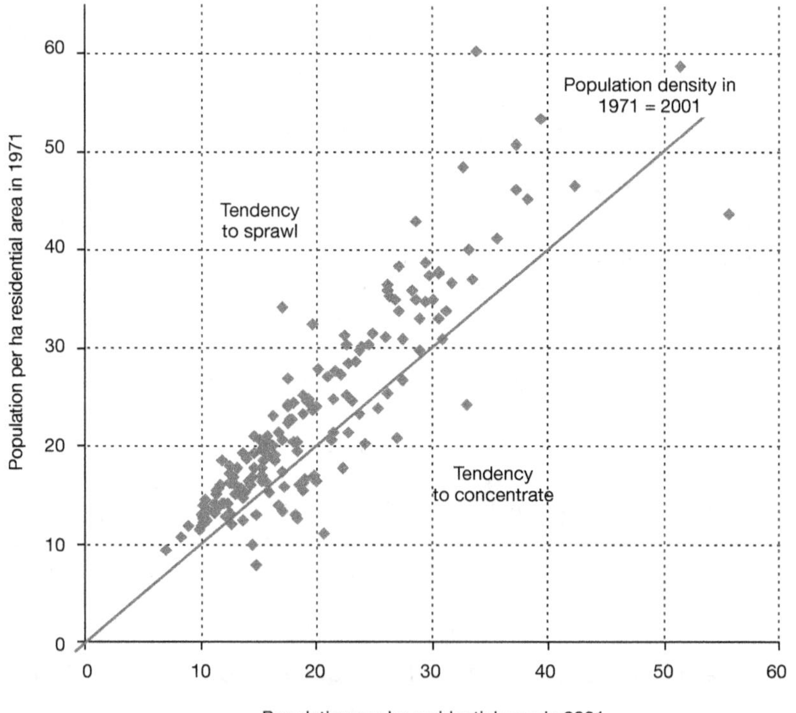

Figure 2.14. Population density in 1971 and 2001.

are shown above the line (population density in 1971 is higher than in 2001). All calculations are based on remote sensing data. This analysis provides a more realistic picture of actual settlement development patterns than densities based on administrative units (Steinnocher and Tötzer, 2001).

A new management paradigm

The analyses above reveal that social and economic trends have affected spatial development in the metropolitan area of Vienna much more than planning. The sustainable use and preservation of green space in metropolitan areas is thus a very challenging and multi-faceted topic affecting and being affected by numerous stakeholders. Metropolitan regions need to cope with various pressures, such as the amount of construction and a number of highly complex unknowns, such as economic success or social well-being in the immediate and long-term future. Nevertheless, complexity and potentially conflicting objectives need to be tackled if the current trend of losing green space to other uses in urban areas is to be reversed and if a resilient urban system is to be achieved.

In 1973, C.S. Holling introduced the word 'resilience' into the ecological literature as a way of helping to understand non-linear dynamics observed in ecosystems (Gunderson, 2000). Resilience is measured by the magnitude of disturbance that can be absorbed before a system changes its structure by changing the variables and processes that control behaviour. This we term 'ecosystem resilience' (Holling and Gunderson, 2002: 28). In the Holling and Gunderson definition, resilience is not about the system returning to a stable equilibrium after disturbance. It is about maintaining the system's structure or function, which encompasses multiple stable states or points of equilibrium. Buffer capacity, the capability to self-organize and adaptive capacity constitute the three properties of resilience (Holling, 1973; Carpenter *et al.*, 2001). In the last few years, the concept of resilience has also been applied to social-ecological systems like cities. According to Holling (1986), two conditions have to be met if we are to apply the concept to systems other than ecosystems: such systems need to be describable in dynamic terms, and they must have more than one potential state of equilibrium (van der Leeuw and Aschan-Leygonie, 2000).

This non-equilibrium view with phases of stability, disturbances and reorganization has already been applied to innovation theories (Rotmans und Kemp, 2003) and is also applicable to urban planning (Pickett *et al.*, 2004). This dynamic and evolutionary approach corresponds to actually observed development patterns. In Europe, for example, there are many, once famous and prosperous old industrial cities, which have to deal with social, economic and environmental changes today. For those cities, the transition from 'centres of production' to 'centres of services' had and, in many cases, still has a tremendous impact on a city's image, labour force and the spatial and socio-economic framework (Tötzer and Gigler, 2005). Such cities had a phase of stability in the industrial era, but global structural change disturbed this state and forced them to reorganize. Confronting this transition and dealing with the changes is the only way for a city to survive in the long run.

If a city shows high resilience, it is able to incorporate changes in its way of functioning without changing qualitatively, which is a major characteristic of a resilient system (van der Leeuw and Aschan-Leygonie, 2000). Attributes of a system that enhance resilience include redundancy, diversity, modularity, spatial heterogeneity, rapid feedbacks, and ecological and social 'memory' (Quinlan, 2003). Transferred to urban structures, resilience can be enhanced through diversification of functions, for example, through mixed uses (Gigler and Tötzer, 2005) and maintenance of agriculture and open space within the city. For politicians and planners, promoting resilience means changing institutions and the nature of decision-making, which includes recognizing the benefits of autonomy, self-organization and new forms of governance to promote social goals and the capacity to adapt (Adger, 2003). An adaptive capacity is essential for a city and refers to the ability of a social-ecological system to cope with novel situations without losing options for the future (Folke *et al.*, 2002). If such systems are to be managed successfully, they require flexible governance with the ability to deal with uncertainties and to respond to internal and external changes.

Adaptive management is a promising approach for managing such uncertainties. The instrument was conceived and first applied to environmental assessment and resource management to develop more effective and resilient policies (Holling, 1978). Resilience and successful adaptive management are intertwined, because if there is no resilience in the system, then one simply cannot manage adaptively (Gunderson, 1999). Gunderson (2000: 434) stated that "most policies are really questions masquerading as answers. Since policies are questions, then management actions become treatments in the experimental sense". This view encapsulates part of the essence of adaptive management. The basis for this approach lies in the recognition that in natural and socio-economic systems, as in urban systems, we often face incomplete information, uncertainty, unknowns, and unexpected events (Holling, 1978). Nevertheless, we need to deal with uncertainty, which requires the use of experiments allowing those involved to assess successful and failing approaches (Walters and Holling, 1990). Another important element of adaptive management is comprehensive monitoring of the area under observation. Crucial, system-relevant indicators should be applied that demonstrate how different aspects of an urban area might be developing or changing over time (Grumbine, 1996; Lessard, 1998). Both the use of experiments and monitoring enable those responsible, for example, urban planners or city managers, to receive feedback from the system thereby determining whether any progress was made. This feedback in turn allows planners to learn about the system and incorporate what was learned into the management strategy (Figure 2.15). Another key feature of this approach is that all stakeholders should be involved in the management process. They should remain informed, provide input and be involved in decision-making (McLain and Lee, 1996). The approach thus requires a much more flexible, adaptive management and planning style that is capable of incorporating frequent changes in environmental and socio-economic conditions, and that can be more responsive to citizens' requests.

Why should this novel approach be relevant to the management of green spaces in urban areas? Conventional planning dictates that planning guidelines, directives and laws are written with the understanding that they will be effective and in place for a specified amount of time and only changed periodically, not regularly and as needed. Planning laws for the suburban areas of Vienna provide a case in point. Laws applying to the area tend to be years or even decades outdated and changes only occur after planning practice has demonstrated repeatedly that the laws are no longer being abided by.

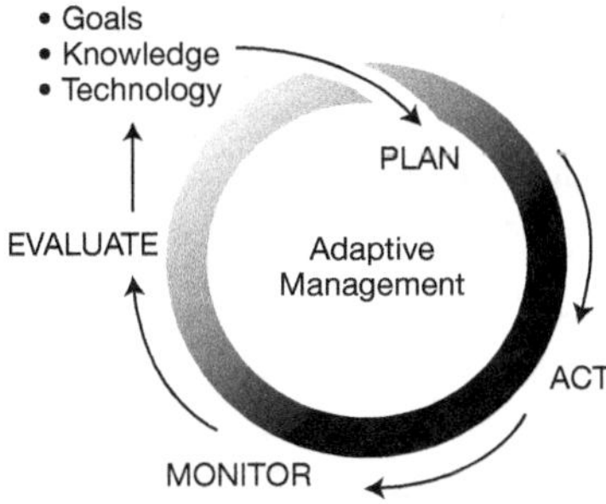

Figure 2.15. Adaptive management cycle.

Thus, policy-making is far from being abreast with actual spatial, social or economic developments and is not responsive to changes because feedback loops are lacking. Handling green space in urban areas, however, requires a more flexible and feedback-oriented approach, because pressures to convert green space to other uses are often immediate and thus occur in time-scales that are much shorter than those for considering policies and laws. At the same time, planners have to design long-range policies in order to curb short-sighted influences from different stakeholders. A good example is the initiative of nine cities which participated in a unique international Sustainable Urban Systems Design competition sponsored by the International Gas Union (IGU). They developed staged 100-year plans supposed to lead to urban sustainability. The winner of the prize was the Greater Vancouver region, which conceptually based its cities^{PLUS}-approach on an adaptive management framework. For them, adaptive management means that long-term plans become part of an ongoing planning process. They follow the principle: "a vision without a plan is but a dream, a plan without a vision is sheer drudgery, a vision with a plan ... can change the world" (The Sheltair Group, 2003: 47). To implement a vision successfully, it is necessary to strike a balance between short and long-range planning interests as well as market developments and, above all, remain flexible and aim at incorporating needed changes to policies.

Other case studies have shown that applying the adaptive management approach leads to the following outcomes (see Olsson *et al.*, 2004; Tompkins and Adger, 2004; Gilmour *et al.*, 1999): the scope of local management is widened to a broad set of issues across scales; management expands from individual actors to multiple-actor processes; participants reach a better understanding of management issues, structure and dynamics of the system; organizational and institutional structures evolve where knowledge of system dynamics is incorporated and social networks develop where information flows, knowledge is mobilized and social resilience is enhanced. All this increases the ability to deal with uncertainty and the capacity to deal with future change.

The adaptive management instrument thus attempts to move away from applying solely top-down, regulation-oriented approaches, but instead encourages citizens, non-profit-making organizations, and other interest groups to also voice their opinions, for example, build and maintain community vegetable gardens and thus self-organize, become stakeholders and participate in decision-making (McLain and Lee, 1996). This creates a broader, more diverse network of interested parties which increases the likelihood that the green space is highly valued and better understood and, therefore, less prone to be converted to other uses. Adaptive management – a new management paradigm – that combines a variety of regulatory and bottom-up approaches for cities and their surrounding areas might be more promising for future green space preservation than current instruments.

Conclusion

About 70% of Europe's population lives in urban areas, which makes green space preservation in these areas a key issue to deal with. Development of transport infrastructure, and construction of industries and residential areas exert multiple pressures on the environment and urban landscapes. Although urban areas are very heterogeneous, there are two characteristics that unite them – dynamism and growth. Cities face change constantly and need to find a way to handle demographic shifts or technological changes that potentially shift spatial patterns and change the entire urban fabric. Evidence of that is the loss of farmland and natural areas due to unprecedented sealing rates in and around cities. The European Environment Agency (2002) identified loss of green space in urban areas as one the most important issues facing cities at the moment.

The Vienna case study demonstrates that multiple pressures affect green space and in particular, agricultural land in the metropolitan region. These pressures include demand for housing, commercial development, non-binding legal instruments, and economic pressures on farmers. The city has a number of green space preservation tools that are to some extent successful in preserving green space in general. However, agricultural land is more prone to conversion because of poor protection mechanisms, an increasing gap in land prices between farmland and land zoned for building, and low income for farmers. In contrast to the city, the situation in the suburban area is even more critical because of less stringent instruments. Zoning laws alone are insufficient to protect green space and agricultural land in suburban areas. As a result, green space and agricultural land continue to be converted to built-up areas at high rates.

As a result, cities need to find ways to curb conversion of green space and to cope with the high level of uncertainty facing future development of urban areas simultaneously if cities are to become more resilient. The concept of resilience was developed for ecosystem management, but has recently been applied to socio-ecological systems as well (Carpenter *et al.*, 2001; Quinlan, 2003). Attributes of urban areas that enhance resilience include diversity, redundancy, spatial heterogeneity, rapid feedbacks and ecological and social 'memory'. Planners and politicians who want to promote resilience need to engage in new forms of governance, recognize the value of self-organization and be willing to adapt to new circumstances.

The adaptive management instrument is tightly coupled with the concept of resilience because only a resilient system can be managed adaptively. Adaptive management is a promising approach for achieving better management and developing more effective policies regarding green space in urban areas. Major features include using experiments to deal with uncertainty, comprehensive monitoring, incorporating feedback mechanisms, involving all relevant stakeholders, and remaining flexible. This contrasts with the often very rigid planning institutions and instruments that lag behind actual developments and do not incorporate new information into policies in a timely fashion. Several studies applied to urban areas (The Sheltair Group, 2003; Tompkins and Adger, 2004) have demonstrated that the approach is useful in developing, for example, long-term plans for urban areas because the scope of local management is widened to a broad set of issues at multiple scales, multi-actor processes develop and organizational and institutional structures evolve. This new management paradigm might enhance the capacity to deal with future change and offer more successful means for attaining sustainable urban development.

Acknowledgments

Empirical work for this paper was funded through the MURBANDY project, a research project of DG JRC-SAI Italy, and the STAU-project, an Austrian Landscape Research project of the Austrian Federal Ministry for Education, Science and Culture.

Landscape policies: from conception to implementation

Chapter 1

The implementation of the Landscape Atlas of Flanders in the integrated spatial planning policy

Marc ANTROP and Veerle VAN EETVELDE

This paper discusses how cultural landscape values in the Flemish region of Belgium were revalorized and became a basis for a new policy that aims for more integrated landscape management in the future. Landscape protection and management is a slow and difficult process in the densely populated Flemish region of Belgium. In 2001, only 2.7% of the area of Flanders was protected as landscape of outstanding value. With the federalization of Belgium in 1993, the regions became responsible for spatial planning, environment and landscape protection. This resulted in a divergence of policies and priorities. Only the situation in the Flemish region will be discussed here. An important change in policy occurred in Flanders in 1995, which placed integrated landscape management as an objective. Simultaneously, the cultural, historical and aesthetic qualities of traditional cultural landscapes were revalorized. These values were used in a new integrated inventory of the landscape: the Landscape Atlas of Flanders. The inventory of relics of traditional landscapes carried out between 1995 and 2001 resulted in a GIS-based landscape atlas. The European Landscape Convention was still not finalized when the inventory was started. Consequently, the development of the atlas and the methods used were specific for the Flemish situation.

As formal and legal definitions of landscape were not available when the inventory for the atlas started, the concept of landscape used was based on definitions and concepts found in the international literature. Landscape and land are not considered to be synonymous. Land refers to soil and ground (Zonneveld, 1995), thus to property and territory. Landscape, on the contrary, is considered to be a common heritage and a collective identity. This concept was included in the Flemish decree on landscape management in 2004. Landscape covers many properties, which means that no one really owns it, but this also makes it confusing as to who then should take care of it (Antrop, 2000a). A landscape is bound by views and not by administrative borders, unless it has received special status, such as legal protection. Nevertheless, the etymology of landscape not only refers to the scenery of the countryside, but also to a homeland and a collectively organized territory, which is given symbolic, cognitive and aesthetic values (Claval, 2004; Olwig, 2004, Groth and Bressi, 1997; Cosgrove, 1993). Thus landscape is an important source of information or forgotten knowledge (Austad, 2000). Understanding the relics is essential for their preservation and integration in future landscape design (Fry and Sarlov-Herlin, 1995). The growing integration between research in landscape ecology, spatial analysis and archaeology illustrates this (Fry *et al.*, 2001). Landscape research and planning becomes increasingly transdisciplinary (Fry, 2001; Tress *et al.*, 2003).

This paper first discusses the societal and legal context in which the atlas was realized and how this influenced the choice of methods used. Then, a summary of the results is given and these are compared with the district zoning plans and biological evaluation map. Finally, the implementation of the atlas so far in policy and landscape management will be discussed. The Atlas was published in 2001 and fitted perfectly in some recommendations of the European Landscape Convention (Council of Europe, 2000). Although Belgium ratified the convention only in October 2004 and it entered into force on 1/2/2005, this stimulated the implementation of a new legislation that aims an integrated management of all landscapes.

Landscape management and protection in Flanders

The legal context

In Belgium, the first law on the protection of monuments, sites and landscapes was issued in 1931. Only 2.7% of Flemish territory was legally protected in 2001 (Van Hoorick, 2000; De Borgher, 2002). Also important was the legislation on spatial planning in 1962, which defined the district zoning plans for future land use. These zoning plans indicated valuable landscapes, based on a first national survey of landscapes (Delaunois, 1960). No strict rules were given; a clear description was lacking on what was valuable in these areas and which criteria were used to define them. In 1993, the three regions, Flanders, Brussels Capital and Wallonia, became responsible for spatial planning and landscape protection. Successive laws and degrees were important for policy of landscape management in Flanders, such as nature protection in 1973 and adapted in 1997, management of forests in 1990, and protection of the archaeological heritage in 1993. The policy on structural planning was introduced in 1997. The decree of 1999 regulated the gradual conversion of the zoning plans into structural plans with much broader vision of landscape management. In the 1980s, the Flemish policy on nature conservation introduced the concept of a 'landscape park', which was replaced by 'regional landscape' in 1990. The law of 1931 was replaced by decrees relating to the protection of monuments and city and town sites in 1976 and to landscape management and protection in 1996, which contained the first legal definition of 'landscape'. This decree was amended in 2000, 2001 and 2004, partly to be appropriate for an international context, the European Landscape Convention in particular.

The interference of different legislations in the context of integrated landscape management is obvious. The procedures for landscape protection and management are consequently slow and cumbersome. Until the 1990s, most of the areas protected as landscape were selected for their natural qualities. One of the reasons for this is that during the 1970s and 1980s, the public became increasingly aware of the ecological deterioration of the environment. As the nature conservation lobby became more powerful, natural landscape values were easily accepted and helped in the protection of such sites. In many cases, the cultural, historical and aesthetic qualities were included as non-vital complementary values. The descriptions of the cultural, historical and aesthetic qualities were rather vague, lacked consistency and theoretical background. Meanwhile, devastating changes were happening at an accelerating speed, particularly since the 1960s. Often autonomous changes occurred faster than could be enforced by law.

Defining landscape

The Flemish decree of 1996 defines landscape as "a confined area with low density of buildings and possessing an internal coherence which has an appearance and coherence that is the result of natural processes and social developments" (in Dutch: "een begrensde grondoppervlakte met een geringe dichtheid van bebouwing en een onderlinge samenhang, waarvan de verschijningsvorm en de samenhang het resultaat zijn van natuurlijke processen en van maatschappelijke ontwikkelingen"). Although the basic elements found in the definition given in the European Landscape Convention can be recognized, two typical aspects of the Flemish situation appear: the extremely high degree of urbanization of the countryside and the necessity to delineate the area where the specific regulations will be applied. The latter refers more to land management than real landscape management.

Criteria for landscape evaluation

Three main groups of qualities are generally used for evaluating landscapes, i.e. natural, cultural and aesthetic. The natural qualities often refer to the natural sciences, particularly biology and ecology, but also include geology, soils and landform. The cultural group includes history, social and economical values as well as religious, symbolic and linguistic aspects. The aesthetic values are often restricted to visual or scenic qualities and sensory ones, such as tranquillity. According to actual Flemish law (the decree on landscape management and protection), four groups of values can be used for selecting landscapes for protection: the natural-scientific value, historical value, socio-cultural value and aesthetic value. Very different criteria are used to evaluate each of them (Antrop, 2003).

References to natural values are mainly supported by lists of rare or endangered species and habitats. Detailed inventories and maps exist already for the biological values. Inventories of archaeological findings are still fragmented and differ in detail according to location. Archaeological inventories are not yet available to the public. Direct indications to archaeological findings are rare as well. Both natural and archaeological values are defined and protected by other legislation. Consequently, the aim of the Landscape Atlas was to collect information complementary to the ecological and archaeological information. Although different values are described according to the legal system, it is their combination that gives the exceptional quality necessary for their protection. A content analysis of the terms used to evaluate cultural, historical and aesthetic qualities, revealed a set of concise criteria, which were used in the atlas (Antrop and Van Damme, 1995). Relics were selected using a set of rules (Tables 2-5) and values were ascribed to each of them, using internal cross-references and a consistent approach, enhancing the holistic approach.

Traditional and contemporary landscapes

Concern about the degradation of our cultural landscapes refers mainly to 'past' rural landscapes and not to the new emerging, modern cultural landscapes. The concept of traditional (cultural) landscapes was introduced in Flanders in 1985 and became gradually accepted as a framework for the new regional planning and landscape assessment (Antrop, 1997). Traditional landscapes refer to the rural landscapes that existed before the important and rapid transformations of the revolutionary age caused by political, social and technological changes. These transformations started at the end of the 18th century and are characterized by changing patterns in mobility, urbanization and globalization (Antrop, 2000b, 2004). The traditional rural character of the countryside is one of slow transition, great stability and an inherent sustainability. Grandparents and grandchildren lived in the same environment and knew and spoke about the same landscape, which was a stable reference point in their lives (Antrop, 2003). Modern landscapes are continuously changing at an increasing pace and are characterized by a polarization between more intensively and more extensively used land (Vos and Klijn, 2000).

The map of traditional landscapes of Flanders shows 77 landscape units consistently grouped according to natural regions and history. The great diversity in the small area of Flanders is obvious and the classification was found to be a helpful framework for the new spatial planning policy (Antrop, 1997). However, there was a need for a finer delineation and more detailed descriptions of the ideal characteristics and condition of these pre-industrial landscapes. A more extended adaptation including criteria and case studies for mapping landscape types and relics was reported in 1995 (Antrop and Van Damme, 1995). This formed the basis for the Landscape Atlas of Flanders, which also allowed the refining of the classification and delineation of traditional landscapes on a map at a working scale of 1/50,000.

The Landscape Atlas of Flanders

The background

The inventory of relics of traditional landscapes in the Flanders region started in 1995 on a provincial basis. Although at that time no GIS facilities were available in the Flemish administration and digital data were very limited and not standardized, the concept of the atlas was conceived entirely as an open GIS-database. The initial plan was to carry out the inventory for the entire Flemish region in an extremely short period of 4 years. There were several practical reasons for this. First, the new spatial planning policy developed in parallel and urgently required basic information about landscapes to be included in the overall more integrated planning scheme. In particular, the discrepancy between the available information on nature and ecology and the cultural and historical assessments became unacceptable. Second, the increasingly faster changes in the landscape and environment demanded a rapid response. Third, the short term of the project was politically attractive and could feasibly be achieved during the course of one legislation. However, the scientific and methodological consequences of these constraints were important (Antrop, 2003). Only existing information could be used and the datasets were very different in nature and quality. In order to fill the many gaps in knowledge, the whole system had to be open and flexible for further extension. Fortunately, Belgium possesses a rich collection of detailed topographical maps of high quality dating from the end of the 18th century. Also several sets of aerial photographs were very important. At the start of the project in 1995 most of these data were still analogue maps. During the realization of the atlas, all data were gradually digitalized and integrated into a GIS.

The method

The historical map of Joseph-Jean-François de Ferraris dating from about 1770 shows the Austrian possessions in the Low Countries on a detailed scale. This map was used as the reference for the traditional rural landscapes. The classification and description of these landscapes served as a descriptive reference base for the ideal landscape types and character. The analogue orthophoto maps of 1990 were used as the reference for assessing the actual landscape. Several editions of topographical maps for dates in between allowed an assessment of the transformation of the landscape. Selection of relics was based on a visual interpretation of patterns on these documents to detect ancient structures that were still recognizable and coherent in the orthophoto maps of 1990.

All relics had to fit into a classification scheme that could be entered into an open GIS-database. New information was likely to emerge and it had to be able to be integrated easily into the system. Also many cultural layers and time periods were combined in most of the relics. In order to create a 'neutral' classification, which would be useful in a multidisciplinary way, the main categories were based on the spatial dimension of the relics selected (Table 5). Relic zones (R) referred to larger areas and were represented by polygons with fuzzy borders. Particular unique complexes were classified as ensembles called anchor places (A) and delineated as polygons. Linear elements such as roads, canals, rivers, ditches and military defence works were classified as line relics (L). Discrete objects such as buildings, solitary trees and landmarks were classified as point relics, coded as P when a description or name was available, otherwise coded as X. Each of the relics received a unique identification code starting with the letter referring to its type, followed by the statistical code of the province in which it is situated and a serial number. This allowed the descriptions of each relic to be stored in the database and linked to the map.

During the inventory process, GIS facilities were increased such that most thematic datasets became available in digital form and the administration's capacity for using electronic information also increased substantially. The final realization of the atlas was achieved entirely using GIS co-ordinated by GIS-Flanders (GIS Vlaanderen, 2001). Technically, the atlas is based on the Arcview GIS 3.2 shape format linked to a Microsoft Access database. However, the files can be consulted using Arc Explorer or a stand-alone viewer in Windows 2000 (Figure 3.1). The atlas was widely distributed on CD-ROM (GIS-Vlaanderen, 2001) accompanied by a book (Hofkens and Roossens, 2001). The Support Centre for GIS-Flanders offered technical support and made the atlas also available online (http://geovlaanderen.gisvlaanderen.be/geo-vlaanderen/landschapsatlas).

Criteria and selection rules

The criteria used for the selection of relics and their evaluation were based on a preliminary study of the criteria used in previous landscape protection cases (Antrop and Van Damme, 1995). The most important were coherence, legibility and preservation condition (soundness) of landscape structures, such as settlement patterns and types, field systems, road networks, landscape type and land use. Additional criteria for the evaluation were scenic quality and information potential. For each of the types of relic, a procedure and rules for selection were described (Antrop, 2001, 2003) (Tables 3.1–3.4). Relic zones were mapped at a scale of 1/50,000 using fuzzy borders to indicate that a more detailed delineation should be done on a larger scale and after field surveying. Selected anchor places were checked on the field in 2000 and were delineated sharply on the map using material landscape features that can be recognized easily in the field.

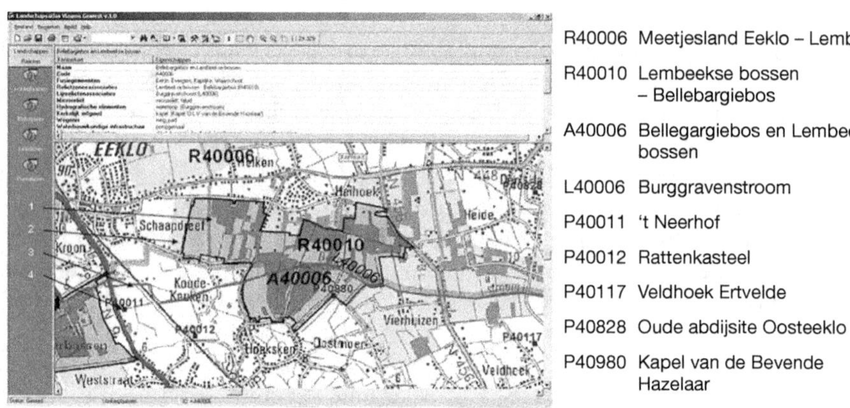

Figure 3.1. Screen copy of the stand-alone viewer of the Landscape Atlas on CD-ROM with the description of a relic area at the top (Source: OC-GIS-Vlaanderen, 2001): (1) anchor places; (2) relic zones; (3) line relics; (4) point relics (background: topographical map N.G.I. 1/100,000).

Table 3.1. Characteristics used to determine the selection of an area as a relic zone.

– The existence of recognizable structures and objects dating from the end of the 18th century up to the Second World War.
– The occurrence of geomorphologic features that structure the landscape or have a physical or natural monumental value.
– The occurrence of concentrations of typical objects or linear elements.
– The occurrence of known or expected concentrations of important archaeological sites.
– The occurrence of particular viewpoints and sights offering a good scenic view over the relic zone.

Landscape with high scenic value and not disturbed visually by modern constructions or infrastructures.

Table 3.2. Characteristics used to determine the selection of an ensemble as an anchor place.

1. Ensemble situated outside a relic zone:
− a complex of clustered heritage elements,
− elements having a distinct historic, morphological or functional coherence,
− elements that are representative of a particular period, style or type,
− in a sufficient preservation condition such that the ensemble can be used as an ideal example.

2. Ensemble situated within a relic zone:
− a complex of clustered individual or linear elements,
− elements having a distinct historic, morphological or functional coherence,
− elements that are representative of a particular period, style or type,
− possess typical characteristics related to the relic zone,
− part of a larger geographical structure,
− forms an aesthetic and undisturbed whole.

Table 3.3. Characteristics used to determine the selection of line features as relics.

− The element possesses a cultural or historical meaning as a connection or border that can still be recognized in the landscape,
− or the element is an ecologically valuable corridor,
− or the element is a natural ecotone or forms a visual gradient or transition between different landscapes,
− the element can be clearly recognized in the landscape and forms a visual and structural element.
The whole element is selected and disturbed segments are marked.

Table 3.4. Characteristics used to determine the selection of point relics.

− Objects that have an intrinsic natural, scientific, cultural, historical or aesthetic value,
− and have a good condition of preservation,
− and are situated in a undisturbed spatial context.
Any legal protection status is indicated.

Table 3.5. Results of the Landscape Atlas of Flanders.

Relic type	Code	Number	Total area or length	% Flanders area	Average size or length
Relic zones	R	515	530,000 ha	39.0	1,029 ha
Anchor places	A	381	221,051 ha	16.3	580 ha
Line relics	L	544	4,851 km		
Point relics	P, X	4,607			

Results and discussion

The results of the inventory were surprising (Table 3.5). Indeed, in the highly urbanized and severely fragmented landscape of Flanders one easily gets the impression that no valuable heritage and attractive cultural landscapes remain. The Landscape Atlas revealed that, on the contrary, many areas still exist where the character of the traditional landscape is preserved. Clearly the selected relics refer mainly to the rural countryside and they complement the existing biological and ecological assessments (Tables 3.6 and 3.7). Many relics of the cultural landscape are located in areas of low biological value. The comparison with actual land use also indicates that not only extraordinary and special landscape types were selected, but ordinary landscapes are also included. However, the maps with information on the number and size of relics, show that the relics form many fragmented and disconnected patches. Most of the relic zones are situated at the periphery of the former municipal territories, which indicates they are more representative of landscapes formed during the late 18th century on former outfields (Van Eetvelde and Antrop, 2005). Most disturbances and degradation of traditional landscapes are caused by urban sprawl around even small centres and follow the dense road network.

Table 3.6. Relations between landscape relic zones and other classifications in Flanders.

Land use*	%	District zoning plans	%	Biological Valuation map	%
Wetlands & water	1.7	Agrarian	60.1	Less valuable	65.9
Arable	34.4			Valuable	20.5
Grassland	27.0			Very valuable	8.2
Heath land, dunes	2.5	Nature, forest	26.5	No value	0.1
Woodland	19.2	Other	12.8	Not mapped	4.5
Built land	10.1				
Other	5.1				

* based on the Biological Valuation Map
Source: after Tack and Van den Brempt (2001)

Table 3.7. Relations between landscape anchor places and other classifications in Flanders.

Land use*	%	District zoning plans	%	Biological Valuation map	%
Wetlands & water	2.5	Agrarian	53.2	Less valuable	11.1
Arable	20.0	Nature, forest	39.4	Valuable	25.2
Grassland	12.3	Other	7.4	Very valuable	59.0
Heath land, dunes	2.3			Not mapped	4.6
Woodland	23.7				
Mixed land use	31.0				
Built land	4.4				
Ohter	4.1				

* based on the CORINE Land Use (1995)
Source: after Tack and Van den Brempt (2001)

Implementation in the planning practice

The minister responsible for landscape protection launched the Landscape Atlas in 2001 during a meeting where a revised, broader and more integrated policy for the landscape was also announced. As this fitted well with the initiatives proposed in the European Landscape Convention, an important means for distributing the atlas was provided. The atlas was available on CD Rom almost free of charge for administrations at the different institutional levels.

The Landscape Atlas was accompanied by a book describing the new policy goals, the methodology used for compiling the atlas, the results and examples of its application in planning, environmental impact assessment and landscape protection. This proved to be a successful strategy as a wide use of the Landscape Atlas was predicted. New amendments to the Flemish landscape decree of 1996 in 2001, 2002 and 2004 anchored the Landscape Atlas in the legislation. It became a legal reference document not only to be used for landscape protection, but also in a general context to conform to the suggestions of the European Landscape Convention. Integration with the spatial structure planning was enforced, particularly for the ongoing development of structure plans at the local level and their implementation in local spatial planning. The anchor places in the atlas can be selected to become heritage landscapes for implementation in local spatial plans by formulating appropriate long-term landscape quality objectives. Besides spatial planning, the atlas is also frequently used in environmental impact assessment. It forms the only description of landscape qualities covering the whole of Flanders in consistent detail and format, and thus can be used as a basic reference. The atlas also became the subject of further scientific research in the landscape character assessment, ecological planning and countryside planning (Antrop *et al.*, 2004; Couvreur *et al.*, 2004; Van Eetvelde and Antrop, 2005). Updating the atlas and how to use it for landscape monitoring are currently under discussion.

Conclusions

The devastating transformation of the landscape since 1960 resulted finally in a growing awareness of the natural and cultural values of the landscape and lead to a gradual change of policy during the 1990s. The Flanders region more or less follows the general trend in Europe in this matter. Nature protection in Flanders was better developed than the protection of cultural and aesthetic landscape values and a renewed interest in these was initiated almost simultaneously with the growing international interest in the cultural landscape. The Landscape Atlas of Flanders rapidly and successfully became integrated in the new planning process because of its speedy and wide distribution and the growing need to be in line with international initiatives. It is now generally used in landscape protection and management policies, environmental impact assessment and structural countryside planning.

Chapter 2

History, time and change: managing landscape and perception

Graham FAIRCLOUGH

Landscape needs to be studied in an interdisciplinary manner and from many perspectives; it cannot be the concern of a single discipline. Nor is it solely an expert concern because it has the potential to be one of the most democratic ways of seeing and looking at the world, and of interpreting and understanding the natural and cultural environment. The European Landscape Convention places people at the very core of its definition of landscape as "an area, *as perceived by people*, whose character is the result of the action and interaction of natural and/or human factors" (authors' italics). It also proclaims that landscape is everyone's common heritage, and that "the participation of the general public, local and regional authorities, and other parties with an interest (in landscape)" requires a full sharing of expertise, knowledge and decision-making (Council of Europe, 2000, 2002).

Yet landscape is more often than not seen as the domain of landscape architects, and sometimes ecologists. It is usually located in terms of government policy in ministries for Environment or Nature rather than also in people-centred ministries such as those for Culture or Spatial Planning. The place of landscape in government policy is, however, more complex than this allocation of responsibility might suggest.

In England (the principal focus of this paper), for example, formal responsibility for landscape sits within the Department for Environment Food and Rural Affairs, and its agencies responsible for the natural environment and the countryside (i.e. English Nature and the Countryside Agency, soon to be combined into a single organization to be called 'Natural England'). Yet, the Government's agency for the historic environment, English Heritage (EH), working for the Department for Culture Media and Sport, also has a strong interest in landscape policy, especially in the field of characterisation and understanding. EH was initially concerned mainly with particular types of landscape (e.g. designed parkland surrounding the country's aristocratic houses, the settings of archaeological sites, or extensive areas of archaeological remains on England's uplands and chalk downlands). Over the past 10 years or so, however, EH has promoted the view that all landscape has historic character and it has, therefore, become concerned with the wider landscape (Fairclough *et al.*, 1999).

In particular, EH is carrying out a national programme of Historic Landscape Characterisation (HLC), part of EH's wider characterisation work that is designed to produce generalized statements about the character and significance of the whole historic environment (English Heritage 2005; www.english-heritage.org.uk/characterisation). It is motivated by the requirements of sustainable management and planning, and the need to understand the past in order to manage the present sustainably. By moving from knowledge to action, HLC aims to provide a strategic framework for spatial planning, agriculture and heritage management decisions (Herring, 1998; Macinnes, 2004; Clark *et al.*, 2004; Aldred and Fairclough, 2003; Fairclough, 2002).

The HLC programme has a diverse agenda. It broadens the scope of 'heritage' in terms of form and function, fabric, and age or recent-ness. It uses landscape as a unifying concept within the historic environment sector to complement traditional site-based 'monument protection'. It also seeks to bring into the cultural heritage sector a greater interest in the semi-natural (but still historic) fabric of the landscape, such as hedges, the pattern of land cover or anthropogenic habitats such as heathland (Berry and Brown, 1995).

HLC also helps to introduce different ways of valuing the historic environment, for example, the differences between public and private perspectives, national and local scales, expert and personal viewpoints, and value in social, economic, amenity or associative terms, and in terms of parallel but different specialist knowledge systems. More importantly, it brought new ways of thinking that involve holistic, integrated views of landscape, the concepts of sustainability, and the goals of integrated planning and management, all central planks of the European Landscape Convention.

HLC also promotes an altered awareness of the historic dimension of landscape in neighbouring disciplines (Countryside Commission, 1996; Fairclough, 2003b). Landscape architects have long-regarded landscape as the 'sum of its parts', as having ecological and cultural, as well as scenic and aesthetic attributes and components, and as having historic depth as well as current significance (Countryside Agency and Scottish Natural Heritage, 2002). Landscape ecologists are fully aware of the central role that human land management makes, and made in the past, to the appearance of the land and to the systems that maintain it. HLC builds on these existing perception by presenting an archaeological and historical view of landscape in formats and languages common to all three disciplines. This was achieved in England at a low level of detail in the mid-1990s in the 'Countryside Character Map' and its associated descriptions (Countryside Commission/Countryside Agency, 1998–99). HLC was developed to provide the raw material to amplify this at more local and detailed level, in a format that could more easily be integrated with other disciplines' perspectives.

HLC is designed as a bridge between disciplines, and as part of the 'common meeting place' that often describes landscape. This is perhaps first and foremost what the concept of landscape is all about: a bringing together of different ways of seeing, and a unifying concept to integrate methodologies and professions (Palang and Fry, 2003). Within this range, HLC offers a particular view of landscape to other landscape specialists and to the public, a view that emphasizes the historic and archaeological dimensions of landscape, its rich time-depth and the central role played by time and change. The great extent of continuous and continuing human-driven change that can be read in landscape's time-depth can be a new insight for some people, which can lead to new ways of responding to, or accepting, present and future change.

Perception and characterisation

Perception of landscape continually change with the passage both of individual time, and of social (collective) time. As new things (buildings or forests, cities or farms, wind-farms or motorways) grow older, they attract increased respect and value, as part of an acceptance that grows from familiarity, or perhaps simply from reconciliation and assimilation. Electricity power lines of the 20th century, once deemed unsightly "blots on the landscape" (Figure 3.2), are by some people now regarded fondly as part of a world they know and cherish because it is familiar. Their eventual loss to new technology will probably meet with some resistance, because certain generations have grown up with them as part of 'their' landscape; they were always there and are, therefore, part of the natural order of things. For some people (adopting a version of a general British preference for tidy, manicured, cultivated landscape), an attractive landscape is one with a sound economic function. For others, as long as the landscape is green, it matters little

Figure 3.2. Electricity power lines once deemed unsightly 'blots on the landscape' – Sheppey.

if it is a centuries-old cultural pattern or merely abandoned wasteland in the early stages of regression.

This paper will argue that people's perception about landscape are changed by knowledge as much as by sentiment. Perception of landscape can change dramatically by learning more about the history of the land, about why it looks as it does, and about the role of time within our present-day landscape. Recognizing that human and historic processes are important characteristics of landscape, and that landscape was not created at any single time, affects how people regard landscape, and thus how they wish to see it managed. At the simplest level, merely recognizing that landscape is a humanly made artefact creates new perception, and thus new mental landscapes. Hoskins' book *The Making of the English Landscape*, for example, first published in 1955, had a tremendous effect well into the 1970s on popular perception of landscape.

The first lesson of HLC is that nature and culture are not separate (rejecting the idea that one piece of land is a natural landscape, whereas another is a cultural one), but that all landscape is the result of the long-term reciprocal interaction of nature and culture. Other types of landscape assessment recognize this as well of course, but to varying degrees (Countryside Agency and Scottish Natural Heritage, 2002). HLC emphasizes that we live with the results of thousands of years of human modification of the world, sometimes suddenly and on a large scale (e.g. mining and forest clearance), sometimes gradually and individually (e.g. modifying biodiversity through breeding, culling and hunting). Simply in crude physical terms, the whole of Europe's environment can be seen to be cultural as well as natural because it contains archaeological remains. In terms of perception, however, the landscape is even more thoroughly cultural. High mountain tops that may scarcely have been reached by people – as close as we get to pristine environments – are culturally modified by perception when viewed as landscape. The act of perceiving them as landscape changes their significance and meaning.

Landscape management needs to reflect this and it can, therefore, usefully be seen as a two-fold process. The first is the most familiar, the protection of parts of the physical fabric of the land that are most valued on one criterion or another, such as hedges, roads, woodland, buildings or archaeological remains. The second may be less familiar, and is certainly more difficult. This is the management of landscape itself, a more complex task concerned with managing change to the complicated interactions between the things that make landscape in our minds. It includes managing perception as well as fabric.

HLC, like other types of characterisation, is more suited to managing change than to protection, for which conventional inventories and databases remain the best approach. It is a method that generalizes and seeks to reduce real-world complexity through adopting small scales and simplification to provide a broad framework for understanding change. It uses modern GIS techniques, but uses them to capture interpretation rather than to record facts and thus to create a partly subjective narrative rather than objective scientific data. HLC creates a GIS-supported model of selected aspects of landscape, as much perception as fact. An HLC 'model' may be incorrect in detail, but the general picture holds together and, more importantly, allows us to reach back beyond detailed documentary history and maps towards the deeper, late prehistoric and early historic, periods. This brings a greater richness to landscape appreciation (although not a new one in terms of landscape archaeology, on which O.G.S. Crawford or W.G. Hoskins pioneered similar approaches throughout the 20th century).

The rest of this paper focuses on three key questions that arise from HLC.
– What can archaeologists and historians uniquely offer to the interdisciplinary understanding of landscape (what in essence is the historic character of landscape)?
– What are the broad implications for landscape management of a more archaeologically informed understanding of landscape?
– What are the repercussions for landscape protection or management (or indeed planning, in the forward-looking sense used by the European Landscape Convention, i.e. spatial planning) of seeing landscape as perception ('an idea not a thing')?

The historic character of landscape

What is the particular contribution of archaeologists and historians to landscape understanding? This is best answered in abstract terms, by putting aside local or regional patterns, or the dominant date of particular places, which are the obvious, immediate contributions of HLC, and focusing on more fundamental issues, namely time, process and change.

Time

Landscape is not only a product of geography. It is more than simple topography, or the lie of the land; it is more than the view from a hilltop or another vantage point. It is also the (current) culmination of human history, our 'story so far'. Not all of the past is visible in any piece of land, but much of it can be 'read' in one way or another, as part of a long chain of cause and effect, and of consequences and impacts, which have made today's landscape look as it does. Thanks to memory, imagination or knowledge based on special techniques, the past can be perceptible in landscape even if its remains are largely invisible below ground. Absences can also contribute to the historic character of landscape. The lack of visible signs of medieval open fields in parts of England emphasizes the wholesale nature of 18th and 19th century enclosure or the regional distinctiveness of areas where open fields were never made; the absence in a territory of semi-natural woodland (except at the edges of parishes) testify to the energy and land management methods of our medieval and later predecessors.

In many places, however, the past is still very conspicuous, and to one degree or another creates a highly visible aspect of landscape. These are the areas most commonly labelled as 'cultural landscapes', but other areas are no less cultural, or less rich in time-depth, except in perception (which will often change with increased knowledge).

The embedding into landscape of the passage and effects of time is one of the defining characteristics of landscape (Macinnes and Wickham-Jones, 1992). The palimpsest of landscape, its layers and rich texture, gives the opportunity to write our stories about both the past and the present (e.g. Clark *et al.*, 2003). The sense of roots, origin and identity that this brings can have as powerful an effect on popular imagination and perception – in other words on landscape – as nature, green-ness, tranquillity or aesthetic feelings.

Process

The passage of time, especially across the very long durations studied by archaeology, is inevitably linked to the historic processes (and human actions) that have created the current environment and landscape, in other words to cultural forces. The European environment (as with most parts of the world) has for thousands of years been the stage on which humans have acted out their lives, and has consequently been greatly modified. Both deliberate human 'agency' and the unintended consequences of action or even inaction (the distinction called 'designed' and 'organic' in the UNESCO world heritage landscape criteria) have created our environment and thus, our landscape. This is so whether that creation has been partly or wholly (supposedly) unconscious (such as the great medieval practice of open field agriculture, or the even older 'prehistoric' farmlands of western England), or fully deliberate and conscious (e.g. 17–19th century designed landscapes, and the deliberate acts of re-planning in England that were concerned with the privatization of common land). Furthermore, people in the past modified soils, flora and fauna so thoroughly that biodiversity itself can be seen as a cultural phenomenon, or as a human construct.

Landscape is also a result of another set of cultural processes, those that frame and influence the way that people perceive it, in relation to ourselves and in the context of current social structures and political views. The act of perceiving (imagining) landscape, the creation of mental landscape, is first and foremost a cultural act, specific to its time and place. Landscape is thus inseparable from human agency and processes.

One result of the link between landscape and process is that neither the fabric of the environment, nor the perception that make landscape, will remain unchanged if the historic processes which created them cease to operate. This has obvious implications for the management of landscape. Formative processes have already ceased to operate in many western European areas – in the deserted sheep-runs of central France, for example, or in northern and western UK where improved farmland has reverted to moorland (Figure 3.3). Similar changes have happened in the more distant past, as with the change to private and pastoral farming in late medieval midland England, or as they did in 20th century England (and will in central Europe) as smaller traditional family farms disappear under the pressures of the Common Agricultural Policy (CAP) and the global market. Perception change as time passes and as the world changes. Peoples' views of landscape alter, for example, as they live more urbanized and globalized lives, and start to regard the countryside as a playground more than as a factory for food. In the UK, a very small proportion of the population has worked the land now for the best part of a century, with consequent changes to how the rural landscape is perceived by the great majority.

Historic processes that create the landscape, whether physically or through perception, are therefore central to landscape management. Understanding them as fully as possible, and thus understanding how landscape works, is a prerequisite for landscape management. Further, if these processes change, or cease to operate, landscapes will change. It might be possible to introduce new processes that create a similar type of landscape, such as new ways of farming based on environmental stewardship, but it will be either subtly different ('inauthentic') or a museum piece, or a sort of 'facadism'. More probably, however, landscapes that were once fully cultural – upland landscapes created by now vanishing sheep grazing, with their drove-ways, walled fields, open aspects, for instance – will retreat into a state of apparently natural woodland, cultural in a very different way. History will become invisible, or foreshortened, and people will identify abandoned cultural landscape as untouched natural wilderness.

Figure 3.3. Deserted sheep-runs – Causses of Gramat.

Change

The third aspect that most characterises the historic landscape is the concept of change itself, the study of which is central to archaeological research. Any study of landscape reveals ubiquitous, successive, almost never-ending change across very long time-scales. Change in the past is easily demonstrable (Figure 3.4), and by and large it is accepted as the reason why we have the landscape that we admire and value today. In contrast, continuing change in the present is mistrusted, feared and often opposed by many sections of the population, usually those for whom landscape values are social, aesthetic or environmental rather than economic. This difference is of course partly because historic change has been neutralized by the passing of time. It also appears to derive from a lack of confidence in the present, and from the slowness with which perception of landscape change (it takes time, a generation or two, for new landscapes to be assimilated into a collective perception). Perhaps too it arises from a misunderstanding of the character of landscape itself, such as a frequent assumption that landscape is natural, unchanging or timeless, all of which are difficult ideas for an archaeologist to agree with.

For the future, change is unforeseeable in detail, but our knowledge of the past tells us that there will be much of it. Perception of landscape will be modified as a result of this ongoing physical change, but perception will also change independently of physical changes. There will be new aesthetics (the environmental aesthetic of the Green movement, for example, or the tendency to regard the countryside as a recreation area and tourist attraction). People will learn more about the history of landscape. They will also start to value recent changes to the environment. Eventually, many recent, present-day and future changes will be absorbed into freshly imagined new landscapes (Figure 3.5).

A tendency to resist present-day change sits uncomfortably between the general recognition on the one hand that historic change has created what we value today, and on the other that present and future change will eventually be accepted as part of future perceived landscape. People remain uncomfortable about new changes in the landscape. There are many pressure groups and campaigns to preserve the countryside as it is. In England, the government sponsors research to monitor and measure change in order to try to control it, using landscape character (which itself contains a reflection of change through time) as a starting point. In other words, we define the effects of past change to try to guide further change, but in the knowledge that landscape has always changed (Countryside Agency, 2004). A definition of landscape character at a particular date is only a starting point for management, a piece of information, not a prescription. Other objectives and policies are needed for this to be converted into action; simply trying to keep or recreate landscape character as it existed at any given point in time is not sustainable because future change must be accommodated.

Managing landscape

The risk in regarding landscape change as inevitable is that landscape professions can simply become witnesses of change, rather than actors influencing it. To avoid this

Figure 3.4. Effects of agricultural changes in Northumberland.

Figure 3.5. Recent landscape changes due to the implementation of wind-farms – Halland (SW).

requires a clear agenda for future change, not simply for reactive preservation. Different sectors in landscape research, however, have different objectives for the management and character of future landscapes. Many landscape architects treasure harmony, and value landscapes that coincide with pre-defined aesthetic codes that are themselves often historic, such as the sublime or the picturesque. Ecologists often interpret 'landscape management' as a tool for enriching or enhancing biodiversity; archaeologists can see it as a tool for protecting individual sites, often extensive complexes which they describe, slightly confusingly, as 'archaeological landscapes'. As well as integration and interdisciplinary collaboration being called for at the knowledge and characterisation stages, therefore, they are even more necessary at the management and planning stages.

Public perception of landscape change is a further factor affecting landscape management. Countryside campaigners in England often put amenity on a par with, or in advance of, aesthetics – they seek to protect 'open' countryside, 'green-field' land, as opposed to built areas, irrespective of the quality or character of the landscape in these areas. Somewhere in this mentality, in UK at least (the question of national cultural difference in landscape perception across Europe is a separate interesting issue), is a strong theme of nostalgia, almost itself now a 21st century aesthetic. This takes the form of a more or less unconscious treasuring of a non-existent golden age of the recent past (precisely when depends, of course, on the age of the person concerned), a time when the landscape was "good", "finished", "traditional", "natural" and since when most changes have been deemed to be negative.

A view of landscape that puts time-depth, historic processes and past and future change at the forefront of appreciation and perception of landscape has repercussions for how landscape protection, management and planning are approached. It can affect our attitudes to new changes and to the future landscapes that are being made all the time. Equally, the holistic treatment of landscape, as a seamless and ubiquitous entity ("all parts of the territory", as the European Landscape Convention phrases it), requires a changed perspective on management compared with previous protectionist approaches. The various methods that European countries have developed since at least the 19th century for protecting, first archaeological monuments, and later buildings, depended for their effectiveness on selectivity. The best examples of something (whether a single building or a whole 'natural park') would be found and kept, the rest being allowed to change with time, or to vanish if necessary. This type of approach cannot be expanded from a relatively small number of discrete sites to the whole landscape.

To achieve landscape management as envisaged by the European Landscape Convention, with landscape at the core of a social policy relevant to everyone, emphasis will need to change from primitive sector-based protectionism (which was an early reaction to high levels of unmediated threats and losses) to a more positive, flexible and holistic management of change as part of spatial planning (which would be a substantive engagement with the process of change, rather than an opposition to it). This is also closely related to sustainable development. Landscape is usually said to be part of the environment leg of the sustainable development tripod, the others being economic and social needs; in truth, however, it can contribute to all three legs. It does this, however, not by insisting on protection or preservation, nor by trying to look backwards to vanished landscapes, but by being clear-sighted about the future landscape that we would like to see. In short, managing landscape requires a designing, forward-looking perspective to create future landscape, but one that is fundamentally informed by knowledge of the past, and particularly of the "past in the present" (Fairclough, 2003a).

Past change has given us our landscape, so there is little logical reason to fossilize and protect it at this stage purely for the sake of doing so; to avoid this, however, our ambitions for the shape and character of the future landscape have to be clearly defined. There will always be a case for the firm protection of individual components which, without supporting uses, might otherwise vanish. To manage the overall patterns and general character of landscape, however, it is more sustainable, and more in keeping with the cultural nature of landscape, for the starting point to be a presumption that further change will happen, rather than a presumption in favour of keeping what exists (or, in the current nostalgic mind-set, trying to put back what is thought to have recently existed, to be more 'natural').

Debates about future landscape can then move on to more positive questions, such as deciding what sort of change should happen, on what time-scale, with what mitigation or modification. The definition of current landscape character is not an automatic guide to what should happen in the future, but it is a knowledge-based starting point. Many countries in Europe are already in a position where conservation, often hand in hand with tourism, sits alongside the economy and politics as a driver for change. In no part of heritage management, nor indeed environmental protection and nature conservation, is this more the case, and more rightly so, than in the field of landscape.

A recent EH booklet *Using HLC* gives many examples of how the results of HLC are being used as part of an engagement with the future, in the areas of spatial planning, landscape management, agricultural policy, and research and communication (Clark *et al.*, 2004). On a large scale, using both management and planning, HLC has been used in England as the starting point to frame responses to government plans for constructing massive new urban communities in south-east England. This work examined the problems of sensitivity and capacity, rather than the concepts of importance or significance, that traditionally drive heritage management (Went *et al.*, 2003; Croft, 2004; Green and Kidd, 2004). This also allows 'condition' and 'character' to be studied separately, for example, to consider whether hedgerow loss over the past half century is a matter of poor condition compared with past character, or an indicator of a new current character.

Landscape as "being in the world": how perception grow and change

Just as the environment itself changes all the time, so too does the perceived landscape that we construct from it. Changes in perception are partly a response to physical change – views of landscape will change if roads and houses are built, if farming practices change, if forests are planted. Perception, however, can also change independently. They have their own reasons for change, linked to the passage of time and changes in fashion, understanding and attitude.

Some examples are relatively straightforward. Various 'rules' for aesthetic appreciation, such as the sublime or the picturesque, have succeeded each other over the last few centuries, each altering peoples' perception. There is the famous example of the English writer and journalist Daniel Defoe who found the mountains in the Lake District frightening and repellent when he passed by in the 1720s, well before the 'discovery' and promotion of the romantic ideal of wonderful scenery by William Wordsworth. Defoe called them "mountains high", with "a kind of unhospitable terror in them", a "terrible aspect", "frightening". Wordsworth's poetry made people see differently: mere words – poetry – can change the landscape. Thomas Hardy's treatment of valley-land and especially heathland in 'Wessex' provides further examples of perception being altered or created by the attribution of new ideas or associations.

Another example is the way perception change with knowledge. A field is a field, for example, until it is pointed out that beneath it lie the remains of a medieval or Roman abandoned city, or a deserted village – then we can perceive it differently. Woodland is simply a group of trees, until the remains of earlier fields give it a place in history, and a new-found sense of artificiality or perhaps transience, or until a landscape ecologist explains that history can be read in the shape of individual ancient trees. Landscape, in other words, is created by the stories that are told and heard about the land (Clark *et al.*, 2003).

Finally, perception will change simply because of the passage of personal time. This is linked to familiarity and personal memory, and involves a person's way of being in the world. We can all think of examples of recent architecture (19–20th century) that were once disliked but are now cherished – the buildings have not changed, other than to get older, but the people making judgements about them have changed. They are older themselves, perhaps, and have readjusted their world-view, their landscape, to the existence of these new things. There will also be younger people for whom these new buildings have always been there as part of their landscape, without the 'stigma' of being new. Such buildings might indeed be cherished parts of youthful memory, part of the 'givens' of a particular place.

The concept in this paper of *'gestion des perceptions'* refers to this aspect.

Whilst continuing to manage physical change to the landscape's material basis (e.g. restoring dry stone walls, coppicing woodland or guiding development in location and design), it might also be possible to manage changes to public perception as well. This is not a matter of trying to tell people what to think, or what not to think. It is a matter of offering new information and making suggestions for new ways of seeing, not because any earlier perception was wrong, but because perception can be multiple, and can be enriched and deepened by new ideas and knowledge.

A recent EH leaflet considered the additions and subtractions to and from landscape made very recently during the later 20th century (Bradley *et al.,* 2004). It proposed that archaeologists should consider the past 50 years of environmental and landscape change for its positive contribution to landscape, as well as for its destructive effects on past landscape. The leaflet's aim is to encourage people to look differently at what is familiar (and often disliked), in order perhaps to understand it more clearly, perhaps to see it differently. The result may well be a desire still to sweep away all the recent additions and to plan a new landscape or retreat to copies of older landscape fondly remembered or imagined. There is a responsibility to understand what we destroy, however, and a more considered, informed and reflective response might well be different in any event. Sooner or later our children or grandchildren will begin to value even what we think are the worst aspects of modern landscape. In an age when the conservation ethos is self-consciously a social driver of what we do to the world, we can easily look ahead, or rather put ourselves in the shoes of a descendant, and consider what from our age we should pass on as part of future heritage.

Managing changes in perception is largely a matter of communication, dialogue and sharing. We do it every time a landscape assessment is published or put on the internet, every time a TV programme shows its viewers images of landscape or wildlife. Archaeologists contribute their own specialist perspectives to this process, perspectives that illuminate the contribution of the deeply ancient as well as the very recent past, whether through visible or invisible remains. They can provide stories in the shape of human processes and actions – cultural explanations for a cultural phenomenon, ones that focus on society and human agency. Other landscape disciplines will contribute their own 'stories' and viewpoints. Spatial planners can contribute a sense of the economic and functional values that landscape holds. The public at large (everyone, including specialists in their personal lives) can contribute personal perspectives to landscape – on value, on amenity and use, on memory.

Out of all these views, and others, will emerge a changing concept of 'landscape'. It is this inclusiveness, whether interdisciplinary or democratizing, that is most notably promoted by the European Landscape Convention. It promotes this in its preamble, with the primary assertion that landscape is Europe's common heritage, an essential element of well-being both individual and social. It promotes it throughout the text through its insistence on democratization and participation, collaboration and exchange. It promotes it in its final substantive article which institutes a European Prize, not for the qualities of an area of land, but as recognition of good examples of peoples' collaborative actions in managing change to the landscape and in shaping the transition from past to future. Landscape is only ever a cultural phenomenon. It is peoples' (and society's) physical interaction with nature, and it is nature filtered through cultural perception. As a result, it always subject to continuous change, and always a product of people and their history.

Chapter 3

From tree-lined banks to hedge landscapes: the dynamics underlying the ideas about the landscape that have inspired public policies[1]

Monique TOUBLANC and Yves LUGINBÜHL

The root of all public policy involve situations that entails a problem for society that must be resolved. The identification of the problem and its formulation are related to the way in which social groups, linked in some degree to government or public institutions, perceive the situation: a perception that depends on both the practical and the abstract relationships that they have with the issue and that they will extrapolate to society. In this particular case which deals with public policy related to the restoration of the hedge landscape, the object is the hedge and the present situation is the major and even radical transformation that this landscape has undergone over the past 50 years. The hedge landscape refers not only to a particular form of organization and structuring of space, but also to the plant life that is a major part of this organization. Perceptions of this object, plant life – and trees in particular – have evolved over the last decades under the influence of different ideologies including those related to ecology and the landscape.

---

[1] The reflection that we propose here is within the framework of the call for research proposals, 'Public Policies and Landscapes: Analysis, Evaluation, Comparisons', launched in 1999 by the Ministry in charge of the environment. It is primarily based on the results of a multidisciplinary research study whose goal was to analyse and evaluate hedge landscape restoration policy. From 1999 to 2003, sociologists, geographers, agronomists and ecologists joined forces to study the ecological, social and agronomical effects of hedge landscape restoration policies initiated in the Côtes d'Armor department in France. It is within this context that Y. Luginbühl and M. Toublanc analysed hedge planting policies as such and their evolution since the end of the 1960s. To do this, they conducted a qualitative survey of the different institutional, political, associative and professional participants, as well as planters and non-planters, among others, involved in these public actions. The results of the survey were compared, on the one hand, with an analysis of administrative and technical documents, produced or not within the framework of the public action studied, and on the other hand, with the observation and recognition of 'concerned' landscapes.

These changes have their own specific language and generate public policies that are themselves transformed with the emergence of new social concerns, particularly those related to the question of nature, identified by objects said to be 'natural', with specific forms, expressing symbols that are not only aesthetic but social and cultural as well. These symbols may refer to professional groups, types of economies or even to ecological processes. When these symbols are attached to objects, they give them a meaning. In the debate on the evolution of the landscape and especially the hedge landscape, both supporters and opponents of the maintenance or restoration of the hedge landscape use their own expressions to refer to a view of the landscape, agriculture and even the world as a whole.

The evolution of these policies, from the 1960s to the present, demonstrates the transformations in prevailing thought trends concerning nature and landscape resulting in modifications in the way they are organized. The *'bocage'*, landscape typified by hedges bordering fields, recognized for so long by famous geographers, is not an exception to this rule. This example is particularly useful in explaining the evolution of ideas about both nature and related public policies. Through this double movement, which at the same time meets with objections and even leads to animated social conflicts, we can witness contemporary landscape transformations as well as deep social change. The duality of aesthetic and political evolution that can be seen in the different positions taken on the issue of hedge landscape is marked by three successive periods that express the parallel evolution of scientific approaches and representations of tree and landscape. The first period is that of the negation of the landscape that could be that of the necessary 'starting from scratch' required by technical and economic progress. The second period begins with the emergence of the landscape on the social scene and a first approach to the issue of hedge landscape; this period is fragmented and centred on the tree, either individually or in groups. The hedge landscape in this case is just a series of rows of trees, without consideration of its wider role. However, within this second period, it is possible to identify different phases corresponding to a series of transformations in the development of these policies. The last period introduces the systemic role of the hedge landscape, and the tree disappears to be replaced by a more complex group where ecological processes are influenced by social ones.

First period: the hedge landscape (or *'bocage'*), an obstacle to the modernization of the landscape

This heading warrants two comments. First, it is voluntarily biased since it implies that the landscape is in the domain of agriculture, whereas it was out of the question in the agricultural sector at that time to imagine the landscape as the work of farmers; the landscape was still a specific concern of the bourgeoisie. Second, this heading expresses the representation that society had at that time of the hedge landscape, i.e. a landscape structure considered as a mark of the archaism of the countryside that contrasting with the rationalisation of agriculture.

In the 1950s, hedge landscape regions, like other landscape regions such as terraces, wetlands, etc., corresponding to 'traditional' means of developing space through agriculture, were either the object of radical transformations (linked or not to agriculture), or abandoned – left fallow or reforested – since they no longer corresponded to the expectations of French society. In this context, the value of hedge landscape regions depreciated and they became synonymous with archaism, to the point of being totally eliminated in many regions where agriculture had become production-oriented and intensive. At that time, ordinary landscape regions were not considered as 'landscapes' and the hedge landscape was not an exception; the landscape as such was still associated with picturesque spaces, out-of-the-ordinary and exceptional, not influenced by social activities and agriculture, in particular.

Agricultural modernization policies, which led to what is sometimes referred to as the second agricultural revolution, bludgeoned anything that got in the way of their implementation. The terms used are associated with the idea of progress and have a positive significance for agriculture: 'modernization', 'rationalization', 'mechanization', 'improvement', 'efficiency'. In the same way, language and ideas related to space are used: 'rational consolidation of plots', 'special remodelling', in other words, 'land consolidation'. For its protagonists, the only goal of land consolidation was to encourage the profitability of farm work. This was not done during this initial period for the purpose of protecting the natural environment but instead, for redistributing land in relation to its agronomic quality, in other words, its capacity to produce. The term 'land consolidation', which was not new, was undoubtedly the one that gradually polarized its opponents and contributed to the creation of a new trend of thinking with a particular vocabulary. This movement was obviously not completely new either. It already existed. Basically, members of the first landscape protection associations had begun to denounce the effects of land consolidation, but this trend was fairly marginal and its arguments essentially aesthetic, like the position of the SPPEF (French Society for the Protection of the Landscape and Aesthetics), founded in 1901 by Charles Bauquier, deputy of the Doubs department, which included artists and writers among its ranks. It is referred to in the arguments of the protagonists of the law of 1906, concerning the protection of natural monuments and sites, that was repealed in 1930, to be replaced by the law concerning classified and registered sites. Until the end of the 1960s, this position was the only one associated with landscape protection. Built on an aesthetic nostalgia of tradition, and in accordance with the ideas of 'Picturesque France' movements in the 19th century, it did not go beyond these formal limits and was not based on an ideology about the landscape that takes ecological, social and economic issues into account.

Opposition to the agricultural modernization policy by criticising the effects of land consolidation is as radical as the policy which was based at that time on an ideology of starting over from scratch that can also be found in other areas, such as urbanism and regional development in general, thus constituting a political priority[2]. The work contingent on the expansion of grouped plots, the goal of land consolidation, lacks any concern for the natural environment and its faunistic and floristic wealth: looking back, land consolidation that took place at that time is considered by farmers today as being "exaggerated". The ideology of agricultural progress that was massively adopted in most of

---

[2] The decades, 1950s and 1960s, were those of regional development with a double priority: agricultural development and the development of rural community facilities. This period was witness to rapid expansion in land development (land consolidation, etc.), the development of water projects (drainage, etc.), forest resource management, wasteland rehabilitation, etc.

Figure 3.6. Typical pollarding : *'coupelles'* and *'ragosses'*, Rennes area (Vieux-Vieil).

the hedge landscape regions was based on certainties, and did not include notions, such as those of precaution or doubt. It is true that many farmers began to express discontent with the shrinking and dispersion of their plots in the 1950s, as well as with the necessity of maintaining trees in hedges, banks and ditches. This maintenance was a role left over from the traditional tasks of medieval times, and the arrival of agriculture in an era of economic and social progress implied the eradication of archaic customs (Figure 3.6). This allowed the peasant to become a farmer.

Nevertheless, this ideology of starting over from scratch was not general and effective in all hedge landscape regions: some of them maintained a fairly dense hedgerow pattern. Regardless, it was certainly not for aesthetic or 'landscape' reasons that some landowners and farmers opposed land consolidation and/or attempted to maintain hedges, but most likely for other reasons, such as an attachment to inherited lands or a feeling of being dispossessed of the land, or perhaps even the fear of seeing a neighbour with more land or a member of the family with whom relations were strained, cultivate this land. It was of little importance that the surface area recovered after consolidation was identical to the initial area. Conflicts between supporters and opponents of land consolidation, even among farmers and inhabitants of rural areas, were undoubtedly more numerous than it previously believed[3], but the preservation of the hedge landscape was certainly not the

---

[3] These conflicts were observed when different surveys were conducted in Brittany (Côtes d'Armor, Finistère and Ille-et-Vilaine departments) as well as in the regions of Champsaur (Isère) and Boischaut (Indre). These conflicts were sometimes violent, and law enforcement was required to restrain the inhabitants of a community opposed to land consolidation from lying down in front of the bulldozers that had come to clear the hedges. However, the issue at stake was not one of landscape quality.

issue. These conflicts set groups of rural inhabitants, landowners and farmers against each other, in battles between 'big' landowners and 'small' landowners or former estate owners and their former sharecroppers.

Finally, in this first period, which prepared the way for the arguments and new policy in favour of hedge landscape restoration, the landscape is almost absent from the arguments and does not seem to be an issue; it appears to only be accessory and has no impact on the transformations of space or the content and evolution of policies. The semantic symbols used here remain quite poor and radical, opposing modernity with tradition or archaism.

Second period: setting up a hedge landscape restoration policy

This period, which can be divided into several stages, is much richer in its multiplicity of arguments and symbols; it reveals the emergence of an ideology concerning the landscape and nature that mobilises a wide variety of words and expressions relating to more sophisticated ideas and deeper reflections. It is not really worth discussing the period at the end of the 1960s when a new ideology about nature – obviously waiting for its time to come – and especially the landscape, appeared. We will only recall that those who supported introducing landscape into regional development projects participated in spreading this new ideology that, on the one hand, affirmed the break with the starting over from scratch ideology, and decided to work with the existing landscape and, on the other hand, promoted the development of qualitative methods based on in-depth analyses. However, this group, that was made up of only a few scientists, landscape artists and architects in the beginning, was not the only one to try to subvert this movement. At the same time, the associative movement, picking up on the momentum of the trend begun in the preceding period, adopted part of the arguments developed as a result of the new social awareness of the environment to support its position in favour of the modification of agricultural policy, focusing particularly on its negative impact on rural areas. The land consolidation policy was stigmatized because of its effects on the hedge landscape. In response to this movement, the Ministry of Agriculture and the agricultural production sectors often insisted that land consolidation consisted strictly of an exchange and redistribution of plots that were not themselves responsible for the disappearance of the hedge landscape; instead, they claimed that it was the contingent work, that is, work whose purpose was to make it possible to cultivate the new field pattern, which led to the elimination of hedges, roads, the levelling of banks, the creation of new road systems and ditch networks for drainage and irrigation. It is totally symptomatic that supporters of the modernization of agriculture emphasize this aspect of land consolidation, as if to exonerate it from its role in landscape transformation. On the contrary, opponents perceive land consolidation as a threat to the hedge landscape.

In this respect, the first landscape arguments in favour of the defence of the hedgerow landscape were based on the idea of an open landscape, such as that of the Beauce, for example, that presents fewer landscape qualities than the hedgerow landscape characterized by the presence of trees. The latter is highly symbolic and makes the ideology of starting from scratch look particularly bad. From the point of view of the majority of the population, a landscape with trees is obviously more pleasant to the eye than a bare landscape: "if there are no trees, it's a real desert" (an inhabitant of a hedgerow region, not involved in agriculture).

Effectively, social arguments that took this new awareness of environmental issues into account very rapidly equated the tree and vegetation, in general, with nature:

"Hedges – a breath of nature", "an island of nature in a streamlined countryside", "nature, we should try to preserve it a little; it would give us some peace and quiet" (quotes from inhabitants not involved in agriculture).

Hedges and trees, therefore, constitute elements of nature with a symbolic value in the collective subconscious. However, during this period characterized by the birth of an environmental collective conscience, this symbolic value extended beyond the strict and immediate meaning of 'landscape' (in the restrictive sense of the term reduced to its aesthetic dimension). Even if some social groups considered that the hedgerow landscape was more pleasant and "prettier" than an open field landscape, it was for other reasons that the question of the meaning of the term 'landscape', was raised, mainly because open field landscapes were symbolic of intensive agriculture (probably not always justified) closely linked to agri-food markets. Some social groups saw economic systems that they rejected operating behind the open field landscape, systems which they would rather avoid. The hedge landscape, from an exaggerated point of view, evoked a system of agriculture that provided better quality, more 'natural' food products. Its disappearance signalled the end of a secular agrarian society.

These symbolic assimilations can clarify and provide a key to understanding the content, at least in the beginning, of 'compensation' policies to counteract the negative effects of land consolidation. As a result, at the beginning of the 1970s – and we can consider this as an initial 'subperiod' – we began to see modifications in the design and conduct of land consolidation aimed at taking environmental impacts into consideration by reducing them or 'compensating' for them. Nevertheless, modernization remained the priority. Preserving nature was not a project in itself; we were content to just limit the damage or, if this was not possible, to correct after the fact, the negative effects of consolidation by compensation. Beginning in 1970, the Ministry of Agriculture initiated environmental studies before proceeding with consolidation practices, but these were essentially based on an ecological approach. The law of 1976, concerning the protection of nature, made impact studies obligatory within the framework of land consolidation. Once again, however, even if some of these studies took the landscape into consideration, it remained a minor issue and rarely resulted in proposals that were followed up. This first phase of the initiatory period of hedge restoration was also part of the ideology of rural development symbolized by the Rural Development Plans (PAR), launched in 1970 with great success and completed in 1975 by regional contracts. Within this context, a 1975 law transformed land consolidation into a rural development tool. The 1970s could be considered the golden age of rural development, and land consolidation followed suit.

This new legal and regulatory framework that corresponded to the emergence of a new social awareness of the environment had an influence on the evolution of land consolidation practices and contributed to the advent of the hedge restoration policy, although it did not have a direct impact on the very first plantations that we observed in the Côtes d'Armor (1972) and that preceded the two laws mentioned above. Even if these 'hedge restoration' practices must obviously be seen within their historical context, it seems to us that they are, first and foremost, the direct result of the 'traumatism' that the unprecedented transformations of rural space resulting from the land consolidation of the 1960s represented for the rural population: "It was necessary to reassure those people who had been shocked by the extent of the destruction of hedges and groves" (the words of a DDAF technician).

This explains why the main objective of hedge restoration policies for a long time was not the choice of species or the establishment of hedges but, instead, the replanting of trees or plants. During the first years of policy implementation, the Government, represented locally by the Forest and Agriculture departments (DDAF), functioned on the basis of planting subsidies; however, what is obvious here is the convergence between planting policy and forestry policy. Beyond the constraints inherent in the administrative system of that time, we must speculate about the limits of the DDAF's 'landscape' knowledge, which appeared to be limited to forests, at least in terms of replanting. The first replanting operations (beginning of 1970s), carried out in departments particularly affected by land consolidation, such as the Côtes d'Armor, were linear plantations of softwoods (Figure 3.7), conforming with forest policy and forest planting techniques at that time, for example, replanting of cedars, Leyland and Lawson cypress and Sitka spruce, species subsidized by National Forest Funds (FFN). The terms used, 'windbreaks', 'shelter belts', 'full planting', are indicative of the collusion between replanting policy and forestry policy, the objective of which was to produce wood. Pretending to be compensation for the effects of work linked to land consolidation, replanting policy

Figure 3.7. Coniferous hedge hiding a farm shed (center of Britanny).

only applied to communities where land consolidation had taken place, as if the razing of hedges and banks had been limited to these communities, which obviously was not the case. It recommended the use of fast-growing species and flat planting, i.e. without banks.

Arguments developed during the first phase of this period focused on objectives that had little to do with 'landscape' (in the strict sense of the term, in other words, aesthetic); to convince farmers about the policy of replanting, arguments focused on wood production and wind protection, with further justifications, such as erosion and runoff protection, after the meteorological event in the Saint-Brieuc bay in 1973. A housing estate located at the bottom of the Langueux hill was flooded during a storm and engulfed by mud caused by runoff water at high tide, preventing the water from draining off. Landscape was still not an issue, or only indirectly, in the motivations of the policy architect. As a result, these plantations were completely disconnected from the existing network and conspicuous because of the differences in their composition and morphology compared with hardwood hedges (pruned or not).

This first phase ended in 1979 when, on the one hand, those involved in replanting policy at the national level, and the Institute of Forest Development (IDF), on the other, converged towards a new form of planting. Hedges were now planted on a plastic film (mulch) consisting of several strata (bushes and trees), with a wide diversity of ornamental and rustic species (Figure 3.8). This hedge model, developed and tested by an agricultural engineer, D. Soltner, satisfied the criterion of softwood hedges, but was not what would normally be expected to be found in the hedge landscape, which is composed almost entirely of hardwood species. Essentially – and this is undoubtedly the originality

Figure 3.8. New diversified hedge in center Britanny (Plemy).

Figure 3.9. Thuyas hedge in a housing estate (Trégueux, Côtes d'Armor).

of this replanting policy – it was almost always based on outside policy, at least in its initial stages, and was not really autonomous. In this case, hedge replanting policy in peri-urban areas (Figure 3.9) was at the basis of the experiment. To integrate housing estates created on a massive scale throughout the country and which modified the landscape of rural villages and communities around the big cities, the IDF proposed a hedge model that could be easily put into effect using mechanical methods (e.g. machines to roll out the plastic film and, today, machines that plant young forest seedlings). This new model replaced the preceding one that had led to the development of cedar hedges in the housing estates built at that time, but also those planted within proximity of agricultural buildings and farms. Experiments were conducted in peri-urban communities (e.g. Trégeux, near Saint-Brieuc), with the mobilisation of the housing development inhabitants to encourage their geographic and social integration[4].

Hedge restoration policy was, therefore, based on this model that was highly successful for several years (it is still 'active' in some regions and operations). It was implemented in the beginning by the DDAF and was reinforced by landscape arguments: softwoods were not considered to be plants adapted to the hedge landscape and were progressively abandoned. The landscape argument was not only about respect for the hedgerow landscape but also for 'landscape diversity' that softwoods could not provide. Softwood reforestation operations that were highly developed from 1960 to 1970 were criticized for this reason: the "all softwood" approach hindered landscape diversity,

---

[4] We can also advance the hypothesis that the softwood species corresponded to the landscape model of the time as can be witnessed by the development of cedar hedges in housing estates built at that time.

"vulgarized landscapes", "led to the uniformity of the region", an expression that was also used to criticize the impact of "urban sprawl" on the landscape by tract houses, single-family dwellings and "postage stamp" woods, built within the framework of forestry policy. Landscape justifications can be added to these, for example, the benefits of hedges for game or wildlife. The 1976 law concerning the protection of nature and that of 1975 discussed above, making land consolidation part of rural development, were part of this trend and made it possible to reinforce the landscape arguments of opponents of replanting projects.

In the name of landscape diversity, the multiple composition model consisting of trees and shrubs was easily accepted, particularly where flowering species are used, as in urbanization operations or the establishment of infrastructures.

A new phase began with the 1982 decentralization law. On the one hand, replanting operations from that time on were implemented by local institutions (general councils and regional councils, in particular) but, on the other hand, these institutions reinforced their arguments with research focused on the rehabilitation of the image of agriculture in French society. An attempt was made to counteract the claims of those who said that modern and intensive agriculture was responsible for the drastic elimination of hedges and banks, by improving the image of the modern farmer that would appeal to the rest of society; this aesthetic argument was thus proposed to farmers, mainly for the purpose of making farms more attractive and to hide livestock buildings. At the same time, organizations like the Chambers of Agriculture undertook the task of promoting these replanting projects and establishing a 'farm beautification' policy: by showing a more positive aspect of the farm where the farmer's concern to improve his living environment was obvious, he expressed his desire to be part of the existing landscape and to contribute to the improvement of the countryside. A major effort was also made at that time to promote rural tourism, country lodges and bed-and-breakfast accommodation. Another idea also introduced at that time that remained popular for years, particularly focused on farmer's wives who were essentially responsible for implementing rural tourism, for example, planting of shrubs and fruit trees to produce fruits for jams or preserves to sell to tourists. Moreover, the necessity of planting a "range of regionally specific plants" began to make itself felt although this idea had yet to come into its own.

In any case, these different arguments had their effects and many farms were enhanced by ornamental plants and access roads bordered by flowering hedges. When planting took place within the proximity of the farmhouse itself, farmers' wives played an important role in the choice of plants; they chose flowering shrubs to obtain a "colourful and bright" landscape effect. Some of these farms became demonstration sites for the farm community, such as a farm in the Côtes d'Armor where the farmer was one of the first to plant vegetation around and at the entrance to his farm buildings to create an image of the modern farmer, sensitive to urban concerns and anxious to contribute to the preservation of nature. The DDAF and the Chamber of Agriculture brought many farmers and their wives to visit the farm so that they could observe landscape architecture considered to be at its best.

As of 1995, the landscape aspect of the hedgerow restoration policy took yet another step: that of 'planting locally'. It was part of a wave of ideologies in favour of renewing regional identities. The species that were recognized and selected by the institutions responsible for these planting projects were referred to as "local species", inspired by the vegetation found in the region concerned. It was also the end of flowering 'urban' species, at least for hedges planted on agricultural land, although these ornamental species continued to be used to 'beautify' farms. We can obviously question the significance of these 'local species' since species considered to be 'local' in many regions were introduced many years ago, such as the Lawson cypress in Brittany or the fastigiate cypresses in Provence. This argument for 'local' vegetation actually demonstrates a new approach to planting by the technical groups that promote it, i.e. a desire to take into consideration the existing landscape and to adapt to it, and to respect its specificities, whereas the softwood species planted at the beginning of these operations were generally different from the naturally occurring vegetation or hedges that already existed. This model recently underwent a change towards a simplification of hedge composition. Although these hedges originally consisted of a fairly wide diversity of species, their composition has been simplified today to only a few species. However, it is ironic to note that it is only now that the argument of contributing to biodiversity has been added to those that already exist.

The 'plant locally' approach was strongly supported by the development of promotion initiatives carried out by associations and, especially in Brittany, by associations that initiated a re-training programme for building earth banks and pruning hedges. However, what is of particular interest is the double movement that took place between the associations and the first research programmes dedicated to the evolution of hedge landscape structures[5]. During this period, when few people were interested in the local forms of maintenance of hedge landscape vegetation, the research programme on the evolution of wooded linear structures developed by INRA-SAD Armorique, in partnership with the CNRS and the ENSP (1994–95), initiated new interest in hedge trimming forms and their maintenance, the maintenance of earth banks, etc. It is obvious that the interest shown in these different local forms of social production of the hedge landscape had an important effect on the perception of the aesthetic aspect of both it and of the forms of the pruned trees themselves. A local vocabulary arose to describe them, depending on the region. In addition to the list of terms for the types of pollards, there are also names for an enthusiast of these pollard forms, a member of an association and an ecomuseum in the Perche region. This new concern for the evolution of local pollard forms and for the earth banks on which they are planted, reveals the aesthetic conflict between those who defend pollards and attempt to justify these strange shapes on the basis of social, ecological, technical or aesthetic (or visual) arguments, and those that oppose them because they represent either an archaism left over from feudal times or a form of "mutilation" of the tree. This conflict concerning shapes brings to light a major difference in opinion on the concept of nature and the type of social organization between pollarded trees on earth banks and hedges recently replanted by institutions to make up for the levelling of vegetation that continues; this difference is clearly indicative of a transition from agricultural to rural. It is significant that the architects of these operations and those involved at the local level qualify hedges planted according to the 'plant locally' model

---

[5] Research programmes had already existed in the 1960s: they led to the symposium in Rennes (1975) on hedge landscapes, where research was presented and discussed on the advantages of hedges for agriculture and the protection of the environment. However, the analysis of the dynamics of maintenance practices was not at the core of the debate.

as 'hedgerow landscapes', whereas they do not use the same expression to designate old hedgerow networks.

At the end of this first period that ended around 1998, hedge restoration policy did not take any consideration of the old forms of this landscape; it was as if this policy had confirmed the end of a long antiquated era, symbolized by these outdated pollard forms on their earth banks – considered a thing of the past – and that it instead decided to promote a new era, that of "hedgerow landscapes", with the complicity of the agricultural sector[6]. Lessons learned from this period nevertheless deserve to be mentioned: until the time when a new concept of hedge restoration was established, the landscape was only seen as a series of formal elements, with no consideration of its social or ecological 'role'. The important issue for the institutions that implemented this policy was to plant a sufficient length of hedges each year that would impress the political or public community by its quantity (even if it was much less than the total length of banks and hedges destroyed, which was the case). With regard to the objectives of agriculture itself, little importance was attributed to the fact that these hedges were planted without any concern for the impact of surface water runoff or the ecological role of the landscape. Agriculture continued to become more and more intensive, despite the fact that the quantity of nitrates and pesticides in drinking water continued to increase.

Third period: towards a territorialized hedgerow landscape?

A new period in the design of hedgerow restoration policy began around 1998. The landscape progressively acquired a new social legitimacy and a new meaning that was no longer reduced to a few aesthetic figures. Until that time, policy focused on the replanted hedge. It often overlooked the banks, ditches and roads; nevertheless, the tree is not the hedge and the hedge is not the hedgerow landscape. In a sense, it is the tree that hides the hedgerow landscape. Hedgerow landscape policy underwent an evolution at the end of the 1990s with a new awareness of the 'hedge network' and the functionality of the 'hedgerow and field pattern'. The 1993 'landscape' law and the SRU (Urban Solidarity and Renewal) law of 2000 that introduced the PLU (Local Planning Programmes) already emphasized the need to "protect hedges or hedge networks" in certain landscape situations. They gave rise to the emergence of a different way of thinking that took into account not only the overall territory but the different roles provided by the elements or groups of elements as well.

We could then add the terms 'linear structures', 'network continuity', 'overall planting scheme', 'coherence', 'global diagnosis' and 'territorial analysis' to those of 'hedgerow and field pattern' or 'hedge network'. The model developed by the hedge replanting policy included the desire to organize space and to make the landscape 'coherent', by including different ways of thinking, particularly those of geographers and ecologists. Policy no longer promoted just planting rows of trees; instead it focused on

---

[6] This declaration does not mean that all farmers agreed with the elimination of the old landscape of earth banks and pollarded trees. Some still continue to prune the oak and chestnut trees and to maintain the banks. However, professional farm organizations (at all levels) do not support this practice.

more global roles and functionalities. Operations were no longer limited to tree planting; instead they attempted to rebuild planted and unplanted banks (e.g. the Côtes d'Armor in 1995), to create groves and to open new roads (or reopen old roads that had disappeared under vegetation) with hedges on both sides. Some planting operations corresponded to the restoration or development of roadways whose purpose was not necessarily limited to agricultural practices. Old hedges were restocked. Planting operations attempted to constitute a hedge and field pattern that could be qualified as 'inter-plot' and that followed the transportation network. This model focused on the interdependence of the elements of the hedgerow landscape and biophysical phenomena. In reaction to the scattered development that characterized planting operations up until that time, this policy attempted to establish a certain unity. Replanted hedges within the framework of this last model continued to be considered as part of the hedgerow landscape.

More than ever, policy objectives were multiple: ecological (to maintain the fauna and flora, expressed forever after as 'biodiversity'), hydrological (to control water runoff), climatic (to protect crops and habitats from wind), economic (to produce wood for heating, especially wood chips), cynegetic (to maintain game animals), landscape oriented (to improve the quality of the landscape), etc. This plurality is affirmed more strongly today. Admittedly as of the 1970s, with the emergence of 'ecology' and the 'environment', the scientific and technical literature has stressed the multiple functions of the hedge – agronomic, hydrologic, cynegetic and, in a more minor and vague way, its landscape role. Nevertheless, the range of functions today is wider, more specific and more integrated, as can be seen, on the one hand, by the evolution of the vocabulary (see preceding paragraph), and on the other, by the use of the master word, 'multifunction-ality', linked to the hedge as soon as it became part of the vocabulary. It is clear that within the dynamics of this policy, that landscape ecology, environmental geography and landscape theories, developed within the framework of different ideologies, all played an essential role.

In the Côtes d'Armor, in Brittany, where a specific research programme devoted to an analysis of hedgerow restoration policy was carried out, different experiences demonstrated the changes in policy design and, through its new manifestations, that of the landscape. We will mention the experience of a local development association, the Rural Initiative Mission (MIR) of the 'Pays du Mené' region that embarked upon a review and action across the whole catchment. In this case, the issue was not to plant hedges here and there but to integrate them into a broad vision for the territory as a whole to attempt to find a solution to the problems created by excess manure production. The approach was based on a territory that was functional from the hydrological point of view and on discussions with farmers. Each farmer was given an aerial photography map of his or her farm, showing existing hedges; the reflection thus consisted of determining the sites where new hedges should be planted to reconstitute a network and thus control water runoff. This involved a global perspective that integrated the ecological role into the territorial management of a farm, on the scale of a small tributary basin. Little by little, action was focused on another tributary basin and, in the long-term, should cover the entire watershed targeted in the MIR's zone of action.

The other experience observed was that of the CIDERAL (the syndicate of a community of communes of Loudéac) that attempted to integrate the role of the territory in a demonstration project involving farmers, the objective of which was "the improvement of the hedge and field pattern[7]". A local co-ordinator monitored the farms in order to try to convince the farmers to plant hedges and maintain them in order to sustain the network.

In the end, these experiences differed considerably from institutional policy – General Council, Regional Council, Chamber of Agriculture and DDAF – even if they used the same financing sources (e.g. the CIDERAL experience was financed by DDAF and the Regional Council, in particular). Institutional policies were transformed by the multiple contributions of experiences in the field, from the associative sector to the scientific community. Those policies, whose main objective was to distribute "kilometres of hedges", changed and instituted global diagnoses and analyses of communal territories or communities of communes by attempting to map planting initiatives (the mapping of planting projects was nonexistent during the first stages: it was impossible to pinpoint planting operations). Despite this evolution, the maintenance of hedges planted is still not resolved, planting sites are rarely found beyond the edges of transportation networks, and the construction of banks is still in its early stages (probably because its cost is higher than that of the planting itself).

However, the evolution of the policy marks a change in scale that is highly significant: by shifting from government to an '*intercommunal*' structure, the object of the policy also changes scale – from the tree to the territory that is a space with a social and political functionality (commune, community of communes) or an ecological one (watershed).

Conclusion

In this overview of landscape policies, the role of the relationships between participants is of utmost importance: transfers of words, expressions and symbols occur between the scientific community, the associative community, the technical community and the political community, which modify the policy and the design of its object. The landscape changed from having a strictly productive role in the beginning, to a decorative and then a multifunctional one that expresses the complexity of its role and that of nature at the same time. These transfers can be seen in the landscape itself. They are the results of individuals themselves and social relationships established during meetings, chance encounters or within the framework of research programmes or political actions. It is difficult to declare that this political dynamic is simply the result of the will of elected officials. It is also sometimes due to chance.

This political dynamic also includes climatic events; the floods of 1973 in the Saint-Brieuc bay and, more recently, those in different watersheds in Brittany, also influenced the way that policy was both implemented and even designed. Nevertheless, it still lacks links with other policies, especially agricultural policy, which is dominated by research whose goal is to increase agricultural production with indisputable effects on the landscape against which landscape policy is still incapable of defending itself.

---

[7] CIDERAL, Assessment of the programme, 'Improvement of the Hedgerow Landscape', CIDERAL 1999/2001, August 2001.

Even though landscape policy has undergone an evolution that takes social and ecological factors into account, it is still necessary to stress the importance of co-ordinating political action with other policies. This is a difficult task and one that also involves deep changes in the way landscape is taught and landscape practices that are still too closely linked to a formal concept that does not sufficiently emphasize social and economic aspects.

Chapter 4

New hedgerows in replanting programmes: assessment of their ecological quality and maintenance on farms

Laurence LE DU-BLAYO, Didier LE CŒUR, Claudine THENAIL, Françoise BUREL and Jacques BAUDRY

The purpose of this study was to analyse public policies that support hedgerow planting and, more particularly, to describe the development of species planted, their contribution to the ecological quality of hedgerow landscapes (*'bocages'*) and their incorporation into farming practices. Our analysis is based on two different viewpoints: the first is related to the objectives laid out in the documents dealing with the implementation of these policies; the other is related to our interpretation of the ecological function of new hedgerows within landscapes and their socio-technical sustainability within the farm unit. Research was conducted in the Côtes d'Armor, a department of France very involved in hedge replanting policies at the national level and where we were fairly well acquainted with the status of the hedgerow network (Le Du-Blayo. *et al.*, 2000).

In documents dealing with the implementation of planting policies, we can observe the progressive shift from a purely quantitative objective in terms of lines of hedgerows planted, to a qualitative objective in terms of hedgerow structure and network organization (Le Du-Blayo, 2004), from a schematic point of view. Their clearest formulation appeared in documents beginning in 1998 with a list of issues justifying the allocation of public financing: closing off fields, erosion control, wind control, floristic and faunistic value, integration of man-made structures into the landscape, etc.

For this study, we chose to develop a 'reading grid' for the maintenance issues and the promotion of biodiversity, with reference to the *floristic and faunistic value* mentioned in these documents, as well as, from a more general point of view, the management of natural resources and the ecological functioning of the landscapes. Many studies have revealed the important role of hedges in the maintenance of the biodiversity of hedgerow landscapes. This biodiversity consists of arborescent and shrubby herbaceous plant species, as well as the animal species that use them as a habitat (i.e. for food resources, reproduction site, protected movement within the landscape, etc.). We cannot study this biodiversity without also taking into account the links between these hedges and the different elements of the landscape (hedges, woods, plots, etc.); in reality, each element by itself is too small to maintain the species and it is the overall network at the core of the landscape mosaic that enables this maintenance by permitting movement and adequate resources in terms of quantity and diversity (Bolger and Scott, 2001; Fleury and Brown, 1997). This landscape mosaic must be considered from the viewpoint of its different elements (hedges, woods, grasslands, crops), as well as from that of the state of its plant cover (cover rate, vigour, etc.). We can thus understand the importance of the addition, removal and organization of the different elements in the space in question, as well as the management of cultivated plant covers and the maintenance of uncultivated plant covers (Baudry and Jouin, 2003). Therefore, the nature, diversity and sustainability of hedge management practices and their context are the key components in the sustainability of hedges, particularly in relation to biodiversity. This brings us back to their integration within rural and agricultural systems.

On the basis of these principles, we conducted two complementary analyses. We analysed the data in hedge planting requests from the General Council of the Côtes d'Armor department in order to interpret them in relation to the issues laid out in the documents available to us and the principles described above. We particularly focused on planted spaces, the hedge structures built there and their location between the periphery of buildings and open fields. We also conducted a field assessment using surveys and observations in a sample area in the Côtes d'Armor department. We evaluated the movement of animal and herbaceous plant species resulting from the connection between new hedges and the pre-existing network. Sustainable maintenance of new hedges implies that one or several of the managers have taken responsibility for planting, shaping, maintenance and guarantee of maintenance conditions. We attempted to evaluate whether or not farmers took on the responsibility for these new hedges and in what way.

Materials and methods

Gathering data on planted species and their location from planting requests

In the Côtes d'Armor, hedge replanting policies are linked to many different interest groups since different organizations have established planting incentive measures: the DDAF, the Chamber of Agriculture, the General Council, regional and local agencies, etc. Nevertheless, since 1988, the General Council has played the major role – that is, of financer – and it is, therefore, for that reason that we conducted our analysis of planting programmes in relation to this organization. A statistical analysis was carried out on replanting demands filed for the overall Côtes d'Armor department with the DAE (Agriculture and Environment Administration) of the General Council of the Côtes d'Armor, the agency then responsible for public planning policy.

The total of 5547 replanting requests that received financing during the period from 1988 to 2000 were analysed. These demands contain information for each hedge and planting entity: the name of the planting entity, its status (private/public), the location of the hedge (building, open field, side of the road, plot margin, etc.), and the length and sequence of species that make it up. Our work consisted of establishing a computerized database from the filed hard-copy requests and, therefore, choosing data and formalizing them. Out of the 67 species used, only 26 were chosen (and organized into 21 codes) – mainly they were the species most often mentioned, to which were added several species considered to be significant by ecologists. Only the occurrences of species were noted, in other words, the presence of a species in a sequence used to plant a hedgerow, i.e. by year, commune, public or private planting entity and location (building or open field).

We would like to emphasize here that the results of this analysis concern only information from planting requests and that there is no doubt, a gap between planting declarations and what was finally planted, that we did not evaluate.

Sampling approach for communes in the Côtes d'Armor department on the basis of cartographic data and agricultural surveys

Field analysis (surveys and observations) was built on a multicriteria selection based on the typology of hedgerow landscape units (Le Du-Blayo., 1995, 2000), the evolution of agricultural production, and types of agricultural organizations within the department. The choice of seven communes representative of these two axes made it possible to establish a common nested study approach where researchers could apply their own investigative methods in terms of geography, ecology, agronomy and sociology. It should be observed that the field approach implemented in the areas of agronomy and ecology took account of the different new hedges present on farms and in the zone studied, without distinguishing them from those included in the General Council's planting requests.

Method for assessing floristic and faunistic value in new hedges, and colonization from the pre-existing hedgerow network

The assessment of the impact on biodiversity of new hedges was carried out on a site of about 100 ha chosen from within the commune of Saint Péver (one of the seven sampled communes), taking two biological models into consideration.

(i) A group of insects, 'forest' beetles, known to be indicators of the density of the hedgerow network and hedge vegetation. These species are characteristic of stable environments because they are wingless, and move slowly from place to place using preferential routes throughout the territory. They use hedges with arborescent and dense herbaceous vegetation as corridors to move from the forest, the source of species, towards farm space. In Saint Péver, we compared beetles in old and newly planted hedges, varying in age up to 20 years.

(ii) Flowering plants of the herbaceous layer that we knew from previous studies are characteristic of the ecological environment of hedges (Le Cœur *et al.*, 1997). We compared the plant biodiversity of three hedges of softwoods at the edge of a garden, 10 new hardwood hedges adjacent to farm plots, planted flat on a plastic tarp and, finally, the edge of a small grazing wood, considered an old structure in this landscape.

Method for assessing the management of old and new hedges

The study of old and new hedge management intersects surveys with farmers (47) and observations in the field. Farms were surveyed within a series of Côtes d'Armor communes chosen because they were representative of the diversity of production systems and the variety of landscapes in this department. Other farms were chosen in the area of Ille-et-Vilaine, where we had information about the technical management of the field margins, facilitating the assessment (Baudry and Jouin, 2003). Field observations were carried out at four study sites (including Saint Péver, mentioned above) each consisting of 100 ha and located within the four Côtes d'Armor communes chosen for this project. Management of a field margin or hedge implies that there is not just a casual relationship between it and the adjacent agricultural activities (e.g. the overlapping of technical operations conducted in the field), but rather the establishment of an integrated technical management system within the framework of the farm unit operation (Gras *et al.*, 1989). The survey method initially consisted of identifying, with the farmer, the types of field margins (structure of the soil and vegetation) on the basis of a 'photographic catalogue'. The types of tasks carried out were then qualitatively recorded by type of field margin. We then analysed the organization of these tasks with the farmer, within a space and over time, taking into consideration the level of the farm unit, material and labour sources, and the production system. Field observations at the four study sites consisted of a description of the maintenance and sanitary status of the hedges, using a data profile.

Results

The objectives underlying the diversity of species planted

The first element to be considered was the large number of species cited in the planting requests: 67 species were planted in subsidized hedges for the period studied, implying a wide diversity of plants (Table 3.8). Nevertheless, as we will see, this list evolved over the course of the programme and was differentiated into hedges planted near buildings (on average about one-third of the subsidized hedgerows) or into hedges referred to as 'open field' types.

It is possible to classify the listed species into five major groups, according to a reading grid based on the ecological functions contributed by these species (resources, functions related to their structure), their type of introduction and their maintenance capacity. Their presence and their different combinations over time and depending on whether they were planted on the periphery of a building or in an open field, indicate the different objectives of planting programmes that we will attempt to identify.

– The first group includes species traditionally used at the time to form monospecific 'curtain hedges' around gardens, for example, Leyland cypress, English laurel, cedar and escallonia. Hedges primarily made up of these species are not particularly multifunctional: their impermeable nature makes them poor windbreaks and their monospecific and limited composition limits their interest for fauna. This group of species was particularly present in the first planting period, 1978–88, with the emphasis on conifers and cedar, in particular. The planting programme then managed by the DDAF was still little

Table 3.8. Species mentioned in the hedge planting requests for subsidized hedges for 1988–2000 Species mentioned in the hedge planting requests for subsidized hedges 1988-2000.

Abelia	Dogwood	English laurel	Plum
Gorse	Cotoneaster	Lilac	Winter currant
Serviceberry	Leyland cypres	Lonicera nitida	Locust
Sea buckthorn	Golden chain	Horse chestnut	Rose
Atriplex halimus	Deutzia	Birch	Willow
Hawthorn	Escallonia	St. John's wort	Redwood
Alder	Elaeagnus	Medlar	Mock orange
Birch	Spruce	Hazelnut	Mountain ash
Alder buckthorn	Maple (field, sycamore)	Walnut	Elder
Boxwood	Forsythia	Tree aster	Tamarisk
Buddleia	Ash	Elm	Thuya
Chestnut	Spindletree	Osier	Linden
Beech	Broom	Poplar	Aspen
Cherry	Currant	Pine	Privet
Oak (pedunculated,	Hornbeam	Apple	Guelder rose
sessile, red, etc.)	Holly	Pear	Wegelia
Choisya ternata	Yew	Sloe	
Kerria japonica			

known to the public and hedge planting was rare. No planting restrictions existed until 1988 when confers were removed from the list of subsidized species.

– The second group consists of ornamental horticultural species, for example, cotoneaster, deutzia, forsythia and broom, originally used as shrub cluster plants and then massively adopted from the 1970s to create low hedges. From During 1988–92, they represented, on average, 40% of the total occurrence of species used in built-up areas. Then, there was a considerable decrease in the use of these species around farm buildings as well as around houses, from 40% to approximately 30% in 1996, and as low as 20% in 1997 (Figure 3.10). Several measures were taken by the hedge restoration service in 1998 that radically modified hedge planting in built-up areas: subsidies were discontinued both for plants from the ornamental group and for hedges smaller than 100 m, limiting planting requests around buildings. This resulted in the end of plantings of wegelia, English laurel, elaeagnus, forsythia and serviceberry. In open field hedgerows, the occurrence of ornamental horticultural species was insignificant, even at the beginning of the period studied.

The third group contains species whose inclusion in the list can be justified by their tolerance to severe coastal conditions, for example, tamarisk, privet, elaeagnus or atriplex. Tamarisk and privet are traditionally used in the department for low hedges along the coast: elaeagnus and atriplex were more recently adopted.

– The fourth group is made up of shrub species (e.g. gorse, spindletree, sloe, medlar, etc.) and trees (e.g. English oak, chestnut, birch, willow, etc.) growing either spontaneously or cultivated (e.g. apple, pear and plum) at the time in old hedges in the region. Hedges planted with these species resemble hedges of the past as they looked after several decades. They are traditionally dominated by pollarded trees, provided that they are managed in such a way as to limit early competition between opportunistic shrubs and future tall trees. However, the immediate impression of this mix of species in an old hedge is perhaps the result of a different type of construction. It could actually be the result of the slow colonization of the hedges, which initially consisted of trees, by shrub species which were progressively introduced by birds.

– Finally, the fifth group is made up of species of trees and shrubs more recently introduced that will replace or mix with species of the fourth group, depending on the situation. Therefore, oak and chestnut (Group 4), the main species encountered

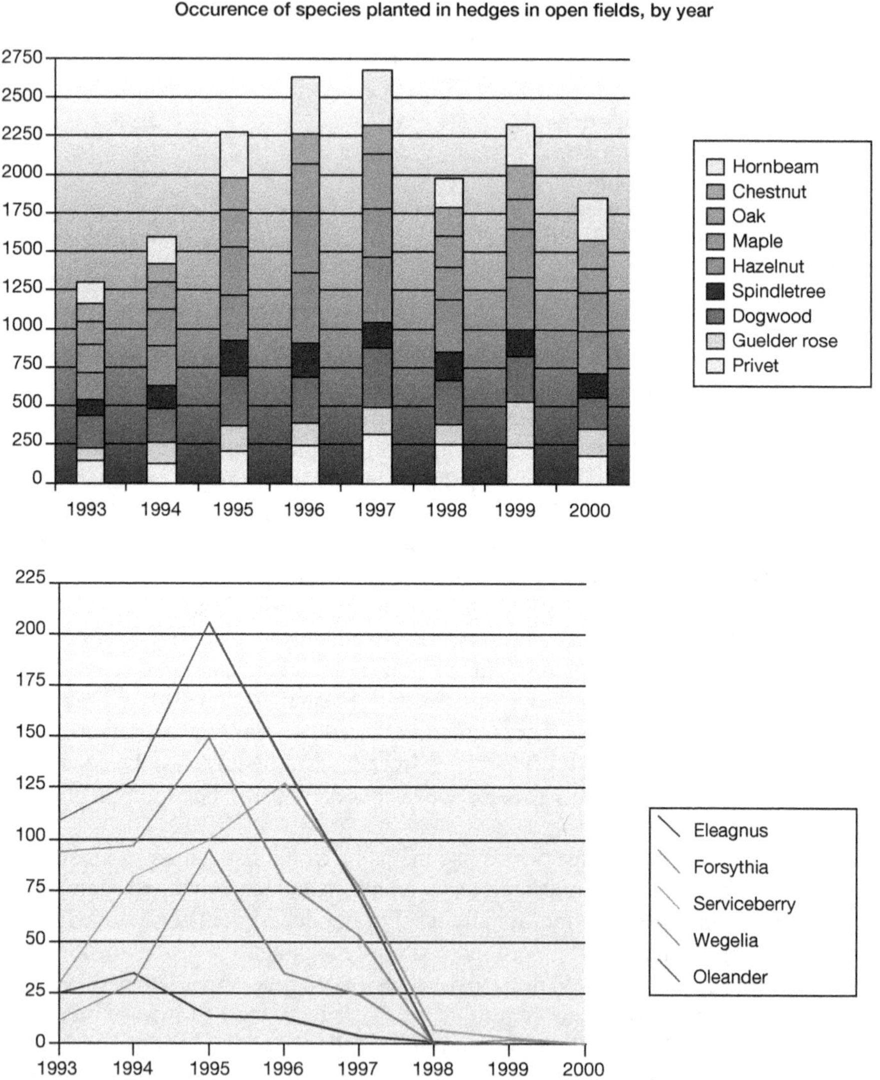

Figure 3.10. Occurrence of species in Groups 1 (monospecific 'curtain' hedge species) and 2 (ornamental horticultural species) planted near buildings.

in traditional hedges, are less frequently used in new plantings than beech or maple (Group 5). Beech is not very spontaneous and not often planted in traditional hedges; maple is a small tree not adapted to the formation of traditional tall hedges. Beech is in great demand for hedges, both in open fields as well as around buildings, and its distribution over the department is proportional to the overall hedges planted. Nevertheless, in this area it is beyond the limit of its species range, whose western limit in France is the Paimpont Forest. Its considerable use in open-field hedges, particularly in the west of the department (Figure 3.11) thus contributes to a significant western extension of

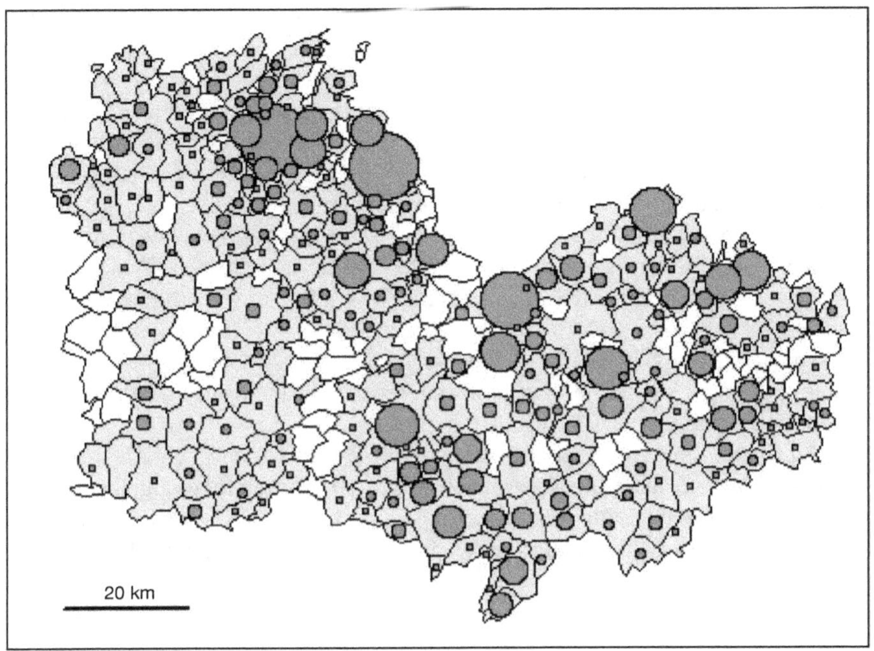

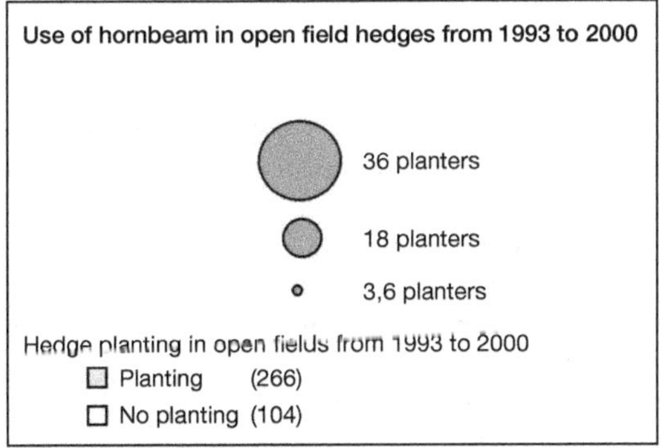

Figure 3.11. Use of hornbeam in hedges in open fields.

its spontaneous area. Other examples include poplar and birch, present until 1992, but subsequently have almost totally disappeared, and the American red oak, removed from the lists in 2002.

In the large majority of the cases (all periods included), whether they be near buildings or in open fields, the hedges planted were plurispecific, consisting of species from Groups 4 and 5 in varying degrees of combinations. In a planting project, the planter proposes a sequence (with a maximum of 7–8 species) that will be repeated over the entire row. The planter chooses from the list proposed by the DAE, with the assistance of consultants. This sequence generally consists of a mixture of tall species, medium-size species, and 'filling' shrubs; these are pluristratified hedges, often referred to as '*bocage* hedges'. Therefore, from 1988 to 2000, 70–88% of trees planted were mixed with shrubs. The choice of sequence, traditional at the time, , made it possible to rapidly produce a wood curtain of young woody perennials (Soltner, 1994). Depending on the purpose of the hedge, shaping and maintenance must be done to create a windbreaker-type hedge in the long-term, or one dominated by tall trees, for example. Our analysis of planting requests did not tell us the aim of the project linked with the choice of species mixture or the means of maintenance projected for long-term management.

During the period 1988–92, 30% of the species in built-up areas were trees such as beech, maple, chestnut or oak; they increased to 40% during 1991–92. Beech and maple were the most frequently planted in these areas, with a higher growth rate than chestnut and oak.

The species planted in open fields were mainly tree species (50–60% of plants), including chestnut, oak, maple, hazelnut and beech. Shrubs were made up the second most common category of species in pluristratified hedges (30–40%), and included dogwood, privet, cranberry bush and spindletree. We also found some fruit trees (e.g. apple, pear, plum and cherry), varying from 5% to 10% of the total, together with other occasional species (e.g. English laurel, buddleia and elm). This overall distribution of species was relatively constant during the period 1993-2000 (Figure 3.11), even if the combinations within the sequences had evolved.

Floristic and faunistic value of new hedges: interactions with the old hedgerow landscape

Beetles

The hedges studied were recent hedges (less than 40 years old) and several old reference hedges similar to the newly planted ones. At the site chosen (Figure 3.12, there is a thicket that could be a source of forest beetles. The very recent hedges were not colonized except for two that are directly connected either to the thicket or an old hedge that hosts the same species. Forest beetles are, therefore, capable of using new hedges, at least temporarily. The proximity of the source of individuals is a key factor for colonization.

It is necessary, therefore, to connect new hedges to old elements of the network to enable species characteristic of hedge landscapes to colonize new hedges and thus to multiply within the agricultural space.

Figure 3.12. Hedges at the ecological sampling site in the commune of Saint Péver.

Flowering plants of the herbaceous layer

Results show that the two types of new hedges, primarily formed of conifers or hardwoods, have a greater floristic value than old elements, hedges or borders. Analysis of the different floristic compositions allowed us to characterize the differences in ecological conditions between new and old hedges.

New hardwood hedges are characterized by a range of light-demanding species within which we find both perennial grassland species (e.g. velvet grass, orchard grass and common sorrel) and annual species such as crop weeds (e.g. cleavers and fumitory). This combination of common species, which are by no means dependent on hedgerows for their conservation, explains the greater floristic value of new hedges. These species, frequently found within agricultural landscapes, easily colonize new hedges and remain there as long as the light supply is adequate. The only species from the forest edge found in new hedges is stitchwort. The area within new hedges, often characterized by a significant layer of humus, is colonized (at the level of planting holes or tears) by nitrophilous species (e.g. nettle and cleavers) or species that can colonize this internal area from the periphery through the development of an underground stem network (e.g. nettle, bramble and bracken).

Old hedges host forest species or those found along the edge of the forest (e.g. germander, dart grass, black bryony and ivy). Eight species, characteristic of old hedges built on embankments, including wild hyacinth, fern and violet, have not been found in new hedges, even those over 20-years old. This suggests that the colonization of new hedges by herbaceous forest species is a long process, hindered by the use of a tarp at planting time. As a result, the area available for spontaneous herbaceous colonization is limited to the edges of the new hedge that remain exposed to light for a long time and are subject to disturbances linked to agricultural activity.

Moreover, the distance to the hedge or old woods, considered a source, appears to be an important factor. In the example of a new hedge connected to an old hedge, we were able to observe colonization by forest species, but only in the area of contact with the old hedge.

New hedges, even if they present a structure that in terms of volume of woody strata is reminiscent of the structure of old hedges and suggests the possibility of establishing ecological conditions close to that of old hedges, do not host herbaceous species characteristic of old hedges that can be considered the 'common heritage' of old hedgerow landscapes.

Technical management and the status of new hedges on the farm

Integration of new hedges in the technical management systems of field margins

Survey-based analysis has validated the hypothesis of technical management systems for pre-existing field margins and reveals the role of several levels of organization within these management systems (Thenail and Codet, 2003). The first level is the differentiation of tools and practices according to the type of field margin (e.g. brush cutters that can be transported on the back for hard-to-reach slopes or chain-saws for pollarding trees). The second level is the coordination of technical operations over a period of time along the field margin, linked to the technical operations carried out on the adjacent grassland or crop. For example, pollarding focuses particularly on developed tree branches bordering corn crops (the shade would decrease yield along the edges) in plots of land not used in winter. Clearing and weeding practices can be integrated during the growing season to keep disturbances of crops or rotation grazing (role of fences) to a minimum. The third level is the spatial differentiation of technical management operations along the field margins that follow the spatial differentiation of crop sequences. For example, the maintenance of pollarded hedges on farms is generally more radical along the margins of cultivated fields with little or no temporary grassland, which are located far from the farm on plots accessible to farm machinery. In comparison, maintenance is less radical along the edges of permanent grasslands more likely located on steep, inaccessible hydromorphic plots. The last level is that of the operation of the farm unit as a whole. For example, all the operations carried out along the field margins represent actual work time that varies depending on the farms studied (e.g. depending on the length of pre-existing hedges), but can be considerable (a month's worth of work for pruning alone) because farms are generally autonomous in this respect. This work implies associated time and labour and the co-ordination of tasks, particularly during the growing season.

As was mentioned above, different types of species were planted in new hedges, with patterns that may include woody species with various structures. The analysis that follows focuses on hedges often referred to as '*bocage*-type', with patterns that include trees (tall and/or average size) and shrubs. As a result, we have excluded monospecific curtain-type hedges consisting of cedar, Leyland cypress, etc., and low hedges consisting of horticultural species. We identified three major types in relation to the integration of these new *bocage* hedges within the framework of the technical systems of the farms (23 out of the 47 farms that were surveyed). In the first case (three farms), new hedges are totally integrated into the management of pre-existing hedges: farmers gave the new hedges a similar shape to pre-existing hedges and they were adapted to the same type of maintenance. Figure 3.13 illustrates this case with a line of pollarded trees. In the second case (three farms), maintenance work on new hedges (mainly pruning of branches) is entrusted to a CUMA[1] or an ETA[2] that provides a complete service, complete with a driver. In this case, a 'windbreaker' shape was chosen (Figure 3.14) and maintenance by lateral pruning is done with a cutter or a bank mower. The third case is the most common (17 farms). It is the status quo, in other words, no shape is given and no management activity is organized (perhaps a few branches are cut). Among these farmers who had not yet done anything (regardless of whether their new hedges were too young or not), seven planned to organize the management of their new hedges (five by lateral pruning and two by pollarding), leaving 10 farmers with no plans for their new hedges (43% of the initial sample).

Observations on the maintenance status of new hedges

Observation of the results of recent maintenance operations confirmed that different technical operations took place on pre-existing hedges. Observations at four study sites were made from April/May to the end of October/beginning of November, successively and at all of the sites and hedges. Maintenance signs included cut branches, chemical clearing/weeding, mowing, grazing and pollarding. These maintenance signs are rare for new hedges. We used indicators to qualify the development status of the hedges (e.g. presence of plants choked out by other plants or vegetation, and the presence of dead plants) and their sanitary status (e.g. signs of disease, injuries or insects). Of the 51 newly qualified hedges, 45% had at least one type of problem.

Discussion

We observed that planting objectives have evolved over the past 15 years. Documents concerning the implementation of hedge planting policies precisely laid out these

---

[1] Farm equipment co-operative.
[2] Farm work enterprise.

Figure 3.13. New hedge planted in the form of a windbreaker (trimmed with a tractor-mounted vertical-blade hedge-cutter (and/or chain-saw).

Figure 3.14. Old hedge planted in the form of a windbreaker (trimmed with a tractor-mounted vertical-blade hedge-cutter (and/or chain-saw).

qualitative objectives for hedges, and we were able to identify multifunctional objectives for the most recent period, even at the ecological level (e.g. erosion control, heritage value, faunistic and floristic value, etc.). Criteria specified in hedge planting requests, such as the species planted or the location between buildings and open fields, revealed the intention to break down the general objectives into implementation practices at the time of planting. Nevertheless, this breakdown is not explicit; for example, what about the choice of species? The reading grid that we proposed for the contributions of new hedges in terms of biodiversity is not sufficient to explain the evolution of the choices made. For example, the elimination of horticultural species (that provided resources for insects and thus participated in maintaining biodiversity) for the benefit of species similar to those found in the old hedgerow landscape can be interpreted as a desire to 'restore' the landscape (and all of its functions?) by directly obtaining the 'final image' of what was in fact the result of a long evolution in terms of development and colonization of species in the old hedgerow network. This is what Yves Luginbühl and Monique Toublanc described as "planting locally" (2006). It is the first issue in the discussion on the necessity, as well as the difficulty, of taking into account the time dimension inherent in these hedge planting actions.

Hedge structure (composition and volume of shrub and arborescent levels) plays a role in biodiversity as a result of the micro-climatic and habitat conditions it creates. The mixture of species planted as specified in planting requests will lead to differences in structure between hedges once they have developed. This is especially the case in the shift from monospecific hedges to plurispecific ones, including tall and medium-size trees and shrubs. Nevertheless, depending on the shape and management of the hedge over time, structural differences are possible, even with the same mixture of species. We previously mentioned the importance of time in the development and colonization of species; we find it here once again in the structuring of hedges and their management. Surveys of farms showed the diversity of management forms on pre-existing hedges and their integration into farm practices, with an organization of work, labour and specialized equipment. Surveys also showed the difficulties involved in integrating new hedges into these technical systems, revealing, in these planting actions, a lack of consideration for the operation of farms and their development and innovation capacities in terms of landscape management.

Information about the location of hedges near buildings or in open fields is essential but provides no information about the spatial aspects of plantations. The link between a garden and the periphery of a building may be an important element in relation to biodiversity. More generally speaking, new hedges, even 20-years-old ones, rarely present the floristic and faunistic value of old hedges. However, their potential greatly increases when they are connected to old hedges. From this point of view, the conservation and restoration of old hedges is even more urgent and the establishment of a network connecting old hedges to new hedges appears to be a central element that should be given serious attention.

The fact that project data were not systematically computerized, similarly the follow-up to planting requests (e.g. plantings effectively done in relation to a project, origin and vitality of plants, planting conditions, maintenance and development, etc.), makes it difficult for the services responsible for these policies to make an accurate assessment. Even if this study contributed to filling in the gaps in the statistical assessment related to planted species, it could still not provide answers to all the questions that arise. For example, it does not allow us to predict the evolution of planting sequences and the total number of plants of each species since a species may have been poorly represented in a project but may be predominant in the sequences when it is used. It should also be observed that the evaluation of hedge plantings in relation to qualitative and multi-functional objectives is more complex to establish and implement than quantitative objectives – because we must go beyond the linear (and monofunctional) aspect of all types of hedges – since more criteria must be taken into account and functions may be antagonistic or difficult to pinpoint.

Conclusion

The establishment of public hedge planting policies is certainly indicative of progress in the restoration of the hedgerow landscape, even if a considerable proportion of the hedge plantings (approximately one third) are on land surrounding public or private buildings and do not contribute directly to the improvement of the hedgerow landscape. We can assume that the evolution of stated objectives in the function of hedge plantings is linked to the evolution of knowledge about how the hedgerow landscape works and, more generally, to the evolution of the social and economic value of the hedges.

Nevertheless, an assessment of this neo-hedgerow landscape is difficult to make because of the many expectations and the lack of tools necessary to carry out a precise evaluation and follow-up. The improvement of public policies concerning the re-establishment or restoration of hedgerow landscapes requires more and earlier in-depth reflection as to the long-term objectives and more effective management through an evaluation based on landscape development observatories. These observatories could be designed at two levels with, on the one hand, a selection of shared joint data covering the entire department and, on the other hand, a local relay to follow the plantations, acquire local experience and validate observations at larger scales on the basis of more in-depth spatial information. The necessity of building information systems is emphasized at this time by those who determine public policy. When we take a close look at the objectives laid out for hedge planting, these systems present many difficulties and, therefore, require particular attention in terms of their construction.

The multifunctionality of hedges in the landscape and activity systems considerably complicates the evaluation of hedges, especially since these functions are not independent of the incorporation of hedges into the landscape, their management and their evolution. Given the time that it takes for a tree to reach maturity, choices made at a given moment do not necessarily correspond to expectations 20 years later. From this point of view, the present enthusiasm in Brittany (France) for individual or collective wood-fired boilers (electric supply) clearly raises the problem of the resource in terms of linear afforestation. 'New' hedges were not necessarily designed or maintained to be logged, and the risk of excessive harvesting is, therefore, real for an old hedge network that is already reduced and weakened. The capacity of policy-makers to maintain diversified forms of management is therefore essential and there is probably not a single solution: the different parties involved (e.g. farmers, private companies, local government agencies, associations, etc.) could each assume different responsibilities to satisfy clearly identified priorities. The redefinition of those concerned and the objectives can only be done by incorporating the hedge network into the dynamics of agricultural production, by simplifying work for the farmer, and by establishing a geographic and functional coherence between new hedges/old hedges as well as between hedges/plots. This involves an investment by local governments in the local dynamics of these planting policies.

Chapter 5

A typology of intercommunal actions related to the landscape

Patrick MOQUAY, Olivier AZNAR, Jacqueline CANDAU,
Marc GUÉRIN and Yves MICHELIN

All public action is developed on the basis of perceptions of its outcome and the mechanisms required to modify it. Intercommunal landscape policies are thus faced with contrasted representations of the landscape and public action. Underlying theories that support and justify proposed actions are rarely explicit. Nevertheless, the representations and hypotheses shared by those involved determine the direction the action will take together with its operational framework. They must be demonstrated in order to develop a true understanding of the intervention approaches of local stakeholders (Muller, 2005).

The action theory of a measure or policy corresponds to the causal supposition applied to the problem to be resolved by policy designers (Trosa, 1992). It covers the hypotheses that underlie policy implementation, hypotheses on the nature and causes of the problem being considered; and hypotheses on the effectiveness and, in particular, the means of production of the actions implemented. This notion thus encapsulates the set of representations and ideas that inspire the designers or participants in the programme as its mechanisms, i.e. the cause and effect relationships between the measures taken and their expected social impact (French Scientific Evaluation Committee, 1997). It is based on an often intuitive perception of the economic and social mechanisms (sometimes referred to as action levers) on which it is necessary to act, in the case in point, those influencing changes in the landscape.

This work intends to analyse these action theories and to propose a typology on the basis of empirical observations relative to the development process for intercommunal landscape policies.

Our analysis is based on three case studies[1] (Box 3.1) focused on rural areas in which an intercommunal landscape action programme was implemented: the landscape charters of the community of communes[2] of Sancy-Artense (Puy-de-Dôme) and the community of communes of Les Cheires (Puy-de-Dôme), and the landscape contract for the region of Les Feuillardiers (Haute-Vienne). Both the landscape charters and the landscape contract are procedures initiated or supported by the Ministry of the Environment for the purpose of developing a landscape action plan. Both of these measures favour a development procedure based on a participative approach, as well as the necessity for a landscape study. The development of these landscape programmes was based on a partnership approach in the three areas studied, bringing together different institutions (e.g. local government agencies, government services at the national level, etc.), and leaving room for consensus-building between the different bodies, both public and private.

Box 3.1. Presentation of the areas surveyed

The **community of communes of Sancy-Artense** (Puy-de-Dôme) is part of an isolated rural area and some of its nine communes are deeply affected by agricultural abandonment. Elected municipal representatives, extremely concerned by the depopulation of their territory over the past 50 years, wanted to take dynamic measures to attract new residents and visitors. Following 5 years of discussions, the landscape charter mainly focused on maintaining agricultural activity and controlling abandonment; urban planning and landscape valorization are the next two objectives.

The **region of Feuillardiers** (Haute-Vienne) is also far from the attraction of city life. Wanting to limit the planting of conifers at the outset in this zone of 17 communes characterized by its chestnut tree coppice, the brand new Périgord-Limousin regional nature park[3] undertook discussions with several partners. An action programme was defined but the landscape contract was never initiated due to the lack of interest on the part of elected officials, in particular.

The **community of communes of Les Cheires** is a peri-urban zone near the Clermont-Ferrand urban conurbation. Its ten communes extend along the valleys of the Rivers Veyre and Monne, at the foot of the Massif du Sancy. The upstream zone is sparsely populated, dominated by agricultural activity, whereas the downstream zone belongs to the Clermont-Ferrand urban centre. The major issue in this case is the canalization of urban sprawl and the management of its impact on the landscape. Although the architectural and landscape charter was primarily created to satisfy a demand of the Auvergne region (for the purpose of obtaining tourist development credits), the landscape finally became one of the major priorities in the territorial project supported by the community of communes.

---

[1] This work was carried out within the framework of the research programme on public policies and the landscape financed by the French Ministry of the Environment between 1999 and 2003 (see final report of Candau *et al.*, 2003).

[2] Communities of communes are public establishments jointly created by several communes. They exercise, in lieu of member communes, the jurisdiction that has been delegated to them, particularly in relation to economic development and land management. One of their characteristics is that they are directly financed by local taxes.

[3] In France, the regional nature parks (PNR) are responsible for implementing preservation and development actions on remarkable but fragile territories – with a view to sustainable development. Although they are always designated by the French Government, their creation and part of their financing has been the responsibility of the regions since the 1980s.

We conducted interviews with people involved in the design and implementation of the public landscape initiative in each of these areas between 2000 and 2001 (40 interviews for the Sancy-Artense landscape charter, 20 interviews for the Les Cheires charter and 20 interviews for the landscape contract for the Les Feuillardiers region). The people encountered belonged to regional and departmental institutions, professional agriculture and forestry organizations, as well as the communes and public intercommunal co-operation agencies for each zone. We analysed the type of arguments given, the landscape perceptions that underlie them and related intervention approaches. For the two zones in the Puy-de-Dôme department, we also conducted a systematic inventory of public measures related to the landscape, even if they were not explicitly linked to the landscape charter: 484 administrative files were thus examined[4].

On the basis of these interviews, we begin by identifying four criteria to characterize landscape interventions. After having deduced a typology of local landscape interventions, the second part is devoted to shedding light on action theories and institutionalized networks mobilized by type of landscape intervention. We conclude with landscape archetypes to which intercommunal policies refer and with realistic ways of combining the different types of interventions identified.

Four characterization criteria for landscape interventions

Four main criteria allowed us to outline a typology of local landscape interventions and the action theories underlying them. The first criterion is related to the concept of institutional network that intervenes in landscape issues. The second deals with the identification of landscape elements that act as intervention supports. The third concerns the means of intervention on landscape processes, in relation to identified means of producing changes targeted by the policy. The last complementary criterion deals with the essential or incidental landscape status of the end-purpose of the policy.

Institutional networks

In the area of landscape, as in other areas, institutional approaches explain that the offer precedes and sometimes anticipates demand by contributing to the modelling of the organization of the stakeholders. This modelling operates through established standards and the channels used to develop and implement them (Callon, 1986). Thus, the public offer leads to the creation of intervention rules related to the landscape, on which local stakeholders[5] will base their behaviour. These stakeholders, however, are generally not

---

[4] The files in question were requests for public assistance linked to landscape maintenance services that had been defined beforehand: maintenance of river banks, restoration of small build heritage, landscape integration of farm buildings, pruning of ornamental trees, creation or maintenance of green spaces, integration of land fills, village landscaping, flower planting, burying of cables, pipes, etc. The analysis covered 10 communes, distributed over the two zones studied in Auvergne.

[5] 'Local players' include all individuals (or groups) locally involved in the action, even if it stems from an organization with a broader scope or purpose, as is the case for representatives of government departments or chambers of commerce.

indifferent to policies and can contribute to orienting their scope by adapting them to local situations: intervene or not, strictly conform to proposed standards or, on the contrary, establish rules or negotiate local agreements. Moreover, policies bring with them a representation of the problem to be resolved (Jobert and Muller, 1987; Faure *et al.*, 1995) that can be reinterpreted at different levels of the administrative implementation.

Thus, public measures, whether they are incentives or programmes, do not only revolve around the financing of interventions. Depending on their institutional network reference, they contribute to the selection of the types of interlocutors, material objects and technical transformations, as well as certain types of management of the uncertainty related to the cost or the quality of the landscape intervention. This notion of institutional network is at the very core of our analysis. Local actors are themselves stakeholders in these institutional networks, generally by sector, that bring together different professional players that share certain specific categories (Douglas, 1986), revolving around specialized institutions. An institutional network thus brings together a group of stakeholders that responds to the same logic, whether it be economic, political or social, and includes institutions responsible for regulating exchanges within this group or between this group and the rest of society (Jobert and Muller, 1987).

Our interviews revealed several institutional networks involved in local landscape action: networks centred on agriculture and forestry, the environment, community facilities, employment and local development[6]. They are made up of local national and regional agencies and the representatives (administrative or professional) of preferentially targeted publics. The first four networks mentioned are highly dependent on decentralised government agencies under the sector ministry in question, to which we can add public or para-public establishments such as chambers of agriculture for the agricultural network. The local development network does not correspond to a given government department but has privileged relations with the DATAR[7]. This network is itself structured by local government agencies and particularly by polyvalent intercommunal structures[8].

These institutional networks each contribute shared visions linked to the context and the interpretation of changes, group interests, objectives and legitimate intervention approaches in their area of activity. They will structure and orient the public action in which they are involved by applying (intentionally or not) the quality standards, rules and concepts that identify them and give meaning to their intervention (Lascoumes, 1996). We can hypothesize that there is a correspondence between action theories that underlie a landscape programme and the stakeholders involved in the development and implementation of this programme (and, therefore, the institutional networks to which they are linked).

---

[6] In the area of landscape, we could have expected to find an institutional network related to culture, but we did not encounter one in any of our three case studies.

[7] DATAR (French Delegation for Territorial Development and Regional Action) was responsible for defining and implementing land development policies, from the time of its creation in 1963 to its reform in 2005. This government agency, under the auspices of the Prime Minister's office, is the equivalent of a central departmental administration.

[8] Multipurpose intercommunal syndicates are public establishments jointly created by several communes to manage a set of services, facilities and benefits of mutual interest. They are mainly financed by contributions from the communes or by payment for their services. Mixed syndicates make it possible to bring together communities (or groups of communities) with different statuses, and not just communes. Mixed syndicates are a valuable tool for implementing an organized development strategy at the infra-regional level.

Defining landscape elements in terms of intervention objects or supports

The landscape can be approached globally or reduced to several flagship elements on which actions (and attention) are concentrated. Obviously, the landscape can always be analytically broken down into intervention objects, and all landscape programmes are generally reduced to a series of more targeted interventions, focused on different means of support that make up the landscape in general. Nevertheless, the approach adopted by the stakeholders may focus on the overall landscape or simply on several isolated landscape elements.

In the first case, the analysis initially attempted to take the landscape environment into account on the basis of a number of interacting elements. The project then aimed at promoting a landscape model in which a space is interpreted in such a way as to qualify it as a 'landscape' (Cadiou and Luginbühl, 1995). It is only later that we focus on the elements of this composition to propose intervention means specific to each one, while maintaining the model of which they are part as a reference. It should be observed that intervention on the overall landscape can thus be broken down into a series of more isolated interventions; the central issue here is that the general outcome of the programme remains within the framework of a wide and global approach to the landscape. The landscape is considered as an overall unit that must be managed as such.

In the second case, on the other hand, the reduction of the landscape problem to specific isolated events is immediate and often radical. The elements dealt with are considered on an individual basis, separated from their immediate environment or their geophysical context, and the landscape as a whole is rapidly forgotten, even if the programme's landscape justifications remain. Interventions are centred on the status of several elements considered to play a key role (e.g. hedges, enclosures, farm buildings, etc.). We expect these isolated interventions to produce a favourable effect on the landscape without necessarily specifying the link between chosen landscape elements and the overall landscape of which they are part (a hedge or an enclosure, for example). Local stakeholders thus can also pursue objectives with which they are more familiar (such as supporting farming).

This slightly simplified presentation could lead to an obvious conclusion: only an assessment of the global landscape can lead to a real landscape policy, relevant to the complexity of dynamics at stake and ensuring coherence between its many separate parts. Nevertheless, this type of landscape policy is particularly problematic in that it integrates many different elements whose interactions are difficult to determine, characterize and subsequently control. The reduction to several symbolic elements appears to be a practical solution at first glance. In reality, this process presents a risk that leads to acting individually on disconnected aspects of the overall system. As illustrated by different examples[9], the individual multiplication of hedges does not necessarily lead to a coherent enclosure.

Intervention levers and the types of changes targeted

In order to produce landscape changes, the action theory can identify different levers corresponding to different intervention targets. Three levers can be distinguished,

---

[9] See Toublanc (2004), for example.

depending on whether the intervention has a direct effect on visible forms that make up the landscape, indirectly targets it by instigating processes that shape it, or attempts to modify the landscape perceptions of the stakeholders.

In the first case, public action defines visible forms characteristic of the landscape desired and proposes intervention approaches that directly modify these visible forms. Actions such as clearing, integrating a building into the landscape and renovation of public spaces can be considered in this category. In this case, landscape policy targets the landscape object and relies on technical acts that make it possible to modify this object to attain the desired results. The technical act in question is related to the final results, which are generally of an aesthetic nature and which may be unrelated to the initial functions of the landscape object. Small rural heritage artefacts (e.g. wells, ovens, communal wash-houses, etc.) perfectly illustrate this shift in function.

In the two following cases, public action instead aims at influencing the production factors of the landscape, whether material processes that shape the landscape or the representations shared by the stakeholders.

Intervention on the material means of production of the landscape concerns all the socio-economic processes, generally of a professional nature, that contribute to the creation and evolution of a landscape, while supposedly targeting other results. Intervention will essentially take place through practices related to the use, development and maintenance of the space concerned. This particularly involves farming and forestry practices that are one of the main factors of landscape change in rural environments. Certain intervention instruments, particularly economic ones (taxation or exemption, for example), only have an indirect effect on landscape improvements – even if they are deliberately pursued. Depending on the situation, the preferred instruments could be economic or legal (e.g. property rights, regulation of activities).

Other instruments focus on aesthetic or technical information for the stakeholders, in order to modify their representations of the landscape. A strong under-lying hypothesis is that these representations influence the behaviour of the stakeholders and that landscape improvements could be the result, in the long run, of a greater sensitivity to landscape issues and subsequent changes in practices. In this third case, we will, therefore, indirectly target the landscape production process with a double intermediation: from representations to individual behaviours, and from individual behaviours to collective landscape production. The communication effort behind such an action theory can be aimed at selected professional groups (farmers, for example, or craftsmen in the building sector), other participants whose impact on the landscape is significant (e.g. people applying for a building permit), or the population as a whole. Schoolchildren are a frequent target of these communication actions, based on an action theory according to which, early socialization and awareness leads to deeper and more sustainable changes in behaviour. In simpler terms, this public is also easier to reach.

The status of landscape objectives

The policies analysed can satisfy principal or secondary landscape objectives. It is possible to imagine a gradient ranging from policies whose only justification is the landscape (e.g. classification of the landscape as a protected space) to policies corresponding to totally different issues and having nothing to do with landscape concerns – which does not exclude the possibility that they will have a considerable impact on the landscape (energy policy, for example). Between the two, we find policies whose objectives are mainly landscape-oriented (but that also correspond to other objectives) and policies with other sectorial objectives but that include landscape concerns.

This criterion makes it possible to complete the preceding and especially to narrow down the field of analysis by defining what would qualify as a landscape programme. In order for an intervention (or more particularly, a set of interventions) to qualify as such, landscape objectives, even secondary ones, must be explicitly attributed to the programme by its promoters. Landscape impact policies with no explicit landscape objective are not within the range of our study, eliminating one of the extremes of the gradient mentioned above.

In the case of interventions on visible forms of the landscape, as in initiatives aimed at landscape representations, such explicit landscape objectives are normally systematic, even if they are not exclusive of other considerations. Landscape interventions can thus be an integral part of a strategy aimed at other objectives, for example, tourist facilities or even the improvement of the attractiveness of a region for potential new residents. Having said this, the landscape aspects of the programme or the policy are evident as soon as landscape objects (representations or material elements) are designated as the main supports of the action, precisely because of their explicit relationship to the landscape.

The ambiguity is even greater for interventions targeting the material production processes of the landscape. In this case, landscape intervention is normally indirect, and the link to the landscape can remain weak within the framework of the official definition of the action. Implicit professional or sectorial objectives are often dominant in the definition of actions where the landscape is only a secondary concern in the end. In the extreme case, the reference will only appear to be a pretext or a justification after the fact to legitimize interventions whose real objectives are elsewhere. This does not prevent us from taking the stakeholders seriously and associating their actions with the final declared landscape justifications, even if we must underscore the differences eventually observed.

Action theories and networks mobilised by type of landscape intervention

The criteria that we have just presented enable us to define different types of landscape actions. We will look at each one to explain the action theories and to illustrate, on the basis of our case studies, the networks on which they rely.

A typology of local landscape interventions

The different types are obtained by cross-overs between the second and third criteria laid out above:
– in the case of intervention support, we take the overall landscape or the concentration of specific landscape elements into consideration;
– in the case of intervention levers, we focus on representations, visible forms and material processes.

Six categories can thus be identified with diverse positions in relation to the landscape and different intervention tools (Table 1). Table 1 provides a synthetic vision of the archetypes underlying conceptions of the landscape, where each archetype corresponds to a preferential intervention lever. We will look at these three preferential levers (i.e. landscape representations, visible landscape forms, and material processes for landscape production), by distinguishing the interventions related to landscape elements and those aimed at the overall landscape, in each case.

Table 3.9. A typology of local landscape interventions.

In each case, the type of landscape intervention considered is given in bold type (with examples in parentheses). This is followed by intervention means in italics (and particularly, public action instruments) that could possibly be mobilised.

Intervention supports \ Intervention levers	Representations	Visible forms	Material processes
Some elements of the landscape	**Awareness by object** (tree, hedge, farm building, etc.)	**Direct management of landscape elements**	**Indirect management of landscape elements**
	Communication actions targeted at schoolchildren Site visits	*Technical acts aimed at modification, maintenance or construction*	*Targeted financial incentives (taxation, exemption, subsidies, etc.) Management of external factors (positive or negative) of production River contracts and territorial forestry charters*
The overall landscape (the landscape as a whole)	**Landscape awareness**	**Overall artefact**	**Generic interventions**
	The same intervention tools as above	*'Ex nihilo' construction Classification procedure*	*Landscape charters Regulatory planning Property rights Use and access rights Special provisions of the mountain law (1985) and the coast law (1986)*
	⇓	⇓	⇓
Archetype	Identity landscape	Scenic landscape	Territorial landscape

Awareness by object

Some interventions contribute to making the inhabitants or local stakeholders aware of the existence and issues involved in the preservation or development of landscape elements, generally seen as historical wealth to be preserved and enhanced (Van-Tilbeurgh, 2002).

The environmental network thus focuses, and has for many years, on awareness actions related to species and flagship or threatened environments. This can correspond to intervention approaches already established within this network, concerning certain species or specific habitats, wetlands, for example. Some landscape objects, such as hedges or even the edges of woods, are perceived as being particularly conducive to the development or preservation of biodiversity, and are given particular attention. In this case, concern with the landscape is only secondary and the purpose of the awareness action is above all ecological, emphasizing respect for nature as the ultimate goal. However, landscape elements are designed as a reflection of certain environmental characteristics of the landscape. The presence of the hedge is, for example, the visible manifestation of the maintenance of biodiversity and thus represents a visual indicator of efforts made in this area.

The relation to territory is more complex, as stakeholders must be made aware of the importance of maintaining the built heritage, for example, low stone walls, small local heritage sites, etc. These types of awareness actions, generally carried out with the aid of communication documents, aim to provide a certain image of the territory, reconciling historical wealth with new uses (through tourism, for example). Landscape elements thus given a historical value are both a reflection (a testimony) to a communal past (Lowenthal, 1994), and the contemporary expression of a new role for the territory, for example, as a place of residence or a tourist destination (Di-Méo and Hinnewinkel, 1999). In this case, it is the local development network rather than the environmental network that is concerned. This could also be a perspective contributed by the culture network which we did not encounter in our surveys: the different stakeholders can be identified as heritage associations, agencies responsible for surveys of historical wealth, protected sites, historical monuments, etc.

Landscape awareness

Awareness of the landscape itself refers to actions aimed at encouraging local stakeholders to interpret the space in which they intervene through a landscape perspective. It includes the dissemination of reference material and images of the landscape or simply forms of animation aimed at modifying how individuals see their familiar space, so that eventually they will change their behaviour to make the landscape evolve in the desired direction. It is generally part of a territorial approach, based on identification with the landscape; in this case, the landscape is a symbol and a reflection of the territory, its history and the population that inhabits it. This reflection can even be imposed from the outside on residents (Kneafsey, 2000), who are not clearly aware at the beginning of the discussions and will (or will not) adopt the expert vision of the landscape that is proposed to them.

In Sancy-Artense, the long effort to draw up a landscape charter was based on many studies and interventions by experts who regularly discussed their investigation leads and results with the directors of the community of communes as the studies advanced. These exchanges were extended to the core of the commission composed of the mayors of the communes involved, who addressed them in their role as elected officials. They were thus able to create a landscape framework and to refer to the landscape to base or justify their land management and development policies. Awareness actions targeted at schoolchildren carried out since 1996 also succeeded in modifying the representations of part of the population. Other planned actions had not been put into effect at the time of our surveys, in particular, training for employees of local construction firms. Moreover, representatives from the CAUE[10] (Conseil départemental en architecture, urbanisme et environnement) are present on a regular basis to assist candidates develop projects that require building permits.

In the community of communes of Les Cheires, awareness is based mainly on the dissemination of information and rules proposed by the landscape charter, through press releases in the intercommunal newsletter and a travelling exhibition. In the Feuillardiers region, the contract development process was not sufficiently participative for us to even consider that it had induced (or targeted) an awareness effect on the local population. Nevertheless, the regional nature park proposed several events focused on 'The tree and the landscape', through round-table discussions, films and workshops on landscape interpretation.

This type of action is conducted by institutional environmental and local develop-ment networks. It extends, on the one hand, a means of action successfully used in the past by the environmental network – environmental education – particularly aimed at schoolchildren and focusing here on a specific dimension, the landscape. On the other hand, it coincides with the specific responsibilities of the communes (and intercommunal structures by delegation) in relation to primary school education and extracurricular activities, in particular. Finally, it responds to the desire of local development militants (and to a lesser degree, technicians) for the personal transformation of the inhabitants via training, participation, exchange and continuing education. Environmental aware-ness operations constantly shift between appropriation and acceptance approaches for standards, and the creation of conditions for adapting standards, or even the creation of new standards. In the cases studied, it was the approach based on the application of standards that prevailed.

Direct management of landscape elements

This case concerns interventions through which different stakeholders directly model certain elements of the landscape justified at least in part by landscape objectives. The range of interventions is extremely broad and many land management practices may be involved, whether initiated by public bodies or private interests. In the intercommunal programmes studied, public intervention is predominant in the form of actions involving

---

[10] The CAUE (Conseil Départemental en Architecture, Urbanisme et Environnement [Architecture, Urbanism and Environment Councils]), are responsible for public advisory missions (addressed at local governments and individuals), in relation to information, awareness and training to promote the quality of architecture, urban-ism and the environment. Established by a 1977 law, these organizations have the status of associations but are created at the initiative of the department.

land management or the maintenance of spaces and public facilities. It also allows for action by individuals, particularly owners or users, eventually assisted by public subsidies.

The maintenance of small built heritage, rural planning, the integration of infrastructures (e.g. underground electric lines, etc.), the integration of farm buildings, the elimination of abandoned cars (financed by the General Council of the Puy-de-Dôme), or even the maintenance of river banks, are all examples of direct management of landscape elements. The community of communes of Sancy-Artense and that of Les Cheires have carried out this type of action relatively systematically.

In this case, it is the community facilities and local development networks that are the most evident. The community facilities network extends its traditional interventions in relation to infrastructures or urbanism by integrating the landscape aspects of these interventions (Graboy-Grobesco, 2002; Menguy and Pernet, 2004). The extension of the operating means of this institutional network to the landscape results in a very normative and technically oriented conception of these interventions (and, as a consequence, of the landscape); reference is made to the idea of landscape integration, seen through standards of visibility and the application of 'good development practices', that can be automatically broken down into a set of similar cases. This systematization of standards can be facilitated by the addition of a section devoted to landscape within the framework of local urbanism plans under the responsibility of the community facilities administration. In the same way, the impact study of engineering structures includes a section devoted to landscape that contributes to the application of these standards.

The local development network is devoted to the traditional interventions of local communities in relation to urban management and the living environment, i.e. management and maintenance of public spaces, green spaces, flower planting, street furniture, renovation of façades, etc. In some cases, the technical standards can be less significant, less impersonal, leaving room for the expression of the inhabitants' wishes or the participation of those residents directly interested. Some of the technical stakeholders with a diverse remit (the CAUE, for example) tend to take on an increasingly important role in local concerns, within the framework of a system characterized by more open competition between government services, research and design offices and even specialized associations. Traditionally, however, local communities depend on the community facilities network to determine these types of developments, particularly in rural areas, thus encouraging the application of technical standards promoted by this network. The increasingly important role in the implementation of landscape interventions by the employment network should also be emphasized, particularly to the employment programmes given the responsibility for maintenance tasks or the restoration of spaces and heritage.

Overall artefact

Intervention on landscape elements can be systematized to the point of wanting to organize and control the landscape in its entirety. This extreme case was not encountered in our surveys; however, it is interesting to analyse an extreme case of landscape intervention. It actually corresponds to an intervention aimed at controlling the landscape as a whole through the direct modification and determination of visible landscape forms. It implies an overall control of the landscape, which is exactly what makes this case just about impossible in reality. Its most convincing incarnation would be amusements parks like Disneyland (Alphandéry, 1996), in which the landscape is integrally treated like a decor, and directly designed, produced and regulated by the intervention of the owner/contractor. Many golf courses built in peri-urban zones were built along this model of the scenic landscape, providing a bucolic vision of the countryside.

We can perhaps find traces of this global artefact in the renovation of villages where the aim is to produce spaces considered to be authentic to meet the expectations of the tourist trade, but reinterpreted and totally artificial. Some of the 'most beautiful villages of France' and other major tourist sites, sometimes criticized for their artificial and, ultimately, lifeless character, could also be included in this category where landscape characteristics are totally controlled. Tourist sites, particularly spas, are often a good example of these artefacts, providing a striking contrast with the surrounding region, as is the case in Auvergne, not far from our study site.

Urban space seems to be more conducive to this type of experience than rural space, still under the yoke of productive processes that are not totally confined to aesthetic considerations. However, the direct remuneration of maintenance services of space provided by farmers (or foresters) could transform some highly patrimonialized rural spaces into these types of global artefacts. The construction of an idealized countryside expected to satisfy the aspirations and representations of urban dwellers, present, either because they own second homes or as a result of tourism (Urbain, 2002), could be a precursor for this type of landscape artefact. These different conceptions could collide at the intersection of local development and community facility networks.

Indirect management of landscape elements

In this case, operators must be provided with assistance or an incentive to modify their practices, in order to improve the status of certain landscape elements. A series of assistance measures for farmers can be included in this category, provided that the immediate aid object is a modification of practices that will have an impact on the landscape, without explicitly conditioning assistance for a landscape intervention (that would be classified as direct management of landscape elements).

The general objective of measures such as the grass premium[11], intended to encourage extensive use of breeding space, is partially landscape-oriented (encouraging the maintenance of space through breeding and, as a result, the maintenance of open spaces), even if it is closely linked to other sectorial concerns (e.g. support for breeders' income). This association between landscape objectives and more classic purely sectorial objectives illustrates why this type of measure primarily emphasizes sectorial institutional networks: those of the environment, agriculture (and forestry, partially distinct) and, to a lesser degree, community facilities. We can consider that this is the result of the marginal integration of landscape considerations in older sectorial intervention approaches, but with some sort of development. Within the range of available agricultural subsidies, the natural handicap compensation (known in France as ICHN), which is often decisive in many regions to ensure the maintenance of open spaces, appears to be outside the

---

[11] The term used at the time of the interviews, before being replaced by PHAE (agro-environmental grass premium).

framework of our study because landscape concern is not explicitly expressed in the basic texts defining this policy.

The landscape effect of urbanism regulations (e.g. regulations concerning roofs that stipulate certain materials and colours, or regulations for the building of new housing estates) are generally within the framework of this indirect management of landscape elements. This effect is obvious in all of the regions surveyed, prompting a strictly landscape-oriented application of urbanism regulations.

By construction, the landscape, in this case, is considered as being the product of the interaction of diverse activities. It is not taken into account as an isolated element within the complex regional system but, instead, as a combination of the bio-geo-physical and human dimensions, particularly socio-economic ones, that shape it. Moreover, this idea of the landscape as a product leads to compromises, through the integration of different rationalities at work in human activity: economic viability, risk assessment, attachment of populations, technical capacity to control processes, etc.

Generic interventions

This landscape concept, seen as the product of socio-economic dynamics, can also refer to other types of instruments that aim at encompassing landscape transformations from a generic point of view, rather than influencing more limited processes that affect certain elements only. These types of generic tools are widespread and common in land-use practices[12], but their effect on the landscape is difficult to measure, particularly because they only offer a general framework for action. The main tools that we describe here are those related to the definition of land use rights: property rights, other rights regulating usage, and all the regulatory documents influencing these rights, particularly urbanism documents. We can also include strategic planning documents when they attempt (among other things) to take into account and give a direction to the overall landscape evolution through a general organization of activities and projects affecting the territory. These generic interventions may correspond to particular dispositions related to the landscape, adopted in favour of environments with specific issues within the framework of the mountain law (1985) and the coast law (1986). The addition of a section devoted to landscape within the framework of local urban planning (already mentioned) can facilitate the relationship between generic interventions and the direct management of landscape elements. Finally, we can add landscape charters that define objectives above all – and not a programme of specific actions – that stakeholders can identify with in relation to the desired evolution of their territory.

Conclusion – archetypes, action approaches and combinations

Three 'archetypes' can be distinguished underlying public interventions aimed at the landscape, i.e. identity landscapes, scenic landscapes and territorial landscapes.

---

[12] Even if they are not necessarily perceived as intervention tools for the landscape, and are even sometimes considered as constraints rather than tools by those involved in land management.

The archetype of the identity landscape is related to the symbolic dimension of the landscape through identification processes and the way in which stakeholders represent the landscape. This identity function can be based on isolated elements, considered to be symbolic, or on the landscape in its entirety, embodying the local community and symbolic of its organization, its activities and part of its history.

Direct interventions on landscape elements or even on the landscape as a whole are related to the idea of landscape as scenery. The landscape aspects of land development are considered as separate entities, in the name of aesthetic references. The purpose of these interventions is to produce, restore and maintain landscape qualities that were presented as being desirable, independently of the main or original functions of the elements in question. The framework thus produced is conceived of as a decor that is to be obtained artificially through actions devoted uniquely to this objective. This archetype of the scenic landscape demonstrates the fact that in their action approach, the stakeholders "unglue the landscape film from the other realities of the territory" (Briffaud, 2001: 52). In extreme cases, the landscape as a whole becomes an artefact, a complete decor, deliberately produced with objectives based on aesthetic values.

The archetype of the territorial landscape is based on the idea that it is necessary to act on the socio-demographic and economic processes at the heart of a given landscape. The landscape in this case is a reflection of activities that take place in the territory and the dynamics that affect it. In a way, it is a product linked to the economic activities and social customs of the people that live there (e.g. their preferences and practices in relation to housing and leisure activities).

These three archetypes are simultaneously organized by the stakeholders, a reflection of the combination of the different action levers. The relation established between economic activities (farming practices, for example) and the landscape can be spontaneously accompanied by work on the identity character of the latter. The enclosure is a product of a certain organization of farms and a regional symbol. Emphasis can be placed on one dimension rather than another and this emphasis can evolve over time. It is possible that the identity landscape takes over when the territorial landscape is no longer valid, and that this recognition leads to the initiation of a scenic landscape!

In the same way, at a more subtle level, action approaches are not mutually exclusive. A fairly logical link can exist, therefore, between the awareness by object and the direct management of landscape elements by residents and producers (as is the case in the Puy-de-Dôme department). The existence of a departmental agreement between the agriculture and forestry administration and the CAUE regarding breeding facilities is an example of a compromise between two forms of action approaches.

However, from a general point of view, these action approaches are not necessarily complementary because the institutional networks, that play a prominent role in our analysis, without being totally impervious, are relatively compartmentalized. Professional representations, often shared and even encouraged by the different administrations, provide solid cognitive frameworks that can even be strengthened by the symbolic aspect of the landscape. This cognitive frequency of institutional networks based on a sectorial approach must be taken into account if we are to go beyond purely sectorial approaches.

If we consider that one of the conditions for improving the landscape is contingent on a high degree of coherence between the initiatives and the representations of sectorial networks, we must then take a closer look at the sites where these actions were applied. Case studies show that the intercommunal scale forms a coherency level of sectorial approaches, particularly in relation to the diagnosis and the follow-up of certain actions (those related to representations, in particular). Nevertheless, the local scale is not the only level of coherence. Our surveys revealed the decisive role of national and regional measures proposed to local stakeholders in the definition and implementation of inter-communal landscape programmes. Efforts to link sectorial policies and networks should, therefore, also be made at the interministerial and regional levels.

Chapter 6

The mayors' polyphonic discourse during a landscape intervention

Jacqueline CANDAU and Patrick MOQUAY

Rural elected officials are becoming increasingly involved in public landscape policies, to such an extent that very often they are the initiators of local landscape interventions. Several studies have shown that they see tourism and the arrival of holiday home owners as factors in local development, a substitute for stagnant or declining agricultural activity (Bages, 1998; Robert, 2001). The conjunction of these two phenomena is frequently interpreted from an economic perspective, according to which the rural landscape is a local resource whose improvement may bring advantages and enhance the appeal of the area for both tourists and residents[1].

This connection between awareness of the landscape and development does not account for all the reasons that prompt local elected representatives to become involved in landscape initiatives (Michelin, 2000). Here it is a matter of a broader examination of how these protagonists appropriate the notion of landscape from a public intervention point of view, although the projects are organized in partnership with other institutional stakeholders. Our hypothesis is that the mayors' scope in the exercise of their function gives them a special role in the implementation of public initiatives to improve the landscape.

Among local elected representatives, especially town officials, we gave particular attention to mayors, who generally have a decisive local influence (Chandernagor, 1993). With few exceptions (usually because of internal conflicts in the town government), the mayor exercises genuine leadership on the town council. The mayor is chosen among local figures and often has recognition and influence prior to taking office (Dorandeu, 1994). In many cases, the mayor determines (at least partially) the composition of the

---

[1] This hypothesis leads in particular to research on land rent as it may be applied to the market for holiday cottages, for example (Mollard *et al.*, 2006).

town council by selecting candidates at election time, which creates a certain compliance within the council. The mayor directs the town's administration, presides over the town council, executes its decisions and has specific regulatory powers (police powers), which bring more responsibilities and a much greater capacity to intervene than the mayor's colleagues on the town council. Lastly, the mayor is a member of numerous committees responsible for co-ordination and decision-making, and consequently has access to more in-depth information than the other elected officials. To carry out his public duties, the mayor must develop a degree of know-how that some have described as 'tinkering' (Le Bart, 1999), which in fact requires genuine social talents, in particular in rhetoric and human relations (Faure, 1992; Fontaine and Le Bart, 1994).

Situations of concrete actions and interactions show the roles played and the protagonists representations, while simultaneously bearing witness to the know-how they employ (Blatrix, 1999). The universe of beliefs, values and representations connected to their function, whether by the elected officials themselves (Ferret, 1996), their partners or the voters, marks the way in which mayors invest themselves in landscape initiatives. If one follows Le Bart (2003), who characterizes the role of mayor as having two purposes – acting to resolve local problems and uniting citizens – one can refine the hypothesis by concluding that this specificity depends on the quality of relations among the various users living in the same area. In other words, intervention for the sake of the landscape does not (only) mean local elected officials enhancing the value of an economic resource; it also constitutes a means to intervene in the relations between users and the rural area.

Our analysis develops a cognitive approach in four steps. First, we describe the interaction context in which projects were developed: the stakeholders involved, the chosen means of public intervention, the evolution of collaborative relations and the localized system of action thus constituted. This makes it possible, secondly, to present the content of the intervention project and the local problems it highlights by specifying, if necessary, the debates between stakeholders to determine whether, and in what way, those points should be discussed in the project. We then analyse the arguments the mayors developed to present the problems they would like to solve, and which they describe as landscape-related. Next, the tenets of the theory of polyphony in linguistics (Ducrot, 1984) allow us to deconstruct the mayors' discourse as proof of our hypothesis. Lastly, we interpret this discourse as the symbolic construction of a community of natives that the mayors would like to promote through their landscape interventions.

This reflection is based in particular on the examination of the landscape and architectural charter of Sancy-Artense (Puy-de-Dôme, France)[2]. The experiment takes place in a zone, which at the time was made up of eight townships that could be considered isolated and rural (Figure 3.15), and was launched in 1996 at the impetus of local elected officials, who chose an incentive-based and participatory system. Our study, carried out in 2000, focused on the process of creating the charter, and we met with the 40 or so people who participated in this process, including the eight mayors. Their input, focusing on the involvement and concerns that each individual attempted to introduce into this intervention project, provided additional material for the analysis that we made

---

[2] This work focuses on certain parts of the joint research project "Local Players and Public Initiatives in the Domain of the Landscape" (Candau *et al.*, 2003) within the 'Public Policies and Landscapes; programme created by the Ministry of the Environment.

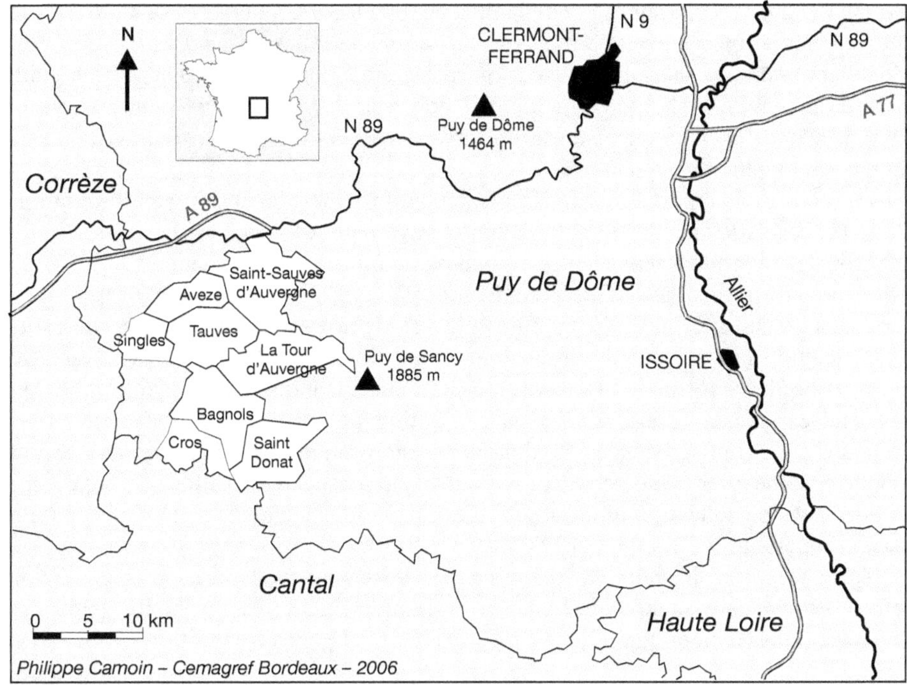

Figure 3.15. Location of the area under study in Sancy-Artense.

of archival records, retracing discussions during the preparation phase: these included the minutes of the meetings of various bodies (e.g. training sessions, steering committee, technical committee, bureau of the Community of Townships, experts' reports, and the personal notes made available to us). This material allowed us to reconstruct the process of reflection (i.e. the formulation of ideas that were then explored or abandoned) and the way the mayors' commitment differed from that of the other persons involved.

The mayors' involvement in a project drawn up with partners

In line with the decentralizing public policy to the territorial level, a landscape intervention project must consider the specific characteristics of the local context (in particular, the system of stakeholders at work there) and the characteristics of the surroundings. The nature of the actions may differ between sites, depending on the issues addressed and the priorities of those involved. This individual treatment is even more necessary when the chosen situations require setting up a participatory approach for the elaboration of the project. This is the case for the landscape and architectural charters.

This means of intervention has no specific legal existence; it is one possible concrete manifestation of the landscape plan (Box 3.2 for regulatory information). In that respect, in addition to the shared nature of its goals, the development of a charter must be based on carrying out a landscape study, given that the approach corresponds above all to a project approach. The project reflects a dynamic vision of the future of an area corresponding to local communities sharing a sense of purpose and a commitment that will be guided by these objectives. It should be mentioned, however, that the charter, unlike a contract, can stop at the formulation of these objectives, without needing to outline a specific programme of actions.

Several factors support the shared implementation of these local landscape programmes among the towns involved. From an administrative point of view, one may point out the connection between landscape concerns and planning and development expertise henceforth exercised at this level (Guérin and Moquay, 2002). A functional approach, moreover, leads certain specialists to recommend working at a scale where landscape entities are sufficiently large. Even if administrative and landscape divisions never agree, the intertownship level seems in this respect to offer sufficient distance to identify the area's main landscape features and, if necessary, to outline coherent priorities (Gorgeu, 1995). It offers a compromise between familiarity on the ground to having a good grasp and awareness of the intervention context, and sufficient perspective to allow a more comprehensive view, the collection of information and aspirations, and a way to concentrate the means of intervention (Moquay, 1998).

As always when it comes to planning and development, landscape issues involve socio-professional stakeholders on whom implementation is partially dependent. Landscape interventions therefore *at the very least* require concertation or, better yet,

---

Box 3.2. Cited regulatory procedures

The landscape and architectural charter
The landscape and architectural charter has no specific legal existence because it is one possible application of the landscape plan. The landscape plan is set out in circular No. 95-23 of 15 March 1995 containing instruments for the protection and improvement of landscapes. The landscape plan "corresponds to a project approach aiming to control the evolution of the landscapes without guaranteeing reflection in a purely legal and administrative framework nor pertaining only to remarkable sites". It resulted in a "reference document shared by the concerned state and local authorities", developed thanks to a landscape study and a common study approach.

The territorial exploitation contract (replaced in 2003 by the Sustainable Agriculture Contract)
The territorial exploitation contract is a means of intervention that is specific to agriculture, established by the Agricultural Orientation Law passed in 1999. It offers public aid to signatory farmers who voluntarily commit to respecting certain conditions. These fall within the framework of the administrative division's agricultural projects and the rural development plan approved by the European Commission in application of regulation (CE) No. 1257/1999 of the Council of 17 May. Landscape maintenance may appear among the objectives of these contracts. The contracts proposed to farmers from a region may be adapted to the environmental issues specific to a given production or territory. Certain so-called 'collective' territorial exploitation contracts may thus contain landscape objectives, such as the one envisaged for the Cros township of the Sancy-Artense area, to combat abandoned agriculture.

---

a partnership. More or less balanced relations will be established among private and public players just as, within the public sphere, they develop among local communities, administrations and para-public organizations. We should note that within intertownship institutions, the town council level remains very present, since these bodies are managed by town representatives (Box 3.3).

In Sancy-Artense, local elected officials played an active role in defining the project. The idea of a concerted landscape initiative took form when, in 1991, the Natural Regional Park (NRP) of the Volcanoes of Auvergne organized training for elected representatives in the Sancy-Artense area, which at the time were grouped together in the Intercommunal Association, a group with various missions. When during one of the training sessions, they saw how interested the participants were in architecture and urbanism, the speakers from the county council in architecture, urbanism and the environment (CAUE, Conseil Départemental en Architecture, Urbanisme et Environnement) and the NRP organized an exhibition and suggested setting up a landscape and architectural charter to preserve the local heritage buildings. At the same time, studies carried out in the area affected by agricultural decline raised awareness of agricultural activities' impact on the landscape. The elected officials mobilized their energies to set up a system to preserve the wealth of the landscape as part of an attempt to prevent depopulation – which abandoned

---

Box 3.3. The stakeholders involved in the preparation of the charter

Public stakeholders
– Territorial authorities: the eight town councils, the Sancy-Artense Community of Townships, the Natural Regional Park (NRP) of the Volcanoes of Auvergne, the general council, the regional council.

– Central government services working through local bodies: departments in charge of infrastructure, the environment, agriculture and forestry, and culture.

– Public and para-public institutions: the chamber of agriculture, the regional centre for forest property, the water agency, the national forestry office.

Private stakeholders
– Agricultural players: associations of farmers.
– Other users: associations of fishermen.

Landscape experts
Three experts employed by three different institutions collaborated: a consultant (*chargé de mission*) from a private firm, an architect from the Conseil Départemental en Architecture, Urbanisme et Environnement (CAUE [country council in architecture, urbanism and the environment]) and a development agent from the NRP.

Main fora for discussion
Two specific bodies were created to elaborate the charter, i.e. the technical committee bringing together the director and the chairman of the Community of Townships as well as the experts, and the steering committee bringing together all the stakeholders mentioned above. The project was also discussed on a regular basis by the bureau of the Community of Townships, made up by the mayors of the eight townships.

agriculture symbolized for them. As this was such a pioneering approach, reflection on the landscape charter, which took nearly 5 years to complete (1992–96), gave rise to a strong mobilization of institutional networks. It is perhaps not an exaggeration to say that the Sancy-Artense landscape became a sort of institutional issue that elected officials seek to perpetuate by finding outside support. Despite the participation of a number of institutional players, the elected officials retained control of the project. The head of the Community of Townships presided over and chaired the technical and steering committees. In addition, each step of the project was discussed at length by the Community of Townships, to which all the mayors belong.

Even if the project to create the charter in Sancy-Artense was largely controlled by the chairman of the Community of Townships, he cannot take all the credit, his discourse suggesting self-attribution notwithstanding (Le Bart, 1992). The way the project developed within a network of players and the collaboration with landscape experts working within the Community of Townships building somewhat breaks up the idea of a decision taken by a single individual at a precise moment in time. Rather, this was a project that took form gradually thanks to the contributions of partners who did not abandon it, but who strengthened their commitment as it advanced. In addition, no landmark event took place to announce the launch or official recognition of the charter through a public signing ceremony under the media spotlight. There were simply articles in the local press explaining the project's goals.

This decision in bits and *pieces* (Barouch, 1989) should not suggest that the involvement of elected city officials is unimportant. When such involvement was inadequate in other areas we have investigated, for example, the Pays des Feuillardiers in the natural regional park of Périgord-Limousin, the intervention project had difficulty reaching the implementation phase (Candau *et al.*, 2003). In this example, the elected officials spent very little time asking themselves about the relevance of a landscape intervention in their area, particularly because they did not have a place where they could meet with their peers to discuss this issue based on each one's specific concerns, as was possible with the bureau of the Community of Townships, composed of the mayors in Sancy-Artense. The steering committee meetings in which they naturally took part did not provide an adequate framework for interaction to develop their ideas, given the diversity of participants present, most of them representing institutions that have more prestige than the townships. What is said and the topics open to debate in a forum for discussion depend, in particular, on who is taking part (i.e. identity and social standing of those present) and the precise way in which exchanges are carried out (who is invited to which meeting and why, who takes the floor) (Candau and Ruault, 2002).

The importance of local elected officials' commitment to a chosen landscape project depends on the specific role they play within the partnership. As we shall see, their involvement is key to 'bringing home' landscape problems to the territorial level, i.e. the transposition of official topics[3], such as closed-off landscapes, abandoned agriculture, etc., into observable problems in the area. During the various interventions we analysed, the official set of themes was generally brought before the partnership by representatives of the administration, here, the regional department of the environment, and they were

---

[3] We are paraphrasing Bourdieu (1992) when he speaks of legitimate problems or problems guaranteed by the state concerning objects analysed by scientific research.

often taken up by landscape experts. They are transposed in two complementary ways. First, each topic is specified by the formulation of territorial problems, which are themselves determined by technical problems whose material dimension provides the basis for solutions which are then included in programming. In Sancy-Artense, for example, the charter's protagonists wanted to make the area "more attractive" and "control the decline of agriculture", and to this end planned to "restore the low walls and hedges along certain paths, while enlarging others", to "save agricultural land from afforestation" and "to start farming certain plots of land again". As a result of this technical transposition of issues, the contents of the intervention become more complex and more precise; "the surface of the problem" (Darré, 2006) widens to give form to a genuine issue. In addition, it must be shown that the area is particularly concerned by these technical problems, i.e they are local problems.

Defining local landscape problems in Sancy-Artense

The first step in order for local players to translate official topics into local technical problems is to make a situation (described here as unacceptable) objective (Padioleau, 1982), so it can usefully be looked at from a landscape perspective in order to find solutions (and access to funding sources). The work of defining the problems thus presupposes that one aspect of the local situation acquires new visibility (Kingdon, 1984), under the effect of accounts told and disseminated by concerned players (Stone, 1989), who consider it to be a problem.

In Sancy-Artense, a diagnosis of all the townships, created by a succession of contributions during the training of elected town officials, revealed an area that was subject to depopulation that was difficult to stop. The distance from economic and urban centres meant it had to depend on its own resources, i.e. agriculture – not very dynamic but active nonetheless – and its rural architecture. It was at this point that the idea of starting a landscape project was born, with the following goal: "Let us save what exists: the charm and character of our rural architecture attracts tourists. It is part of the scenery".[4]

During the preparation of this project, discussions seemed to dwell on the subject of buildings, not on the subject of rural heritage as might have been expected, but on farm buildings and, even more, the building of new homes. The county office in charge of architectural and environmental consultation proposed architectural standards (e.g. the slant of the roof, colours of doors and shutters, roofs and outside walls, materials to be used, volume of buildings, etc.). They were adopted and subsequently communicated to citizens through each city hall, which offered information and advice free of charge. The most controversial point of discussion concerned where to build new homes. Representatives of the DDE (Direction Départementale de l'Equipement, the government office in charge of urbanism) and the representative of the county office responsible for architectural and environmental consultation, expressed alarm about the "moth-eaten patterns" they saw emerging in the area. They felt that such disorderly urbanization damaged the aesthetic quality of the landscape. In contrast, the mayors, who were eager

---

[4] See report presenting an assessment of training for elected representatives, written by the Natural Regional Park of the Auvergne Volcanoes and all the townships of Sancy-Artense, united in a joint association.

to facilitate insofar as possible the arrival of new residents, tended to be more conciliatory and rejected any regulations that were too strict; in terms of project implementation. They did not want, for example, to commit to a project for regulatory protection (by establishing a zone to protect architectural heritage, whether urban or rural – known by its French acronym, ZPPAUP) or to drawing up township maps[5]. While the charter was being drafted, these controversies did not, however, cause strong disagreement among people with different opinions, who preferred to deploy their interaction skills (Cicourel, 1973) to maintain good relations, which were essential for them to be able to work together (Candau *et al.*, 2003).

Discussions also focused on the topic of agricultural activity, in particular following several studies that indicated how the land use plan had evolved. It also showed the financial difficulties facing farmers in this mountainous area, even though it is part of the Saint-Nectaire region where the renowned cheese is produced (bearing the *appellation d'origine contrôlée* label, a guarantee of quality)., Consensus was reached very quickly about the growing scrub encroachment in certain sectors and the need to act, if it was not possible to stop this trend, it had, at least, to be brought under control. The charter thus outlines ways to save agricultural land from afforestation, and encourage collaborative projects tobring abandoned farmland back into cultivation; a collective 'territorial exploitation contract'[6] in connection with this idea was discussed for Cros, the township hardest hit by this phenomenon. The fact that a majority of elected town officials work in agriculture or a related sector no doubt contributed to the increasing importance given to these agricultural issues. More likely, the landscape studies undertaken by an agro-geographer led to a shift in the very conception of landscape, seen not just as part of the scenery, but as the setting and (evolving) territory for people living there. This shift may be detected in the topics addressed during meetings and the introductory terms used by the chairman of the steering committee, who no longer referred to "developing tourism" but to "making the area more attractive", adding that this project was also of interest, and perhaps of greatest interest to, the inhabitants of the Sancy-Artense communities.

The structuring role of the numerous rural paths was also noted by the various landscape studies. The question of whether or not they need to be restored does not appear in the minutes of meetings to draw up the charter, but was spontaneously addressed by the elected town officials during the interviews we conducted for our research. They were taking action in this direction (or planned to), torn between the need to enlarge the paths so they could once again be used for agricultural purposes, and the need to preserve them in their current state, rimmed by low walls and hedges, so that they would be used mainly by walkers and hikers.

This period of discussion and reflection finally led to the formulation of four general objectives that were ranked in order of importance: maintain agricultural activity, bring 'agriculture decline' under control, improve the management of urbanism, and increase

---

[5] The ZPPAUP sets up the strict control of modifications to the sites it covers, subject to the authority of the *architecte des bâtiments de France* (architect of the buildings of France), which essentially removes some of the mayor's prerogatives on questions of urbanism. The township maps delineate the areas on the land register where building is allowed. Once this type of regulatory document has been drawn up, building permits may only be granted for areas where construction is allowed.

[6] See Box 3.2 for a succinct presentation of this system of intervention.

the value of the landscape. While the first observation targeted the rural buildings to be preserved, a wide range of factors, both material (e.g. new agricultural or residential buildings, fallow agricultural lands, forests and paths) and immaterial (e.g. training and awareness), were ultimately identified. What at first seemed to be a simple project to develop tourism had, by 1996, become a larger project covering a wide range of topics, which continued to evolve. One may interpret this evolution in the substance of the intervention project as a collective process of knowledge production based above all on discursive exchange (Darré, 1999; Candau and Ruault, 2002). In this example, the landscape was used to analyse cognitively (Muller and Surel, 2000) what was deemed to be a negative trend in change in land use, whether in terms of an increase in abandoned agriculture, the decline of rural paths, or the architectural quality of new buildings.

Some of these problems were taken up by the mayors who explained their local importance using specific logical arguments.

From spatial issues to social issues: the mayors' polyphonic voices

In order to convince people of the importance of these spatial issues, and even to strengthen their position when dealing with partners with different views, all the mayors we encountered refrained from speaking in the first person and brought users into their discourse. With the exception of one development agent from the Natural Regional Park, they were the only ones among the protagonists of the charter to use such argumentative reasoning. In such a way, they position themselves as spokespersons, not only for their inhabitants/voters, but for anyone else present in their township, which their elective responsibility authorizes them to do. As their legal representative, they can speak in their name. Herein resides the true specificity of the mayors compared with their partners, i.e. using the voice of users while presenting their arguments.

Their testimony reveals personal involvement that in reality is composed of several voices. It resembles a play where characters, or rather stylized figures, appear, such as the farmer, the inhabitant, the hiker..., they are not specific individuals (Mr or Mrs So-and-so). These figures appear because the mayors mention their point of view on a spatial problem to be resolved. The theatrical metaphor based on characters in a play shows that what these actors say is constantly interwoven with the words of others, a phenomenon at the core of Ducrot's polyphonic theory (1984). This theory sets out a fundamental distinction between the speaker and the enunciator. The speaker is to polyphonic discourse what the playwright is to a play, and the enunciator is the equivalent of the characters. Here the mayor, the oral speaker, creates his discourse by borrowing from a series of different voices (masks) in addition to his own. From a methodological viewpoint, an analysis of the content identifies the passages where the enunciator appears. Either the enunciator is named ("the farmers", "the breeders", the hiker", "the visitors", "the owner", etc.) or he is simply referred to ("someone who wants to build a vacation home", "to get by with their farm equipment"), or, lastly, reference is made to a vague figure that is not clearly identifiable ("one", "people"). Traces of the speaker's voice are, however, clearer and more easily identifiable, with sentences in the first person: "I", (possibly "we"), "my", "mine", etc.

The mayors interviewed do not all express landscape problems in this way, but only those which are the most difficult because they are recent or controversial. The restoration of minor heritage buildings, for example, is a problem shared by all the institutions involved in the charter; there is no need to convince anyone. It must also be possible to glimpse solutions to the problem at hand. For example, no mention is made of the management of property belonging to "sections" of the townships (*les sectionnaux*), a problem for which no clear solution exists, and on which the mayors themselves disagree.

This argumentative rhetoric focused on four spatial problems: the spread of fallow land, agricultural buildings, building homes, and the restoration of rural paths[7]. By way of illustration, let us look more closely at the last two. The mayors we spoke to mentioned the applications they receive for serviceable land and building permits, whether for new residents or those with holiday homes. They do everything in their power to give a favourable reply: "Let's say someone wants to build a vacation home in the village where he was born. I cannot see why he shouldn't be able to" (Mayor of Bagnols). "There are one or two new houses built every year. It is our duty as a town to make plots of land available to these people so they can come live here" (Mayor of La Tour d'Auvergne). These mayors are opposed to the sorts of regulations advocated in particular by the administration with regard to the establishment of newly built homes. These regulations would lead them to turn down applications for building permits, which they do not wish to do; for them, welcoming new residents is more important than the harm caused by building a home outside the village. By relaying the expectations of future residents, they consolidate their position when dealing with their institutional partners.

When elected town officials talk about improving or restoring rural paths, they take into account both farmers and hikers, whose needs are different. Farmers would like to have the paths enlarged so they can get through with their farm equipment: "The farmers all complain there are not enough paths to get to their fields. There are many paths that are too narrow, always with a wall running along both sides. They were built to be wide enough for an ox and cart" (Mayor of Cros). By contrast, hikers like these walls or hedges along the paths: "Keeping the old walls is not a problem for the tourists; I believe they like them" (member of the Town Council, Avèze district).

On listening to the mayors, one can identify three main types of users – the farmer/inhabitant, the holiday home owner and the hiker – whose expectations may at times be hard to reconcile. The many meanings of the word 'landscape', at times presented as an obstacle, here appears to be an advantage because it makes it possible to unite highly diverse concerns without necessarily using the term landscape to have to explain their being taken into account. The polysemy of the word landscape is significant but is not the source of the problem being discussed among those concerned (Swaffield, 1998).

Thanks to these argumentative dynamics, the mayors of the Sancy-Artense area defend, of course, the relevance of the technical problem to be solved, but in addition they give it a social dimension. The dramatic art of their account, constructed in such a way,

---

[7] The mayors of the eight Sancy-Artense communities that we interviewed share similar views about the four problems, compared with the opinions that certain other players may have expressed. This does not of course mean that they agree on all points and details, and even less so on issues addressed during the interviews (the management of *les sectionnaux*, for example).

gives new importance to the material problems to deal with in the name of landscape: they are also social issues.

It is interesting to ask how the mayors position themselves in relation to these expectations. In other words, does the speaker share the point of view of the enunciators that he brings into his discourse?

Faced with contradictory interests, the mayors do not choose sides but create a magistrate-like figure for themselves. They play the role of the judge, which is easily identified in their discourse, since they then speak in the first person singular: "If it's necessary to enlarge the path, what should be done with the old wall? The farmer would like to get rid of the old wall. But in terms of our hiking path heritage, it would be good to keep it. So a balance has to be struck between the two. And I'm the one to judge" (Mayor of Avèze).

It is thus possible to see to what extent this social designation gives landscape-related problems greater importance: they become more technically precise, connected to social issues. The territory is alive, described through practices that are in the midst of changing and that lead to potential conflicts among users. The mayors enact this transposition to the territorial level by putting themselves in a position to prevent conflicts linked to changing professional and domestic uses in their towns, for the sake of the landscape. This is how they respond to what appears to be one of the central tasks of all local governments, i.e. conflict prevention and management (Kaufman, 1991).

Through the landscape, building a community of natives

The social dimension the mayors bring to the definition of the projects leads to questions about the concrete effectiveness of such interventions. The argumentative rhetoric that we described above says nothing about the mayors' actual commitment to resolving the landscape problems under debate. Are public policies about landscape above all symbolic policies? They would thus have little effect on the issues of space but would simply be rhetorical constructions. Le Bart (2003) makes this assumption for environmental policies: "This institutionalization of the 'environment' sector makes it possible not only to make clients of ecologist militants, but also to show sincere voluntarism in a context that in reality is marked by powerlessness". Keating (1993) expressed a similar observation in terms of economic intervention, considering that local elected officials have every interest in displaying their voluntarism, despite doubts concerning the actual effectiveness of their interventions.

The interventions undertaken in Sancy-Artense may moderate this evaluation. The town councils have considerably increased spending in this area, if one is to judge by their investments in environmental services (Aznar, 2002). Of course, the landscape charter led to actions that were ultimately rather limited (e.g. free urbanism consulting, raising awareness among schoolchildren, etc.). Preparing the charter, however, provided a way to consolidate partnerships and especially to make progress in a collective reflection on current social issues that leave visible traces on land use planning.

Because the landscape is a combination of material elements whose evolution depends on the social uses that are made of it, the landscape makes it possible to intervene, at least to a certain extent, on these social dynamics at work. The mayors' commitment to a collective project specifically targets these social dynamics. They want to prevent conflicts and make it easier for different users to coexist. When they play the role of public magistrate, they want to work for the sociability of users. This role, more specifically transposed here in rural townships, is identified by Le Bart (2003: 103) for all the mayors: "For him [the mayor], it is a matter of erasing all internal divisions so that the only group that exists is the community of the town's inhabitants". Playing this unifying role, or more precisely the incarnation of a local community, corresponds to an earlier means of expression and positioning used by rural elected representatives (Kesselman, 1967). The work of elected officials evolves, however, in line with the social transformations of rural living (Moquay, 2002).

What then is this community of inhabitants? This communitarian utopia is played out in local fairs and events, such as the 'Bread Fair', started a few years ago at Cros. Bread is baked in an old oven, a minor piece of local heritage restored in the name of the landscape. Those who have their secondary residence in Cros ask to work, for free of course, alongside the 'villagers'. "They are also from Cros – a little bit", says the mayor with satisfaction, noting that they wear the same badge as the locals.

Heritage weaves a link between the past and the present, the landscape allowing elected town officials to create unity among the various residents. Whether they live there year-round or occasionally, as secondary residents[8], they are all natives for the mayors, i.e. they are "from here". Taken to an extreme, it seems that this status may even be extended to the very infrequent inhabitants who are visitors, whose interests the mayor also protects.

Because they are mobilised for the social revitalization of their communities, the mayors invest themselves in landscape projects that help them create this community of natives. This symbolic construction is, of course, largely utopian, wiping away conflicts and social differences. Our study shows, however, that it appears indispensable for the development of concrete actions.

Presenting the social importance of landscape problems at the territorial level: the particular role of mayors

The involvement of town officials turns out to be indispensable in the initiatives that attempt to address the landscape of a given territory. A lack of involvement on their part makes the transition to the programming phase more difficult, due to a lack of local anchoring for the problems expressed in the intervention project. Such anchoring depends on a double cognitive process: building the local importance of the technical problems put forward, and the representation of their importance based on the underlying social issues. Not only are the problems to resolve geographically precise, but they are also explained relative to their social dimension.

---

[8] On this point, Perrot and de La Soudière (1998) consider that "More and more secondary residents alternate between their primary and secondary homes, rather than using their secondary homes as an alternative. This process suggests that in a growing number of cases, they could be considered permanent inhabitants of the villages".

The mayors develop this social intensity by conveying the expectations or concerns of certain users, whether inhabitants or visitors. This role of spokesperson is all the easier to play because it corresponds to their legitimate function as elected representatives. Beyond the issues of aesthetic coherence of an area, in such a way one sees that the landscape makes it possible to reflect and to try to regulate relations among people who live in the same area but whose interests may diverge. The experience in Sancy-Artense shows that the mayors become involved in public landscape interventions particularly for this reason.

This appropriation of the landscape is based on two characteristics of this environmental asset. On the one hand, its use is always localized in space. On the other hand, it is essentially the result of a process of activities (e.g. urbanization, demographics and production) whose aim is not landscape-related (economists speak of a "joint product" or "externality"). The landscape components are subject to localized multiple uses. The co-ordination of uses, therefore, raises *de facto*, the issue of several categories of stakeholders living together, a situation that is the primary concern of the elected representatives, mayors and town officials. The mayors position themselves as spokespersons for the diversity of users and create a judge-like figure for themselves.

In doing so, they propose the 'playing field' that is indispensable in the application of any process of intervention. They demonstrate to their partners, especially the institutional ones, their ability to persuade the city council and citizens to accept the application of the intervention system in their towns. Yet nothing allows us to say that the expectations conveyed by the mayors reflect the concerns of the various users. The fact is that no public debate was organized for people in the towns of the Sancy-Artense community. One may ask whether such a public debate risked challenging the mayors' roles as spokespersons, thereby weakening their central role in the process of elaborating public landscape intervention.

Public participation in landscaping action

Chapter 1

From the landscape perception until landscaping action. How long is the way?

Rosário OLIVEIRA, Milena DNEBOSKÁ and Teresa PINTO CORREIA

New challenges for integrative science

During the last centuries, science focused more and more on ambitious and pressing challenges and goals. Hand in hand with technical progress, however, the importance of humans' understanding of, and empathy with, nature has been vanishing. Between microcosms and macrocosms, man as individual has often become lost. As a consequence of the parallel incredible speed of economic growth in the developed world, man has been progressively driven away from nature and, inescapably, away from himself as well (Nowotny *et al.*, 2002; Lorenz, 1997).

Re-establishing the proximity between man and nature presupposes the formulation of new knowledge, concepts and understanding of the various meanings implicit in this relationship. At the same time, an increased awareness and responsibility is needed from the policy until the individual spheres.

The complexity of this demand requires integrative concepts such as that of landscape, so that nature and society can be considered as a whole, including the various perceptions and interests of different sectors and groups (Kaur *et al.*, 2004; Tress and Tress, 2001). Thus, the integration between natural and social scientific research can play an important role in complementing old fashion planning and management practices, where decisions about landscape were under the domain of the technical, research or political perspective and did not consider any other points of view, such as those of local people (Gobster, 2004; Buchecker, 2003). The landscape approach, considered in this paper, presupposes an in-depth knowledge of what local inhabitants think and feel in relation to their landscape and to what extent they are still engaged with it. Accordingly, research can be seen as a practical scientific field oriented towards providing solutions

Figure 4.1. Location of the two case-study areas (Mértola and Monforte municipalities) in Portugal.

for environmental, social, economical, aesthetic and cultural problems, but it can also make an important contribution to understanding, from a theoretical point of view, the complexity of the concept and all the transdisciplinary connections (Tress and Tress, 2001).

Our approach involved the holistic study of spatial reality and human context. The objective was to understand whether the perception of landscape can act as an interface between landscape planning and an effective management in remote rural areas under specific socio-economic conditions and with uncertain future development perspectives. We use empirical data for the study of landscape perceptions from two case studies in southern Portugal, both with similar characteristics to many others in the Mediterranean region (Figure 4.1).

What does perception mean within the landscape research approach?

In Portugal, and possibly in most other Mediterranean countries, landscape planning and landscape design have a longer tradition than landscape management. Often, there is an abundance of plans and regulations, but weak and unsatisfactory application in the field (Cancela d'Abreu *et al.*, 2004). Planning without management is a sterile exercise. This controversial situation is one of the reasons for the many conflicts and reluctant acceptance of conservation proposals by local people (Bucheker *et al.*, 2003; Scott, 2002). Plans and rules can be handled by an expert in the office, but landscape management is much more complex and requires more detailed knowledge. As proclaimed by Council of Europe (2000) "landscape management means actions from a perspective of sustainable development, to ensure the regular upkeep of a landscape, so as to guide and harmonize changes which are brought about by social, economic and environmental processes". Thus, action involves people and different people have different perceptions about the same landscape and reality (Stewart *et al.*, 2004; Bell, 2001).

Understanding the role of landscape perception seems to be fundamental if we wish to communicate with people about the landscape and if we want to integrate

226

their perspectives, needs and expectations of, and for, their territory into management proposals and involve them in integrated management practice.

There is no clear and simple definition of the concept of perception and it is often used in different contexts ranging from psychology, to physiology, medicine, philosophy, arts, aesthetics, communication and landscape research. Even within the landscape context, perception can mean many different things. In our study, we define landscape perception as the result of the relationship between individuals, groups of the local community and the landscape. This, means more than a cognitive process, but also includes critical and emotional attitudes towards a specific landscape and its changes. These perceptions were assessed through personal interviews.

As a result of investigations into landscape perception, it is possible to discover some important concerns of local groups, which should be considered in management proposals as consideration of these concerns would increase the likelihood of success in the planning and achievement of management goals (Palmer, 1997). In this way, we started with the assumption that there should be a strong interrelation between perception and landscape management.

Insights into remote rural landscape – two case studies from Alentejo

In remote Mediterranean regions, many aspects of rural life remain unknown, developing in their physical, social and cultural remoteness (Vos and Meekes, 1999).

However, the importance of rural spaces has been increasing since new functions beyond agricultural production have been broadly required. More and more people are temporarily or permanently seeking to be closer to nature and landscape, in particular are looking for places to which they can feel rooted. These transformations are rather complex and dependent on multiple drivers, impossible to approach in the framework of one research project.

Both studies were developed under an applied research perspective, in the sense that the final results and conclusions could be integrated into specific policy and planning tools. Nevertheless, only some of results obtained in these projects were used on this paper.

The municipality of Mértola

Context and problems

Mértola is a peripheral rural municipality located in the south-eastern part of Portugal. It is a sparsely populated area (fewer than 7 inhabitants/km^2), with a low level of economic dynamism. As with most areas in the south of Portugal, at the beginning of the 20th century, wheat production was the main land cover in the region and almost all the economy was based on agriculture. In the 1970s, due to changes in national and world policies and markets, a significant exodus of inhabitants started to empty the villages and surrounding fields. By then the poor and unsuitable soils for agricultural production had been eroded and the area was among the most threatened by desertification (CCD, 1997). The functiuns of many agricultural areas evolved and production's ideal started to be part of the past. Agricultural fields were abandoned and scrub encroached. While this dramatic change to the landscape was happening, people had to face this transformation without any alternative and hope for the future. In the municipality of Mértola, by 1985, just before Portugal became part of the European Union, shrubland was the main land cover (60%) (Casimiro, 2002) and the situation caused a very severe human and economic depression.

Since 1986, the Common Agricultural Policy (CAP) has introduced new opportunities for agriculture and afforestation activities and since then, it has been the main economic support to help local farmers to continue their practices (Oliveira, 1998, 2001). In addition, new landowners started to be attracted to the area, particularly for hunting. The introduction of a livestock premium, the application of agri-environment and afforestation schemes, especially Regulation 2080/92, presupposed land use and landscape changes. From 1985 to 1995, approximately 50% of the area shifted from shrubland into productive use, where extensive livestock systems occupied 35% of the land and approximately 15% started to be cultivated again (Casimiro, 2002). On the other hand between 1985 and 2000, afforestation occurred in 14% of the municipality area.

Even though it was not the major quantitative land cover change seen in the municipality in the last decades, the rate of afforestation in Mértola was one of the highest rates in Portugal between 1992 and 2000, and, in interviews, local people refer it as the most significant transformation.

Methodology

The research focused on the Mértola case study, financed by the National Foundation for the Science and Technology, aimed at evaluating and understanding the land cover changes that have occurred in the municipality of Mértola during the last decades and to understand how people perceive these changes in the landscape.

Using satellite image interpretation, land cover changes were analysed for three different periods – 1985, 1995 and 2001. Different users of the landscape (e.g. local residents, landowners and farmers, and visitors), were interviewed in three areas in the municipality of Mértola (i.e. Amendoeira da Serra, João Serra and Fernandes), each of them occupying about 2200 ha. These areas represent different kind of landscapes and various implicit dynamics. Two of them are located inside a natural park, which occupies roughly half of the municipality, while one is outside the park. Technical staff and decision-makers at different levels (i.e. local, regional and national) were also interviewed.

In total 123 interviews were conducted, each lasting about 1.5 h. The interviewees were selected not by statistical representativeness but by maximum variety (Buchecker *et al.*, 2003; Hunziker, 1995). Qualitative analysis was separated into two blocks, local residents, farmers/landowners and visitors in one block and technical staff and decision-makers in the other. There were thus two groups of questions, each including five thematic subjects.

For the purpose of this paper, only two of the themes were considered – the perception of landscape change and knowledge about CAP.

Results

Land cover changes were quantified both in the municipality area and in the three areas for three different periods for 1985–2000. Based on this, we observed that in none of the areas was the afforestation with pine tree (*Pinus pinea*) the main change that occurred. Nevertheless, independently of the quantitative land cover change, the majority of people (74%) perceived pine tree afforestation as the main change to have taken place in the last 10 years. It seems to indicate that this new land cover has had a particular impact on the perception of landscape change.

Some individuals saw afforestation positively, while others had a more negative perception. People were positive when they thought that afforestation had introduced new aesthetic impressions on a dry landscape and also provided an appreciable income for farmers. Negative perceptions of the afforestation were related to considerations such as: it does not fit with the "traditional" landscape; it does not contribute significantly to the local economic dynamics; it reduces/replaces agricultural and hunting areas; and it significantly increases migration and contributes to the human emptiness of the area.

Some people (20%) identified the decrease in cereal production as the main land cover change to have occurred in the area, even though it happened more than 20 years ago. They refer to many positive aspects in addition to the composition of the landscape during this period (e.g. economic activity, social and cultural coherence, more employment, etc.) in comparison with today. At the same time, the education level of the people interviewed was low and most of them (81%), either residents or landowners/farmers, were not able to identify any kind of financial support mechanism, while remaining critical of subsidies as a whole, seeing them as something abstract associated with the national or European level. Many of the farmers (33%) recognized that it is not possible to continue any agricultural or forestry activities in this area without financial support. Furthermore, some (20%) consider the natural park as a major obstacle in the development of the area, because it has very strict regulations concerning activities which could conflict with nature conservation priorities and 38% had no idea about its role. It is not surprising, therefore, that there is some resistance of attempts to motivate and build capacity to address issues of depopulation and loss of attachment to the land and the wider landscape.

Even though people recognized the many benefits of the CAP financial opportunities, they said that sometimes they feel very pessimist about the future, as they do not believe that the depopulation, abandonment and impoverishment trends of the region will be reversed. The majority of locals thought that the agricultural and forestry activities are totally dependent on CAP subsidies and this made them sceptical about the future. As this cycle of desperation gets stronger, these negative perspectives become firmer and of people lose their attachment to the surrounding landscape.

CAP was recently reformed and more policy changes will be considered, at least until 2013 (Cordovil *et al.*, 2003). New schemes and opportunities will appear in the second pillar of the policy. Thus, rural development requires more effort to promote innovation and capacity building to activate many bottom-up processes, which can guarantee the improvement of life, environment and landscapes quality and many other requirements implicit in the new CAP.

Ribeira Grande in Monforte municipality

Context and problems

Driven by the intention of reversing the process of abandonment and increasing decay of the Monforte municipality, the municipal council launched a project aimed at possible valorization of the landscape. This will concern providing improved recreation functions, both for locals and as a way of attracting tourists. The river of Ribeira Grande is the most important natural value of the municipality, with interesting landscape features and heritage components. Research on the cultural landscape emerged because of new appreciation of monuments within their context, with several classified monuments close to the river.

The complex project for the valorization of the landscape (also including biophysical and ethnography research) was made possible thanks to the INTERREG programme on river bio-corridors. The study presented in this paper researched the local landscape memory through interviews with local stakeholders about the river landscape, its history, their relationship with it, and its future management, focusing on the preservation of special features.

The study area is located in the western part of municipality of Monforte in northern Alentejo, close to the Spanish border. It concerns particularly the landscape surrounding the initial section of the small River Ribeira Grande, within the municipality boundaries. The traditional rural landscape consists of gently undulating hills covered mainly with mixed *montado* (agro-silvo-pastoral system of semi-open forests of holm and cork oaks).

The municipality of Monforte has a large area, but low demographic density (8 inhabitants/km^2). There is a weak secondary sector (13%), but agriculture (38%) and services (49%), make up the most important sectors. As with the entire Alentejo region, this area has experienced a slow and permanent loss of population. Monforte inhabitants are predominantly elderly people (35% over 60 years), the middle age generation is lacking, and there is a slight dominance of women over men.

The Ribeira Grande landscape has had an almost uninterrupted human presence since prehistoric times. Until the first half of the last century, the area was marked by an increasing population density and intensity of land use, based on large-scale cereal cultivation. Since then, however, the slow but constant process of human desertification has continued. Now most of the land is used as extensive pastures, mainly for cattle and sheep.

This transformation, together with a shortage of shepherds led to a parcelling of landscape with barbed wire fences, which made access very difficult. Former landscape use along the river is demonstrated by the ruins of various water mills, small bridges and other constructions; other features connected to water, such as, vegetable gardens, wellsprings or orchards still partially exist.

Although the larger landscape pattern and image of Ribeira Grande has not changed much, more detailed observation reveals the diminishing of functions and uses of the territory, as well as loss of more detailed mosaics. Together with the change in land use, there has been a change in the attitude of people towards their landscape, i.e. a growing alienation from it.

It is essential for future management and landscape recovery to know the present relationship of a population to the landscape and its memory structure. The memory structure of cultural landscape is understood in this study as a web composed of essential elements and their connections. These elements, here called 'angular stones', generate the life inside the landscape, and make such an impression on people that they refer to them later in life as places of special importance and affection (Oliveira, 2001).

For a revitalization of the landscape, creation of new uses within it, and especially for the restoration of the relationship, man – landscape, it is necessary to determine the most important elements of the landscape structure. It is assumed that angular stones are a possible gateway into landscape revitalization through the relationship and attachment of the population to these elements within their landscape (Cílek, 2000).

Supposing that the essential elements of landscape structure inherited from the past are still present in today's landscape, the study aimed, through interviews, to identify the most frequently mentioned places and features. These are considered as the 'angular stones' of the current river landscape, with meanings retained in local memory.

The analysis should help in identifying which places still present within local memory, are the 'angular stones of landscape memory structure' and should be used in its recovery.

Methodology

Forty-two inhabitants were interviewed, divided into seven different landscape user groups. Criteria for selection of respondents from among the population were age and variety of uses of landscape, for example, professional or amenity users like farmers, hunters, fishermen, etc.

Four groups of questions were defined, concerning knowledge of the landscape, its past and present uses, and expectations for its future development.

Analysis of the familiarity of the population with the river landscape, was centred on the frequency and variety of places visited at different times by different groups of stakeholders. The following groups were compared because of their contrasting characteristics and relationships with the landscape: elderly citizens vs young students and farmers vs amenity users.

For the analysis in the present paper, two main groups of questions were considered:
– Questioning about the frequency of landscape visits in the past and present, expected to produce an image of the relationship and proximity of the interviewees with the landscape.
– Questioning about the quantity and localization of places visited (in the past, present and suggestions for future visits), expected to obtain the identification and localization of angular stones.

Results

Comparing visits to the area around the river in the past and in the present, the frequency and especially the number of places, strongly diminish. The spatial distribution of the places visited, based on the stakeholders' answers, showed a noticeable decrease in visits to more distant places, and those located far from main roads, not accessible by car.

This reduction in sites visited is especially striking (a decrease of almost 50%) in groups of elderly citizens and farmers, who, ever since childhood, have had the closest relationship with the landscape.

Different reasons were given for not entering into the landscape today. Among elderly citizens, the argument is mainly a need to rest in the village after a whole life of hard work in the fields and by the river, and also the impenetrability caused by fences and brambles. The younger generation does not visit the river landscape because they are frightened of getting lost, or they have no interest in it, or there are too many brambles and no defined paths.

Even though elderly citizens used to be rather evident in the river landscape, some of them even living there, nowadays their frequency has decreased, and many (34%), do not visit it any more. Among students, there are some who, previously, never went to the river (16%), but all of them now visit, at least the nearest places, from time to time. Farmers are another group who used to have an almost permanent presence in the area, and due to their work, they make frequent visits to it (85% visited several times a week). Among other amenity users, the frequency of visits has in general diminished (particularly those who visited the area once a month, down by 35%), except for those who never previously went there (10%) and those who never visit it now (5%).

The quantity of elements still remembered (especially by elderly citizens), together with the lack of variety in the places visited (especially by younger generations), signify that the landscape of Ribeira Grande is now at a 'turning point'. There are many things preserved in local memory, which should be transferred to future landscape users and on which should be based landscape valorization and new functions, like open air sporting and educational, natural and cultural activities.

To attract people back to the landscape, as well as to increase and motivate new users, the landscape must provide new functions. Only then will they start feeling concern for its future and in that way cooperate in true participative management (Buijs, 2003).

A combined reflection and conclusion

In both case studies, people have lost or are still losing the traditional functional connection with their landscape leading to a loss of their emotional relationship with it. Together with this lack of relationship, the lack of interest in its management is connected to current alien landscape management practices, neither understood nor accepted by the local inhabitants.

Former traditional landscape management assured a more effective attachment between people and the landscape because of their direct and permanent involvement. Actually this attachment, as well as knowledge of landscape history, and the meanings of earlier landscape, are disappearing, as clearly shown in the Ribeira Grande case study.

Even though, the new functions that emerged recently are considered to be an opportunity by some landscape users, others do not seem to be sufficiently aware of these possibilities or know how to get engaged with them. This was exemplified by new pine afforestation in the Mértola case study.

When approaching future landscape management, it is important to define functions that assure the attachment of people with the landscape and to stimulate new perceptions that might generate new directions for management.

Landscape is not only something perceived by people, as defined by European Landscape Convention (Council of Europe, 2000a), but also something that has to be managed by people. In this understanding, not only experts, but also the population, should be part of the landscape management process.

But how can a new rural landscape development strategy succeed if people have, for a long time, been losing their attachment to the landscape?

We consider local participation is the fundamental tool for introducing innovative proposals to reach consensus and motivation among all stakeholders. Only in this way is it possible to expect an increase in the intellectual and social capital and thus to cause a shift from identity loss to a more creative and active relationship, where local people and decision-makers can integrate their perspectives within the same building process. This co-operation then, like a raising spiral, supported by a plurality of people's perceptions and landscape management actions, can lead towards more sustainable future landscapes, lived and appreciated by their users. To reach this 'ideal scenario', it is necessary that stakeholders to gradually become aware of the importance of their participation in the decision-making and landscape management processes.

Based on both research experiences, we consider interviews as one of the key steps towards an active exchange of information about the landscape. While researchers are becoming aware of problems within the actual landscape, the interviewees are gaining awareness about the importance of the topic through the concern demonstrated by the investigators.

It is important to let people know not only the reasons and aims of the study, but also to share with them the proposals based on the research results. This should be an opportunity for reaching consensus, reconciling the population's expectations and viable technical solutions based on landscape potentials.

Perception and management should be considered as being part of the same whole. Local participation is a way to deal with changing process. People change their attitudes if they are able to share their visions and opinions with others, trying to reach common solutions. This is the fundamental basis of participation and in our opinion, an important step in finding out how people can reinvent landscape. Understanding landscape perception appears to open a door towards landscaping action. In this way, landscaping action could be considered landscape management in which the human dimension and direct involvement of people are assured. Nevertheless, we cannot forget that there is still a considerable distance between the best theoretical proposal and its concrete application, especially in situations where the public is not sufficiently knowledgeable and motivated enough to get involved as in both of our case studies.

Despite the complexity of this issue, we believe that local level is the right scale to stimulate the public awareness and participation process.

Neither of our studies intended to cover all the steps up to the direct involvement of local people in their landscape management. Their contribution was primarily to understand, through the landscape perception study, the weaknesses of present landscape management models and to present suggestions to be incorporated in the management and planning processes for the future.

The results from the Mértola case study will be incorporated into the final version of proposals that will be compiled. In the smaller study of Ribeira Grande, final proposals for valorization of the river landscape were delivered to the municipality council and presented to the public. The results will be also considered in the municipal plan of development.

Local participation is desirable nowadays, not only because new perspectives on landscape research are developing, but also because during the last decades, we have observed that a technical approach is not enough to guarantee successful landscape management (Cílek, 2002).

Most European rural areas demand creative and innovative management models and solutions. Local participation, in the form of consultative democracy, in landscape management is limited, since citizens' involvement in social and cultural life is narrow and almost extends only as far as the vote. We should proceed towards new models, one of which is participative democracy where active and effective participation is expected.

Local participation can be seen as a practical bridge between research and planning, and between planning and management. Nevertheless, integration between research and implementation is still only beginning (Stewart *et al.*, 2004). At the same time, planning processes are rarely sensitive enough to articulate environmental meanings or represent the plurality of perspectives of the community.

Participatory approaches require persistence and plenty of time and can often be misleading if not all the specificities of landscape and the interests of its users are included. A strategy and action plan for participation should be developed for each case. Even though the studied landscapes have many similarities, it is impossible to define a common action plan. Based on our experiences, we are sure only about some important aspects, which are necessary in order to initiate an active participatory process.

– *Information* for the population, depending on the characteristics of each local community and the subjects to highlight.

– *Stimulation* and *encouragement* of an active citizenship through local initiatives, such as debates for sharing ideas on specific landscape problems and needs.

– Demonstration of the *public willingness to participate in the decision-making process related to landscape management as the result of local initiatives.*

– Co-operation between local communities and authorities in the implementation of identified and viable common solutions.

Thus, through knowledge of landscape perceptions we can improve our research and planning skills in landscape management. But we are only able to reach holism, integration and sustainability if public awareness and participation play as equal a role as the expert perspectives (Scott, 2002; Luz, 2000).

Cosgrove (1985) reformulated the concept of landscape perception, which was strongly natural science orientated until 20 years ago, in order to "allow for the incorporation of individual, imaginative and creative human experience into studies".

Now it is time to include the idea of public participation into the landscaping action and in that way bring back the individual to himself, nature, society and landscape.

The incorporation of public participation processes in three landscape planning projects in the Murcia region of Spain

Santiago FERNÁNDEZ MUÑOZ and Rafael MATA OLMO

Landscape, social perceptions and public participation

Landscape does not only deal with the physiognomy of a territory; it also involves mankind's sensitivity towards his geographic space. The landscape, therefore, constitutes a meeting point between the object and the subject, between the 'being' and his 'visibility', an idea that synthesizes – according to Jean Marc Besse – the tension between "on one hand, the activity of the spectator and, on the other, the fact that there is something to see, something to be taken in" (Besse, 2000: 100). Thus, the cultural content of the landscape is seen not only in the materiality of each configuration shaped by anthropic action in nature, but also in social images and models of preferences[1] based on multiple viewpoints and perceptions that change over time.

For a concept of landscape that is committed to sustainable land management, the representation of groups and social actors is of interest, above all, as an expression of the different ways of seeing and interpreting the landscape reality, and of expressing aspirations and objectives that do not always coincide. The perceptive dimension of the concept of landscape, therefore, refers to the need for public participation in order to learn of – according to the Florence Convention – "the aspirations of populations" and to formulate "objectives of landscape quality". It is not a question of a frivolous proposal for landscapes *à la carte*. Neither is it a question, as Michel Prieur wrote, of "surrendering

---

[1] Here the concept of 'landscape model' (modèle paysager) as understood by Cadiou and Luginbühl (1995), as a "cognitive model that allows a space to be read and qualified as a landscape".

to fashion (…). If the Convention insists so much upon the question of participation, it is in order to juridically express the specificity of 'landscape' in the best manner possible. The landscape only exists through what is perceived. A policy exclusively involving the experts and the administration would produce a landscape *supported* by the people (…). The democratization of the landscape is not only associated with the new scope of action introduced by the Convention of Florence, but is also expressed through this collective and individual acquisition of all landscapes, which requires the direct participation of everyone in all the decision-taking phases, for their transformation, the follow-up of their evolution and for the prevention of inconsiderate destruction" (Prieur and Durousseau, 2004).

Landscape planning and public participation in the Murcia region of Spain

In Spain, much remains to be done with regard to landscape planning and consequently, participation by citizens in relation to landscape. Although significant advances have been made in recent years in the characterization of the landscape in the state as a whole and in certain regions (Basque Government, 1990; Gómez Mendoza, 1999; Mata Olmo and Sanz Herráiz, 2003), there has been little experience in relation to actions taken. Even today, the safe keeping of landscape values is predominantly attributed to the policies and regulations governing protected natural areas. The latest state law on nature conservation, dating from 1989, recognizes, among different forms of protection, that of 'protected landscape'. Most of the regional autonomies that have approved conservation laws (a total of 14 out of 17) have also incorporated the concept of 'protected landscape', with certain nuances in their definitions. The protection of few landscapes, however, has been declared to date, and criteria and dimensions vary according to regions, and furthermore, these are concentrated in a small number of regional autonomies.

Little has been done, either, by the town planning sector, in spite of the fact that the first Land Law in 1956 and subsequent reforms thereof, highlighted the values of landscape as one of the criteria for the classification of protected rural land. There has also been minimum implementation – except in Catalonia and the Balearic Isles – of specific planning measures such as the Special Plan, conceived among other reasons for the defence of landscape areas of great interest, without the need to resort to other nature protection systems.

More recently, in the last 15 years, the noteworthy legislative development in land planning by regional autonomies, which have been constitutionally assigned exclusive responsibility in this respect, has contributed to a certain updating of the landscape, which, in any case, has been promoted by the general interest in the landscape question in other European countries. The generalized model of land planning regulations in all regional autonomies, which gives the planning instruments of regional and subregional influence great capacity for action, endows these instruments with a certain power for the safe keeping and improvement of landscape values, because references in laws, where they exist at all, are always generic and lack any capacity for determination.

The development of the regional autonomies' legislative framework (all regions now have land planning laws, some are even second generation), is in contrast, however, with the few instruments for regional and subregional planning that have finally been approved and applied. Some of these plans, specifically those designed for several regions in Andalucia, or those recently approved for the island of Menorca (Balearic Isles) or Tenerife (Canary Isles), consider landscape to be an important factor, a concept heretofore unknown in Spanish land planning. In these instruments, landscape is considered in a specific manner, for example, in the Menorca Island Land Plan, it is seen both as a way to understand and diagnose the territory, and as one of the main objectives related to planning and awareness of landscape as territorial heritage.

In this context, we must mention the implementation in the last 4 years of three projects relating to landscape study and directives in three subregions of the Murcia region: the Huerta[2] of Murcia, the Noroeste (north-west of Murcia) and the Altiplano (plateau). The 'Dirección General de Ordenación del Territorio' (land planning department) of the regional government has considered these studies with the objective of "implementing analysis and diagnosis of landscape units in order to assess these; to make proposals for territorial action for the improvement of the landscape and the design of regulations for protection and planning in relation to other uses, in particular rural tourism, productive economic activities and infrastructures".

The technical prerequisites of the three projects follow a classical methodology, which consists of a systematic inventory of the 'physical and anthropic environment', the demarcation of homogeneous landscape units, and the drafting of a diagnosis and the definition of certain proposals conforming to specific landscape directives, for incorporation into the territorial Land Plan at regional scale, an instrument established by regional Land Law. Using this as a basis, and following another approach (Mata *et al.*, 2001), the research team[3] proposed, that the cross-cutting social surveys, be incorporated into all phases of the study, despite the fact that the technical specifications made no reference at all to such participation.

What should be the aims and levels of public participation in a landscape project? There is no simple or single answer to this question; it will vary according to the social characteristics of each territory and with the degree of social commitment to public participation, while in all cases the lack of experience in the formulation and application of social survey initiatives in landscape management planning will remain a constraint (Zanchini, 2002). The main difficulty found in such participatory processes is often political, not technical. Participation is not the result of a decision by experts, it is often a political choice that is subsequently added by other means (Rebollo, 2002).

Once the themes included in the social surveying initiative have been defined, it must then be decided who should become involved. The term 'interested parties', often used in the literature on participation, is relatively ambiguous in a landscape plan, because practically the whole population can be deemed 'interested'. The first option should be between participation based on the individual, which collects the opinions of society as a whole in statistically reliable samples, or through consultation aimed mainly at the social

---

[2] Huerta: irrigated and cultivated plain, especially in the provinces of Valencia and Murcia in Spain.
[3] The projects were drafted by a multidisciplinary team of consultants from public administrations, directed by Rafael Mata Olmo and co-ordinated by Santiago Fernández Muñoz. The management of the projects by the administration was the responsibility of Antonio Clemente (Noroeste and Altiplano) and Clemente Pagán (Huerta of Murcia). (Región de Murcia, 2002, 2003,2004).

actors and partners. It is not a question of excluding certain parties, as it is possible, and in many cases, recommended, to combine both methods. In any case, the objective is to establish a sample that represents society, that contributes different landscape perceptions and aspirations, and, using this base, in dialogue with experts, to construct a socially and politically accepted landscape project, which is feasible in terms of management.

It should be remembered that the social surveys were initiated by the research team and were not, therefore, methodologically defined by the regional administration. In this sense, each of the projects was conceived as an experiment with which to test different methods, considering the geographic and social realities of each space, and remembering that, together with the participation, each study aimed to meet the requirements for the landscape inventory, characterization, diagnosis and directives established by the technical prerequisites.

Although belonging to the same region, each *comarca*[4] subjected to planning has different physical and social features, which have influenced decisions on the public consultation methods. The Huerta de Murcia, with an area of approximately 250 km^2, constitutes one of the most representative landscape image of old irrigation systems and intense housing development characteristic of Mediterranean valleys and alluvial plains. It is currently a space undergoing a process of 'metropolitanization', with a population of around 500,000 inhabitants, half of which lives in the city of Murcia (in the centre of the *huerta*), the other half residing in small, traditional *huerta* workers' settlements, the population of which has grown by up to 10 times in less than 50 years, in houses scattered throughout the valley, between *huertas* and citrus orchards. The profound change in landscape and functionality caused by housing development has not prevented the *huerta* from continuing to be a fundamental element of identity in the *comarca* and the region, and a space with a rich hydrological and cultural heritage, which also provides excellent views from its mountain borders.

On the other hand, the Noroeste is a large subregion (almost 3000 km^2), of a predominantly rural nature, with marked contrasts in physical features (e.g. high forested mountain, agricultural plateaux and small irrigated valleys) and a small population for its size (65,000 inhabitants). In this subregion, the population relates better to the reality of the landscape, although the considerable size and internal diversity of the area, and the numerous smaller population nuclei, of great territorial significance, present problems for effective public engagement in the participation process. The Altiplano[5] subregion (*comarca*) is much smaller in area and has a smaller population; it has a clear local identity, with a long tradition of environmental and cultural study and protection. All project areas, however, have little experience of public participation and no processes of any depth or continuity could be identified.

The territorial complexity of the subregions in question and their little experience of the process of participation, together with a lack of specific material resources for participation, only allowed for public consultation processes of limited scope. These processes attempted to glean the opinions of people and social partners on the different aspects of the plan in order to incorporate them into each of the study phases (e.g. landscape

---

[4] Comarca is a subregional geographic area with physical and historical features. Hereinafter it will be referred to as 'subregion'.

[5] Encompasses the municipalities of Jumilla and Yecla.

inventory and characterization, diagnosis and proposals). The ultimate objective was to obtain the impressions of the population on the character, condition, values and actions in relation to landscape, contrasting these with the opinions of the experts and representatives of the local government, and subsequently to formulate proposals for directives.

The questionnaire, which was used as a basis for all the processes, regardless of the method followed in each case, covered the following aspects.

– **The character and identity of the landscape:** elements or aspects that enable us to characterize landscape; local designations of the landscapes (what they are called); most representative and favourite places, spots or environments.

– **The view of the landscape:** itineraries, vantage points and places from where it is possible to have a panoramic views or scenic partial views of the local landscapes.

– **Processes, changes and problems of the landscape:** identification, classification and ranking of the territorial dynamics that cause changes and problems in the landscape.

– **Landscape aspirations and proposals:** definition of elements and landscapes meriting protection, improvement or recovery, and classification of the necessary actions in this respect.

Methods for public consultation and the characterization of the landscape in the subregions of Murcia

The geographic peculiarity of each subregion, the different numbers of citizens affected and the experimental nature of the participation initiatives justified the use of three different consultation methods.

In the case of the Huerta de Murcia, the Delphi method was chosen, the objective of which was to obtain the highest consensus possible from a group of experts on a given subject. The Delphi in the landscape of the Huerta de Murcia consisted of three questionnaires, one for each of the study phases. The first open questionnaire on the characterization and identity of the landscape was followed by two more dealing with processes and proposals into which was incorporated the findings of the group's answers from the preceding questionnaire. In the second round, the interviewee had the chance to situate himself in relation to the opinion of the other participants, thus facilitating consensus.

The group of participants (Table 1) in the consultation was made up of the 'experts' and the 'involved'. The former are people with knowledge and experience in the study of landscapes and in problems related to traditional Mediterranean irrigation systems, and to the specific processes taking place in the Huerta de Murcia. The 'involved' people are those who do not have theoretical knowledge greater than that of the average person, but whose activities will be potentially affected by the proposals in the plan. On this basis, a group of 47 participants belonging to different political, scientific and social institutions, and to irrigation and production organizations from the *huerta*, was created.

In the large and diverse subregion of Noroeste in-depth questionnaires and interviews were carried out with associations and social partners. Participation of associations provides the opinions of a limited number of highly qualified people, with a high degree of social standing, which, in some cases, is more useful than other types of surveys.

Table 4.1. Participants in the Delphi.

Professional colleges
Segura Water Board
Experts
Farmers
Ecologist groups
Cultural institutions
Junta de Hacendados (Landowners' Association) of the Huerta de Murcia
Political parties
Journalists
Trade Unions
Land planning and environmental experts
University departments

The selection of the 294 social partners consulted in the landscape study in Noroeste was based on a criterion of plurality, such that the participants represented the range of social and economic interests of the area. The results of the survey were amplified by a series of in-depth interviews with social partners from the primary and nature conservation sectors, coinciding with the drafting of a Sustainable Development Plan for this subregion.

In the landscape project for Altiplano, a survey applied to the population as a whole was opted for, in order to obtain the opinions of the highest possible number of citizens, thus avoiding the possible bias of participation based on the more informed, committed or involved sectors in society. To this end, a representative sample was defined, making use of a stratified random design to avoid deviations resulting from place of residence, age or gender.

The results of the Delphi in the *huerta* landscape

The Delphi method has shown its effectiveness in public consultations relating to the landscape of a metropolitan area, which is densely populated and poorly structured from the social point of view, and whose population shows an increasing lack of knowledge of the natural and cultural heritage of the environment. In the identification of the character of this excellent example of a Mediterranean peri-urban landscape, the interviewees considered of great importance everything related to irrigation and hydrological infrastructures, the River Segura, its natural dynamics and recent routing, and the relationship between the river, floodplains and *huerta* as an area of agricultural production and residence. One of the milestones unanimously highlighted by participants was the *Contraparada* dam and its surroundings, a hydrological 'project' dating from medieval times, which distributes water through the large channels on the right- and left-hand banks of the River Segura (Figure 4.2).

The opinion of the experts was also accurate and useful in the identification and classification of the changes and conflicts affecting the landscape of the *huerta*, and indicated in the first place the scattered housing and lack of town planning discipline. A clear expression, with regard to landscape, of the housing development process is the disappearance of the unbroken contact between population settlements and the surrounding agricultural landscape, characteristic of the traditional settlement system, the development of the so-called '*huerta* roads', and the proliferation of single-family dwellings within the rural housing. The consensus of both the experts and public resulted in development of clear proposals for priority actions for the safeguard and improvement of the landscape, and for the classification as an element of urban quality and standard of living, and locally, as a tourism resource associated with hydrological features of historic interest. The recovery of wetland landscapes, the control of housing development and the restoration of degraded landscapes take first place on the list (Figure 4.3).

The Delphi, therefore, provided an informed view of Huerta de Murcia's landscape, which was highly valued by the experts and politicians and has been incorporated by the research team into every phase of the study. The local experts made a cultured reading of the landscape that transcends the strictly visible, highlighting the role of the hydrological and agricultural structures in the historic modelling of the landscape and in the management thereof.

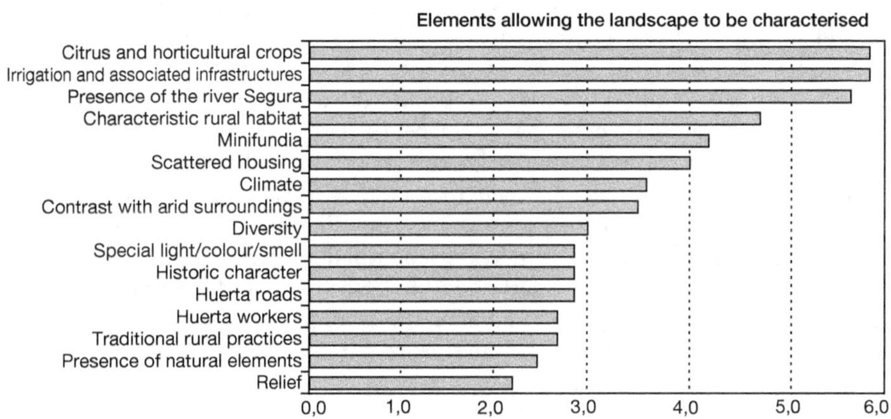

Figure 4.2. Elements used to characterised the landscape.

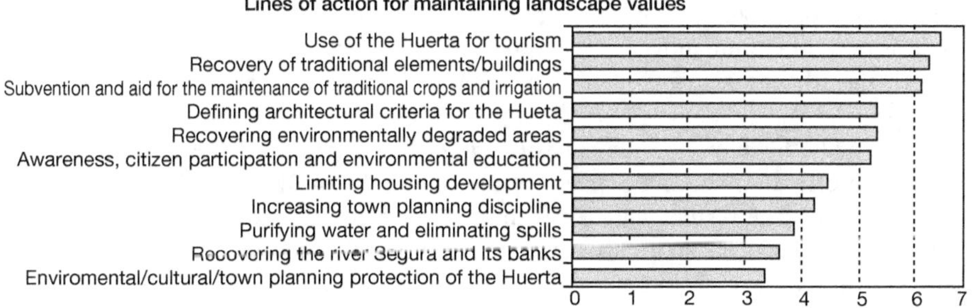

Figure 4.3. Lines of action for maintaining landscape values.

The consensus regarding the great value of the hydrological heritage constituted a good argument to propose – as was done in the study of the Huerta de Murcia – that projects for the modernization of infrastructures in traditional irrigation areas should consider the cultural and ecological value, and, as a last resort, the heritage value, of water distribution networks, instead of the priority and, sometimes, exclusive objective of improving efficiency. This position is particularly pertinent to metropolitan areas in which the irrigation-related landscape plays an important role, not only with regard to production, but also an environmental and territorial role in a context often dominated by types of part-time farming or even recreation.

The landscape surveys in Noroeste and Altiplano

In the projects for the subregions of Noroeste and Altiplano, it was decided to obtain a broader social perspective with the use of questionnaires. In the former, the decision was made to involve a large number of social partners, whereas in Altiplano, the survey was applied to a representative sample of the population. The questionnaires included open questions for participants to answer spontaneously, the objective of which was to record the perception of the population, avoiding a previously drawn up list of characteristic elements, places or sites, or vantage points and itineraries that might condition the result. It was, however, considered necessary to present the interviewees with a broad but limited series of problems in order to identify the territorial dynamics that, in their opinion, had the greatest effect on the landscape. Experience from consultations in other territories also led to the predefinition in the survey of possible strategies and actions necessary to protect landscape values and to attempt to improve and restore these.

Both surveys were useful in the design of the diagnoses and the formulation of proposals; although, in general terms, the contributions were less informative than those provided by the Delphi in Huerta de Murcia for drafting the plans, they did lead to the reconsideration of emphases and strategies and, in any case, revealed the social view of the landscape. Joint analysis of the answers from both consultations allowed for illustration of these perceptions.

The population of Noroeste perceived a broken, forested landscape, with a predominance of natural elements (e.g. mountains, forest vegetation and water), in which agricultural uses were also mentioned as characteristic components, whereas in Altiplano the population interviewed identified the features of their territory through four main elements: i.e. aridity, vineyards, the mountain relief rising above the plains of Jumilla and Yecla, and the physiognomy of the urban nuclei. The social perception, therefore, draws a picture of a predominantly agricultural landscape, with clear topographic contrasts, in which the two main towns also form a part, although at greater scale, of the local landscape.

Along with the lists of elements allowing characterization, the consultation also requested information about the places or areas with the most representative landscapes in the subregion. The results were revealing in two senses. In the first place, the population identifies landscapes in their own district, perhaps for reasons both of proximity and a feeling of not belonging to the *comarca*. In the second place, a clear role is played by mountain formations and forests as an expression of the most characteristic or identifying landscapes of both territories. The agricultural plains landscapes, dominant in Altiplano and also present in Noroeste, appear in second place, in spite of some of their constituent elements, such as the vineyards or croplands in general, being indicated as characteristic of the landscape. People, therefore, tend to identify images of the landscape with landscape milestones, i.e. those which stand out or are unique, rather than the many other components of the area, but which in fact define its appearance. The highest points and places with a significant presence of water, together with sanctuaries and other religious areas, were the places most indicated by the population from the landscape point of view. Next are the town centres, confirming that, in the opinion of the interviewees, these are elements that make up the landscape and, at the same time, constitute representative images.

The survey in Noroeste, and to a lesser extent Altiplano, demonstrates that many landscapes of greater interest, in the opinion of the experts, and some of the best watchtowers are completely unnoticed by the local population, who know, value and frequent only those landscapes and vantage points with good access and recreational facilities. This fact, which could only be established through public consultation, has enabled the Land Plan to be oriented towards the landscape and territorial culture of the population, through the dissemination of information on unknown and unfrequented landscapes, but which, however, correspond to the most socially valued images.

Processes, problems and proposals

Processes and problems related to water are clearly the most important factor in landscape for the populations of Altiplano and Noroeste. The over-exploitation of subterranean aquifers for irrigation and the lack of hydrological resources take first place, together with the disappearance of wetlands (Altiplano) and the pollution of rivers and aquifers (Noroeste). Three of the six most relevant problems, therefore, are related to water quantity or quality (Table 4.2).

Table 4.2. The ten main problems affecting the landscape.

Noroeste subregion	Altiplano subregion
Over-exploitation of subterranean waters	Water shortage
Water shortage	Over-exploitation of subterranean waters
Visual impact of quarries	Lack of town planning discipline
Lack of appropriate nature conservation policy	Absence of forest management with landscape criteria
Lack of town planning discipline	Solid waste outside rubbish dumps
Forest fires	Disappearance of wetlands
Pollution of rivers and aquifers	Disappearance of traditional agriculture
Absence of forest management with landscape criteria	Poor architectural quality of urban growth
Lack of appropriate policy related to treatment of waste and rubbish dumps	Visual impact of quarries
Deterioration and destruction of traditional agricultural buildings	Pollution of rivers and aquifers

Interpretation of the results of the surveys cannot be separated from the debate on the National Hydrological Plan that has taken place in recent years. In a context of water shortage in the south-east, priority was given to water management in the Murcia region. The debate over the River Ebro transfers has been particularly intense in areas like Altiplano, for which the construction of a specific branch is provided, to take water from the main channel of the transfer. Furthermore, the participation processes indicated that the lack of water is mainly attributed to natural causes (drought), as opposed to management of this resource, this being a very deep-rooted interpretation of the problem in Mediterranean regions (Saurí and del Moral, 2001).

Public consultation highlighted a clear tendency to identify territorial dynamics and problems with problems related to the landscape, so that many processes of territorial change are considered to be relevant from the landscape point of view, although they may have little real effect on the characterization and conditions of the landscape. Thus, in Altiplano, the over-exploitation of aquifers is one of the main landscape problems; it is a phenomenon, however, that has few visible manifestations (it only involves a few streams and small wetlands) and little effect on the most characteristic and socially appreciated elements and landscapes of the subregion. On the other hand, in Noroeste, the over-exploitation of aquifers does have a clear effect on the landscape, with a decrease in water courses flowing into springs, and reduced flows in rivers, implying an evident risk for the survival of many river and riparian landscapes, which appear in the public consultation as the most characteristic and valued components of the area.

Along with problems related to water resources, the populations of Noroeste and Altiplano indicated the proliferation of dispersed construction on the outskirts of the main towns, as another of the more serious problems related to landscape. This result is surprising and significant at the same time. It is noteworthy that the population identifies a relevant landscape problem in a type of landscape such as the peri-urban one, which heretofore had not been appreciated. From this public consultation experience in the region of Murcia, we can conclude that, although the population identifies the character and identity of the landscape with notable or unique elements, the people also understand that landscape is their perception of their surroundings when they are asked about problems relating to landscape. The landscape that stands out is not, therefore, the high mountains, forests or sanctuaries, but rather the peri-urban landscape in which much of daily life takes place.

The proposals for the landscape resulting from the processes of participation in Noroeste and Altiplano were similar but not identical (Table 4.3). In both cases, the recharge of aquifers takes first place, followed by different kinds of environmental actions, such as the protection of natural spaces or the control of dumping. On a second level are proposals related to the maintenance and rehabilitation of the rural and historic heritage, and then, actions referring to accessing and viewing the landscape, and to greater town planning discipline.

The population, therefore, mostly opted for proposals of a proactive kind, i.e. those which involve some type of physical intervention in the territory, with rapid, specific and tangible results. The nature and order of the proposals also reflects the relative assimilation of the concepts of territory and landscape; only this can account for the fact that the recharge of aquifers takes first place, in spite of the little direct effect it has on the landscape, ahead of other lines of action associated with more specific landscape

Table 4.3. Proposals for maintaining and improving landscapes.

Noroeste	Altiplano
Aquifer recharge	Aquifer recharge
Reforestation	Improvement of rural and forest roads
Protection of natural spaces	Restoring rural and historic heritage
Control of spills and waste waters	Protection of natural spaces
Improvement of rural and forest roads	Reforestation
Management and clearing of forests	Conservation of traditional crops
Town planning discipline	Town planning discipline
Regeneration quarries	Creation of a network of vantage points of landscape interest
Conservation of traditional crops	Control of spills and waste waters
Control and treatment of waste and rubbish dumps	Limiting enlargement of irrigation systems

problems. Thus, consultation with the population in general, and particularly with the social partners in the case of Noroeste, highlightedthe preference for actions that do not conflict with the activity of any economic sector in the subregion. With regard to water, the proposals agree with the very widespread idea in the population that the problem mainly lies in the scarcity and deterioration of the resource – hence the importance of aquifer recharge – and not in the type of use or management, which might involve some kind of limitation on the growth of irrigation practice.

Some final reflections

Public consultation in the cases studies in Murcia has in the first place helped to highlight the existence of different models of landscape perception and different social representations, not only in each subregion or district, but also among the different groups and social partners within each one.

The identity of the landscape is usually based on the prominent images of nature, and not on the physiognomies of greater territorial significance. When the population is asked about problems and conflicts related to the landscape, however, the scenario of daily life, the peri-urban space and the problem of scattered housing are considered as landscape.

The participation experience also shows limited local knowledge of existing landscape resources in each subregion, even with regard to those resources that respond to the most appreciated landscape images. It would seem that only those landscapes with easy access, preferably motorized, are known and frequented. Thus, one of the main directives of each study is the dissemination and interpretation of available landscape resources, through the compilation of a network of itineraries for access to local landscape diversity.

The public consultation also highlights certain contradictions or incoherence between what is desired with regard to landscape and opinions about certain processes (socio-economic or ecological), which affect the safe keeping of the most appreciated landscapes.

The methodology proposed by the administration did not include any reference to a public consultation, so the decision for participation processes to be incorporated into the landscape planning projects was taken by the authors. However, technical and political project co-ordinators accepted developing Delphi and the consultation as a further enhancement to the project, although they were not personally involved in any of these actions.

Our modest experience of public consultation in the drafting of the landscape directives for three subregions in Murcia enables us to conclude that participation in landscape-related matters should not be conceived as a basis for designing inventories of actions or as a 'suggestion box', but rather as a process incorporating "sensitive and perceptive interlocutors and not mere containers of preferences" (Fin Arler, 2000). It is a process revealing images and desires, but also knowledge gaps and contradictions, a mirror reflecting the 'awareness' and 'conscience' of society with regard to its territory. The role of the expert is not reduced to a mere registrar of requests and aspirations, or of a mediator in a confrontation between opposed images, although the latter is very important. Much can and must be contributed to the knowledge of landscape diversity and the dynamics thereof, to the evaluation of its fragility and vulnerability, to making contemplation and interpretation possible, to considering strategies and actions for the defence and improvement of the landscape, in an attempt to establish convergence, which is not always easy, between social perceptions and aspirations, which often do not coincide. Having said all this, a previous and vital step must be remembered: to highlight the fact that the landscape exists, and that it is something more valuable and profound than mere visible appearance, and that it deserves to be defended and appreciated.

Chapter 3

Landscape at the crossroads
of local development choices:
What knowledge for which issues?
Which tools for action?

Emmanuel GUISEPELLI and Philippe FLEURY

At a time when public action increasingly focuses on local negotiation and contractualization between the different stakeholders in the name of 'territorial governance' and 'sustainable development', landscape is often a consideration in land development projects. However, there are many different interpretations of the meaning of the landscape in these projects and it is necessary to clarify them, from the point of view of analysis and research as well as the mobilisation of landscape as a tool for action.

This paper addresses this question by showing the diversity of landscape status in relation to action. To pursue our analysis, we developed a broad definition of landscape. We based our research on a definition in which the landscape, far from being limited to an aesthetic view of space, is, above all, a social construction, the visible and perceived product of the interaction between a society and its environment at a given time in history. This position, shared by many ruralists in the world of research, offers possibilities that are relatively unexplored in relation to action at this time, since development experiences based on this approach are rare and have not been used to their advantage. If we consider the landscape as the product of social interactions, it is relevant to advance the hypothesis that it can be a tool for negotiating development projects, provided that the appropriate methodology is adopted (Guisepelli, 2001).

The aim of this paper is to show that the way in which the landscape is approached within a consensual process reveals the ways to approach development itself. We also show that the conditions necessary for a clear and explicit expression of the objectives of each of the concerned parties, based on the landscape, are only met under certain methodological conditions. Within this perspective, after providing an overview of the diversity of landscape status in the French Alps, we will draw on the contribution of the analysis of landscape discussion experiences to the construction of land development projects. We will then attempt to show how the production of action tools, based on the landscape, can be achieved within the framework of a research/action exchange.

Our reflection approaches the landscape from the overall perspective of land development. How does the emergence of the landscape within the political arena recast relationships between research and action? What are the development challenges brought to the forefront by scientific and empirical knowledge stemming from the landscape? What are the representations and values that underlie them? Can we design tools that will lead to a collective reading of the landscape?

Diversity of landscape considerations in development projects (analysis in the French Alps)

The landscape question is often linked to important issues, for example, land management, urban pressure, green tourism, new roles for agriculture, such as the maintenance of space, tourist infrastructures, etc. However, landscape appears in debates from different perspectives, depending on the context. In the French Alps, we analysed the diversity of landscape status and its links with development issues.

Diversity of landscape status in the French Alps

We studied the way in which the landscape question is dealt with in local development projects (e.g. development contacts in the Rhône-Alpes region, Leader+ projects, etc.), and particularly in those where agriculture is an issue (e.g. local sustainable agriculture projects). To do this, we analysed different projects to determine the moments during negotiations when the term 'landscape' was evoked, to identify specific issues attributed to the landscape and to characterize the collective consciousness of those involved in the process. Therefore, we were able to identify four types of situations.

1. The term 'landscape' does not – or very seldom – appear in local discussions. This was especially observed in some rural areas far from urban population centres. The questions that we can link to the landscape theme are actually dealt with at the local level in terms of 'space management' and deal primarily with the closing of environments for safety reasons and with the living environment of permanent inhabitants. This is the case of the highly rural areas of the Baronnies (Drôme and Hautes-Alpes) and the valley of the Tinée in the countryside near Nice (Alpes-Maritimes). Space management issues and those dealing with the maintenance of agriculture are linked to the commitment to avoid the depopulation of rural zones that are already sparsely populated. When it is explicitly a question of landscape, this term most often refers to historic monuments or natural curiosities. Space itself rarely has the status of a landscape.

2. The landscape clearly appears. It is the effective and unpremeditated manifestation of a development model. In this case, it is not a question of landscape management but, instead, its use to establish a local identity or to communicate information about a model. Thus, the Trièves (Isère), with its 'wheat mountain' image, the Beaufortain (Savoie) or the Thônes region (Haute-Savoie), with their well-maintained pastures, impose the quality of their landscape as the consequence of a harmonious development to be sustained. The landscape in this case becomes the exemplary figure of an idealized development.

When linked to the notion of 'countryside', the landscape takes on an identity role whose image is the reflection of that which determines the specificity of the site and can make it possible to demand some degree of autonomy at the political level. The local stakeholders[1] who see the situation in these terms are generally elected city and inter-community representatives. On this point, they are generally in agreement with farmers and representatives of the agricultural sector (e.g. elected representatives of unions or chambers of agriculture) and are strongly supported by resident associations where they exist[2].

The landscape here takes on an essentially symbolic function and is not used as a basis for reflection about the meaning of the territory and its future, except when it is implied by a certain status quo.

3. The landscape appears in development projects as an object on the same par as other sectorial actions. This is far from being the most common situation. In these projects, in the Tarentaise-Vanoise region (Savoie), the valley of the Arve and the Faucigny (Haute-Savoie), the landscape is seen as an aesthetic attribute to 'sell' the site. Landscape management consists of aesthetic initiatives in the towns and the treatment of landscape 'eyesores' (e.g. abandoned spaces, electricity pylons, etc.). For projects of this type, we often observed the important role of scientific and technical expertise in the landscape analysis and proposal of solutions (Guisepelli *et al.*, 2000). These experts are most often architects, urban planners and even landscape architects working in engineering and design departments, often in partnership with the DDE (Departmental Facilities Directorates) and the CAUE (Architecture, Urbanism and Environment Councils).

In this situation, a formal and aesthetic approach to the landscape is taken. The aspects emphasized are those of legibility (e.g. the necessity of a clear separation between the different components of the landscape, forest and agricultural spaces, in particular), the opening of new spaces (e.g. the rehabilitation of fallow land), the force line (e.g. the need to delineate the structural elements of the landscape), eyesores (e.g. unsightly elements in the landscape, such as abandoned spaces, electricity pylons, farm buildings, etc.), attraction points (e.g. features that immediately attract attention, such as natural curiosities and landforms), etc. Action proposals arising from this perspective consist of developments aimed at the rehabilitation of force lines while eliminating eyesores (e.g. the abandoned spaces around a village or along the edge of a forest, etc.).

Local stakeholders are often placed in a reaction situation when faced with a 'technical' diagnosis. It is difficult for them, if not impossible, to create and propose their own diagnosis (Guisepelli *et al.*, 2000). Thus, the social and economic questions raised by local stakeholders in relation to the landscape are only partially or not taken into account at all in these types of projects.

---

[1] Defined as any entity – group or individual – involved in the management process and having an initiative or reaction role (Amblard *et al.*, 1996).

[2] These players can make occasional alliances with environmental protection organizations, even if they are generally more inclined to promote protection measures at the local level than to defend the specificity of a territory.

4. The landscape appears as a means for addressing broader development issues. In this case, the group effort carried out on the landscape and its dynamics consists of linking the visible aspects of the landscape to the social, economic and, sometimes, even the environmental conditions of local development. In the Alps, recent collective landscape visualization experiences by stakeholders show that this approach leads them to establish diagnoses that are sometimes very different from those of the experts. The causal sequences referred to by the stakeholders to explain the landscape and its dynamics are not just the expression of social representations; they have their origins in the relationships between stakeholders and the tension that exists between their desire to develop new resources and the advantages and obstacles involved in doing so. The landscape makes it possible to reconcile the visible and the underlying means of operation of the territory with which local stakeholders are confronted on a daily basis.

These different configurations are not impervious; they can overlap in territories, particularly in those where several development projects and approaches exist.

The relationship between the landscape model and the development model

Northern alpine landscapes have been the object of very strong representations for three centuries by the social and cultural elite (Luginbühl 1990, 1996; de la Soudière, 1991). These representations, such as the regional model[3] (i.e. a pastoral vision composed of typical elements such as chalets, snow-covered peaks in the background, pine forests, pastures and cows) or the model of majestic mountain peaks (i.e. the vision of nature in all its splendour with glaciers and mountain pastures) (Guisepelli, 2005), have become reference models at every level of society. They are even the inspiration for some local development operations to make the materiality of the landscapes conform to these representations, for example, fir tree planting in the hardwood valley of the Belleville (Savoie) and the construction of chalets with wood from old barns (to give the impression of rusticity) in areas where stone is the traditional construction material, are just some typical examples in the Savoy region (Guisepelli, 2001). These representations are found side-by-side with or overlying local representations, some of which were themselves transformed into models and have been reproduced within the rural world of the Alps. This is especially the case for the 'Beaufortain' model (Savoie), symbolic for the farmers of the northern Alps.

Ideally, the Beaufortain model represents an exemplary agricultural development for the farmers of the northern Alps and some of the inhabitants of the mountain, for example, open, cultivated and productive slopes, products of sound management run by small family-size units, prosperous and organized around a product of quality and high added value – Beaufort cheese. Although the agricultural profession is aware that this model is technically relatively unreproducible outside of this zone, its evocation refers to a strategy of "making idle land talk". By expressing the necessity of a landscape maintained by agriculture, like Beaufortain, farmers demand a greater recognition of their

---

[3] The landscape model is a cognitive scheme that makes it possible to read a space and qualify it as a landscape (Cadiou, Luginbühl, 1995).

work, their practices and their place in society. Thus, introducing the emblematic model as an ideal landscape model is a means they use to claim a more important place in the local decision-making process.

A new, imported model, adopted by certain elected officials of mountain communities catering to tourists, has been superimposed on the one described above. This model is seen by them as a development path to be followed. The typical ideal is the Tyrolean mountain landscape. It is similar to the emblematic model of farmers in relation to its landscape components, but arguments in favour of its implementation make no reference to an agricultural identity and culture. Assumed to meet the expectations of tourists (which is partially true), this Tyrolean model emerges as a result of questions on the requalification; of mountain territories to develop summer tourism (Guisepelli, 2001). Elected representatives justify the relevance of their landscape ideal by attributing the expansion of interseasonal and summer tourism in the Tyrol to the quality of its landscapes and architecture. The functions attributed to agriculture for the conservation of open landscapes according to this model do not have the same scope as the emblematic model: agricultural activity here is part of the aesthetic value of the landscape in relation to an outside clientele while representing a certain idea of tradition and authenticity at the same time. It is not an essential element for local development and identity, as in the emblematic model, nor considered as an economic and decision-making contributor to the territory. Only maintenance actions on the part of the farmers are important here.

These two landscape models, despite their similarity from the point of view of their materiality, encompass very different concepts of local development and the place of agriculture, therefore, to propose a landscape model is another way of proposing a development model and even of highlighting specific sectorial interests.

To speak of the landscape is often to speak of the socio-economic organization and development of a territory, since a reading of the landscape always involves an explanation of its materiality. Some emphasize chains of ecological causes whereas others speak of economic and social dimensions. However, with respect to action, it is no longer a question of the overlapping of different individual 'readings' of the landscape; instead, it is a group reading that is built up progressively and in different ways, depending on how the different stakeholders interact.

Group landscape readings in the Northern Alps

In a group, depending on how the relationships between scientific knowledge (e.g. researcher scientists and experts) and 'lay' knowledge (e.g. local stakeholders) are established, the evaluation of the landscape and the causal sequences that explain its status and dynamics and, as a consequence, the resulting action priorities, are different. Experts and stakeholders do not have the same approach to the landscape. The stakeholders' diagnosis can even be at odds with the issues identified by the 'specialists'.

Scientific readings and stakeholder readings

This first example shows how landscape architects, agricultural engineers, elected officials and farmers see causal systems and different solutions when faced with the same mountain landscape in the Tarentaise region (Savoie). The portion of the landscape in question is located on a southern slope. It includes a village, an upstream forest zone and an agricultural grassland zone surrounding the village. This area is in the process of being colonized by woody perennials in the steep areas; only a few flat parcels of land have a herbaceous cover with no woody species.

An initial reading of this landscape can be made by observing the dynamics of series of vegetation, from grass to forest. The observer is struck by the decrease in farming but the causes are not taken into account, and often the action proposals recommend a mechanical treatment for the woody species (e.g. clearing, harvesting, etc.), focused on clearing the area around the village and re-establishing 'force lines' that structure the landscape. This is the approach of naturalists and landscape architects. It was also that of urban architects from an engineering and design firm that worked on roads and landscape issues in the Tarentaise region[4].

The agricultural engineer, by focusing on field practices, tries to understand the relationships between different practices, for example, cutting, intensive grazing, extensive grazing and percentage of land covered by woody species. He then refers to farming operations, particularly 'practice determinants', i.e. the reasons that explain why farmers graze or cut in one area rather than another. From this point of view, landscape dynamics are, therefore, linked to changes in agricultural practices, the logic of which can be understood at the farming system level. This approach makes it possible for the engineer to propose a certain degree of flexibility, to determine under which conditions farmers should cut or rehabilitate some of the abandoned plots, leading to a maintenance assistance approach at the plot level.

When a collective reading of the same landscape takes place, farmers and elected representatives refer to causes other than changes in the landscape. While mentioning the abandonment of slopes because of the difficulties involved in mechanization and labour-intensive practices, they speak of the greater decrease in the number of farms compared with other regions. Different reasons are evoked, for example, the drastic subdivision of the land, late implementation of a milk collection system by the dairy farmers' co-operative in Beaufort (having an adverse affect on farm maintenance), a limited community budget in comparison to the budget of communities on the north slope with their ski resorts and a farm assistance policy. These causal chains that explain the landscape are complex and, once established, open the way to proposals for solutions that are controversial rather than unanimous, for example, establishment of fiscal solidarity between tourist communities on the north slope and less privileged communities on the south slope, assistance to encourage young farmers to settle in the area, land assignment planning, milk collection assistance in difficult areas, etc.

We can object to this example that an interdisciplinary research would have revealed the different causal systems referred to by the stakeholders. This is highly probable but rarely possible and are we not supposed to have confidence in the practical knowledge of the stakeholders and thus, approach expertise as a diagnostic element among others? From this perspective, the expert's work is no longer the only basis for action that would

---

[4] Study report: 'Accès en Tarentaise et enjeux paysagers', Schéma de cohérence de la Tarentaise, by Verrier, C. (ASADAC), Fabre M., DDE / SAU 1 Savoie, (1997).

make it possible to 'objectively' characterize and deal with a situation. It becomes an element that contributes to the enhancement of a more overall collective reflection on the issues and that participates in arbitration on the relevancy and feasibility of different options.

The players' refusal of experts' diagnoses

In 2004, all the concerned parties (e.g. farmers, elected officials and inhabitants) in the Albertville Basin met to take a look at the future of the area, 20 years in the future. During a debate on the subject of actual and desirable development, a discussion took place on the basis of a visual aid proposed by the group coordinators. It was a block diagram, accompanied by several landscape photographs representing the spatial distribution of farming systems (e.g. arboriculture, crops, mixed farming systems and dairy farming), in relation to altitude and exposure. This visual aid strongly contributed to the construction of a collective reading of the landscape, not because the group members agreed with it but because they agreed to discuss it.

Basically, rather than a 'vertical' organization of the territory as proposed by technicians, it was the 'horizontal' issues that were emphasized. Participants particularly insisted that two urbanization rings around the towns of Albertville and Ugine be emphasized on the block diagram. The first ring corresponded to a dense urbanization zone and the second to a more diffused area. This classification into two urbanization zones around each city, although general, made it possible for them to discuss the conditions for the existence of agriculture on the basis of varying degrees of urban influence (e.g. real estate competition, the encroachment of urbanization on farmland with the problems it brings as well as new marketing opportunities for local products, etc.). Finally, the advantages and disadvantages of each urbanization zone were discussed in relation to different types of farm products.

The refusal of the players to accept the block diagram representation as originally presented (which should have lead to a discussion on the adaptation of agriculture to each 'stage' of the slopes) as a basis for another diagnosis was motivated by the need for a debate on the relationships between agriculture and urbanization and on the perceived and shared necessity of agricultural 'visibility' within an urban context. On the basis of this discussion, particularly innovative development perspectives were developed. For example, in agricultural spaces in the first ring, farming was seen as playing the role of a 'green pocket' with high added value products. Agriculture thus becomes the means for allowing urban residents to discover the countryside. The creation of an 'agricultural discovery tour' was proposed.

Landscape in this case is clearly a function of the relationship between the city, the farm and the environment. By showing interactions between several spaces (e.g. constructed zones, fields, grasslands and forests, as well as intraurban parks and gardens), the schematization of the spatial organization of the landscape made it possible to discuss new challenges and to give substance to action projects.

This case is almost a textbook example of the difference between an expert opinion, however well supported, and the expectations of a group of concerned members. This group assessment would not have been possible in a situation where the diagnosis was presented as the framework for discussions. This is an example of group dynamic practices that encourage the participatory and active approach of its members, leading to this expression.

When landscape justifies the social selection of space

The following situation, which was just a part of the negotiations, illustrates both the difficulties involved in negotiations revolving around the landscape and the fact that we can justify social segregation of space in the name of landscape management.

As part of this project, a group reflection phase on the landscape took place (in 2003), by highlighting zones that would be of concern over the next 20 years. The group was divided into subgroups[5]. One of the subgroups (composed mainly of farmers) came up with the idea of distributing space management between several categories of service providers, such as professional farmers, landscaping firms and 'small farmers'. Flat land was attributed to the first group, who did not feel that they had to satisfy a demand for maintaining spaces that did not meet criteria of access and flatness. Areas with slopes were left to landscaping firms and "small farmers" or to "professional farmers who had the time", provided that the task was remunerated. The justification of a "social selection of the landscape" was supported by the idea that agricultural sustainability was based on its professionalization. From this perspective, operational and technical tools were to be attributed to professional farmers, including the best plots, i.e. those in flat areas. This is a development concept that, once again, was presented and materialised by highlighting particular zones on photographs.

The critical assessment of these zones by other subgroups made it possible to come back to this vision that was deeply modified as a result. The confrontation between this way of seeing the situation and the one formulated by other involved stakeholders (e.g tourist office, elected officials, resident associations, etc.) made it possible to 'insert' new variables into the explanation of the status and dynamics of landscapes and to formulate expectations of agriculture other than just maintaining an open landscape. This made it possible to reconstruct a systemic approach to land development in which interactions between agriculture, tourism and the inhabitants are approached from different points of view, for example, the relationship between urbanization of farm-land and intensification of remaining land, consequences of changes in agricultural practices, but also of urbanization on the quality of the local environment and tourist infrastructures, etc.

This case shows that the application of a method based on a simple expression of the "stakeholders' word" is not sufficient (especially when groups are dominated by a particular category of stakeholders), and that a collective endeavour must be built on group dynamic techniques that allow the expression of the different interests involved. For the group leader, this implies that he/she has mastered group dynamic techniques as well as the ability to identify different interest groups present and to assess the social

---

[5] Three work subgroups, each one consisting of elected officials, farmers and representatives from the tourist sector, worked on land development scenarios for a 20-year period, based primarily on a reading of the land-scape. A set of four landscape photographs, aimed at reflecting the diversity of land and development configu-rations, was given to the subgroups. They were asked to choose priority zones and to give reasons for their choice and the feasibility of their proposal.

consequences that can lead to one policy or another. His/her independence in relation to each group is of the utmost importance.

Discussion

Using landscape as a tool to instigate action

Negotiation processes provide a privileged context where landscape representations can be approached as a means to set development strategies in motion. Collective evaluation of the landscape shifts the traditional conceptual framework, separating scientific approaches to material aspects from social representations of the landscape. In the case of action assistance, it is difficult to pretend to be able to exhaustively describe the material aspects of the landscape, from the expert's point of view, and then to see which elements of this scientific representation, referred to as 'objective', are mobilised in social representations. Stakeholders' diagnoses link physical and ideal facts based as much on their knowledge of the land as on development opportunities that they would like to put into action. Recognizing this, from the point of view of the research involved in the action as well as from the point of view of expertise, means adopting a position that lends legitimacy to the stakeholders'diagnoses. This implies analysing causal chains and solutions identified by stakeholders to deal with the landscape as a strategic interpretation of their objectives and action leverage tools that they identify. Which development projects implicit in their reading are pushed to the forefront? What are the possible areas of agreement? We must also question the nature of these diagnoses. Which socio-territorial organization concepts do they reveal? Why is one variable linked to another? Analysis of these questions seems vital to us if we are to avoid the dangers of a fragile consensus on landscape forms (like the Beaufortain and Tyrolean models).

Even if many landscape management tools present this as an object of consensus, they risk building projects on the basis of a misunderstanding: agreeing on material forms without understanding what actually lies behind these forms in terms of project and concept of the socio-economic organization of the territory. We think that a sound agreement on the landscape means going beyond the mere form of a model and requires the recognition of the legitimacy of all parties to contribute and to support a land development project. Operational landscape management must identify issues for which there is a true convergence of objectives and means (e.g. recognition of farmers as decision-makers and not as simple executors, as well as recognition by farmers of possible contributors and issues other than those related to agriculture).

To do this, mediation between scientific and 'lay' knowledge is essential. It requires the co-ordinators, on one hand, to have the operational ability to manage participatory dialogue procedures by allowing all the participants to express themselves and, on the other hand, a certain distance enabling them to identify the often implicit values and strategies of the players in relation to development when they express their opinion on the landscape. Landscape photography is a particularly effective medium. Maps and block diagrams are also useful but these objects must be dealt with cautiously because their utilization runs the risk of maintaining a hierarchy between the one who produces them (the expert) and the one who is asked to express an opinion about them (Mougenot, 2003).

An example of a tool: 'identifying landscape-interest zones in a local development project'

By basing ourselves on a landscape approach as a social construction, we have designed a dialogue tool for a group of stakeholders making it possible to identify landscape-interest zones in a local development project. It is necessary to go beyond the simple identification of landscape action priorities and to contribute to a development project with a larger scope concerning agriculture and its role in land development. To do this, landscape is approached in such a way as to make it possible to have a debate on all of the human activities produced by the landscape, particularly agricultural practices, as well as the reasons for which the group is concerned with the landscape, for example, tourism, local environment and/or the cultural identity of the inhabitants, etc. Nevertheless, experience shows that the operational implementation of the tool implies that the group has already conducted a survey of the site, i.e. activity trends, farming trends, definition of major priorities related to the future of the territory and its agriculture. Work on the landscape then makes it possible to test the representations and the intentions of all those involved and to assess the feasibility of different actions. The landscape is an object for effectively visualizing objectives and imagining what we would like to change and how we intend to do it.

From a practical point of view, two or three groups of stakeholders work on different landscapes in the region. It is the group itself that chooses the sites it considers worthy of consideration. Then, on the basis of photographs, or directly in the field, participants identify, highlight and comment on what they consider to be a landscape-interest zone (Guisepelli and Fleury, 2003)[6]:

– the stakeholders discuss the landscapes and defend their position on the issues that they identify, for example, improvement of the living environment, cultural identity, natural risks, value of beautiful landscapes for tourists, etc.

– collective delineation of landscape-interest zones by the stakeholders.

– eventually, the group co-ordinator or expert can contribute another point of view (e.g. the historical analysis of photos, or information concerning the aspirations of people outside the group in relation to the landscape, such as tourists).

– To make the transition from this descriptive phase to action priorities, it is necessary to answer certain questions. Why do you think there is a problem in a certain zone? What are the activity or farming trends at the root of it? What should be done? Are we being realistic? How can we organize ourselves to act more effectively?

It is obviously not in the questions but in the resulting discussion that a collective landscape reading (not necessarily consensual) and its social construction gradually take form. Several elements appear to be important in this collective construction that progressively evolves from a description of landscape forms to social relationships and the evolution of human activities that are at its root.

First of all, these exchanges often lead to an awareness of the diversity of ways of perceiving landscapes and assessing priorities. For example, in the case of a mountain

---

[6] Landscape photography can also reflect the point of view of the person who took the picture and thus introduce a 'hierarchy' between experts and players. Nevertheless, the players, with the exception of technical tools that they control less effectively, can more easily question the choices made in the case of photographs.

landscape, ski slopes are elements considered by elected officials and some farmers to open up the landscape and enhance its quality. However, the majority of farmers see only agricultural land resources, grasslands and pastures as open spaces, of high aesthetic quality. Despite these differences, an agreement was reached on the necessity of keeping the outskirts of the villages open and preserving the least steep areas for agriculture whenever possible. Beyond the reading of physical forms, this technique makes it possible to understand the justifications that stakeholders evoke to defend their landscape choices. Each utilization of this tool makes it possible to declare that landscape is the vehicle for expressing different aspects of the relationships between social groups and local development.

An action management tool built on the basis of the landscape must, therefore, satisfy different conditions: participate in the emergence of the convergences of interests between stakeholders; take account of the means necessary to carry them out; contribute to a mutual recognition of representations, objectives and the legitimacy of each one.

From this point of view, developing a tool to enable us to deal with a collective reading of the landscape for the purpose of development implies approaching the analysis of its materiality by combining expert approaches with stakeholder approaches. All representations of the materiality of the landscape and its determinants, whether from the stakeholders 'or the experts' point of view, provide a group with important information. However, what really counts is that this representation comes to be, not as a unique truth but as a subject of debate and controversy, which can eventually even be rejected. This implies another conception of the approaches to the materiality of the landscape by research scientists and experts. We must abandon the idea that these are the only approaches with which stakeholders can choose solutions. They must be conceived of as 'germs' whose role is to instigate discussion and to give rise to other causal chains to extend the scope of action options.

For these reasons, it seems important to us that efforts to develop tools focus on the consensus-building stage between stakeholders. Landscape has often been used as the means to avoid conflicts and progressively and smoothly arrive at an agreement within a group. The examples that we have developed here show that debate is a necessary step in the construction of a sound agreement. However, taking issue is one thing and arriving at an agreement is another; this is not easy to do. Group dynamic tools based on debate management are necessary and do not exist at this time.

Conclusion: from action to research

These perspectives modify the limit between the production of knowledge and action. The production of knowledge, of know-how, is no longer approached as the exclusive domain of research scientists and experts. This issue can be approached from two points of view. What have the stakeholders learned from the researchers and experts? And, what have the researchers and experts learned from the stakeholders (Callon *et al.*, 2001)? Moreover, applied research concerning landscape is not exclusive of action; it contributes to it, creating what is referred to as 'intervention research' (Hatchuel, 2000). We are no longer in a downward movement of research towards action, but in a more complex exchange process. This also leads to a renewal of research themes revolving around the landscape. It must be approached as a means for expressing development projects. To do so, it is necessary to analyse the stakeholders' diagnoses, not as the expression of biased representations of an objectivable reality, but as the fruit of experience lived on a daily basis whose legitimacy and relevance are the result of this experience, a combination of both reality and desire.

Our paper does not question the relevance or importance of landscape actions focused on the development of the materiality of the landscape. However, it analyses how, in different territories of the alpine range, local stakeholders speak of, or want to speak of, local development in terms of their landscape. This complexity is linked to the polysemous nature of the landscape and to the fact that it is also and above all a social construction. Approaching landscape as a means to address development implies the specific group dynamic methods that we described above. How they can be developed remains to be seen and, in the same vein, negotiation with regard to the landscape constitutes yet another area of research to be explored.

The enhancement of research through action requires a reflexive attitude concerning the operational consequences of its products and its ethical stance. Avoiding this issue by reducing the landscape to a scientific and technical dimension without restoring its political dimension is equivalent to playing a paradoxical game that limits the landscape to a circle of specialists while declaring that it represents a global issue for society. This paradox leads to another: representing the landscape as a fundamental issue while limiting actions to simple superficial problems, without addressing specific political and social issues that must be taken into consideration.

Landscape: a window of opportunity for regional governance? Landscape scenario workshops as a participatory planning tool

Wolfgang PFEFFERKORN and Barbara ČERNIČ MALI

"Since the middle of the 1990s, the role of landscapes as a political issue at the European level has been steadily increasing. Despite the absence of formal, statutory European instruments for landscapes, they have captured the interest of both scientific and governmental bodies alike. [...] A landscape approach offers holistic assessment and planning tools to define and develop the interface between nature and culture. Hence, landscape, as the place of human interaction with nature appears to be the heart of sustainability"[1]. For a long time, the discussion among concerned academic persons and spatial and landscape planners (at least in the German speaking parts of Europe) about landscape development has been dominated by the conservation and nature protection aspects. In the last few years, this discussion has been superseded by a new and more comprehensive understanding guided by the idea of sustainable development, taking into consideration, on the one hand, that ecological aspects have to be seen in close relation to social and economic aspects of a region or a protected area and, on the other hand, that the concept of sustainable development cannot be based only on (scientific) knowledge, but requires a discussion of values in society.

In this new understanding of landscape and landscape research, inter- and transdisciplinary approaches become more and more important. Transdisciplinarity – which means the inclusion of non-scientific persons into research – demands everyday language as communication standard:
– What will landscape look like in Saas Fee/Switzerland in the year 2020?
– Who will run the few remaining mountain farms in Carnia/Italy?
– Will the bears and wolves reconquer the woods of the eastern Austrian Alps?

---

[1] European Centre for Nature Conservation (2000).

These and other questions were discussed in several projects by researchers, public administrators and the local population in the whole alpine bow between 1997 and 2004. Hereby, inter- and transdisciplinary research groups used the method of 'future pictures and future stories of cultural landscape':
– to develop and discuss 'likely futures' for their regions;
– to define desired and non desired trends;
– to identify measures for supporting desired development and to prevent undesired development;
– to find common agreements for local action.

With this approach, the authors wanted to show that the idea of opening the scientific world to non-scientists – which is necessary when seriously considering the concept of sustainable development – is possible and thus research can directly help to support local action.

Scientific background

The results presented below are based on the research project REGALP ('Regional Development and Cultural Landscape Change: the Example of the Alps. Evaluating and Adjusting EU and National Policies to Manage a Balanced Change', 2001–04). The project was co-financed by the European Commission and several national institutions under the Fifth Framework Programme[2]. The method was developed in the following research projects.
– 'EU-Extension to the East: Chances and Risks for Sustainable Development of Cultural Landscapes in the Styria-Slovenia Border Region' (2000–02). Research project financed by the Austrian Federal Ministry for Education, Science and Culture under the Austrian Research Focus 'Cultural Landscapes'.
– 'Sustainable Development of Alpine Cultural Landscapes in the Austria-Slovenia Border Region' (1998–99). Research project financed by the Austrian Federal Ministry for Education, Science and Culture under the Austrian Research Focus 'Cultural Landscapes'.
– 'Future Pictures and Future Stories of Cultural Landscape' (1998–99). Research project financed by the Austrian Federal Ministry for Education, Science and Culture under the Austrian Research Focus 'Cultural Landscapes'.

Methodological design of regional scenarios, future pictures and future stories

The methodology of 'future pictures and future stories of cultural landscape' was based on the following steps (Figure 4.4).
– Analysing macro-trends.
– Defining integrated cultural landscape scenarios.

---

[2] For more information: www.regalp.at

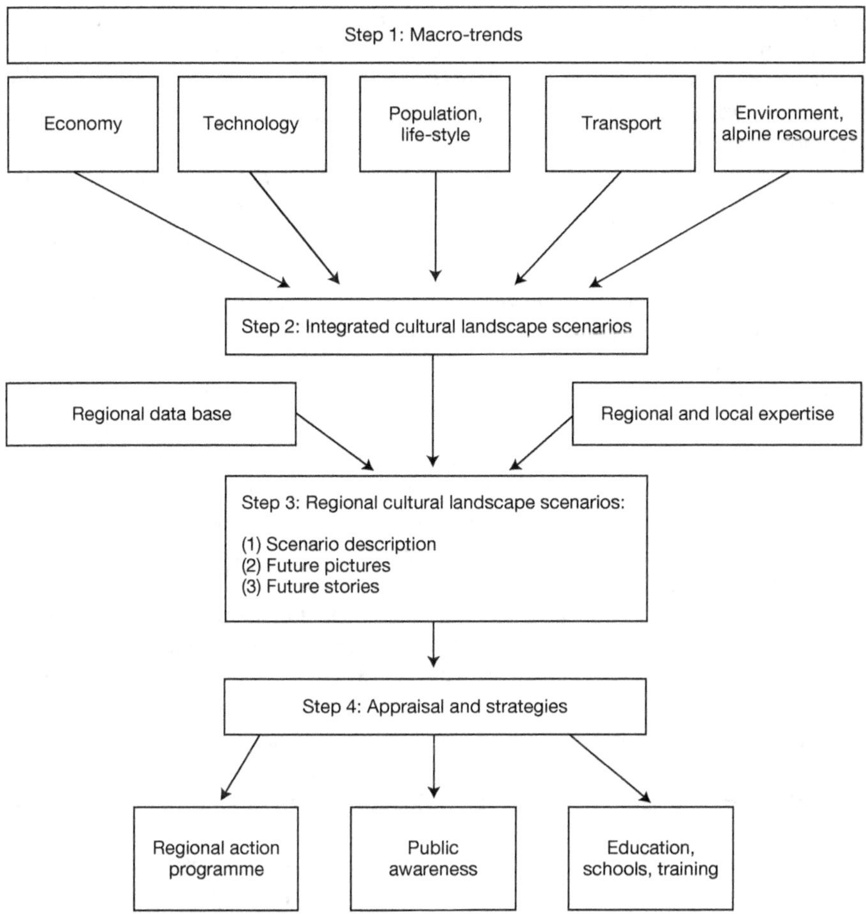

Figure 4.4. Steps towards regional cultural landscape scenarios.

– Detailing regional scenarios by using data, future pictures and future stories of cultural landscape.

– Developing strategies to promote desired and to avoid undesired developments.

In REGALP, the first step was a comprehensive review of research dealing with general present and future development trends and sectoral forecasts for demography, technology, economy, transport, agriculture and environment. Thus, the main development trends at the European level, which will influence spatial development and cultural landscape changes in the Alps until 2020, were identified. In REGALP, these macro-trends were compiled in a 'Macro-trends Reader' (to be downloaded from the project website)

This reader was the basis for the second step, the elaboration of two opposing integrated cultural landscape scenarios for the Alps in 2020. As these scenarios were not intended to be tools for detailed forecasts, but a basis for discussions within the research team, as well as with experts, policy-makers and regional stakeholders, the research team decided to omit extreme and improbable scenarios like a global ecological showdown, a third world war or a 'Garden of Eden'. However, the REGALP team decided to present two 'likely futures', each different from the other in terms of the aims and weighting of public policies governing alpine space.

In the first scenario, policies are not significantly oriented towards sustainability, therefore, it was called the 'Inertial Scenario'. The main characteristic of this scenario is increasing polarization of spatial development in the Alps. The second scenario depicts a situation where policies have a stronger positive impact and basic references of sustainability such as the Alpine Convention[3], the European Landscape Convention[4] and the European Spatial Development Perspective will be implemented at different levels. This scenario was called the 'Towards Sustainability Scenario'.

These two alpine-wide scenarios were the basis for the third step, where the research team developed regional cultural landscape scenarios for the seven REGALP pilot regions. Hereby, the researchers used a detailed analysis of regional data as well as the knowledge and expertise of regional stakeholders gained through interviews.

The regional cultural landscape scenarios consist of the following three 'products', which were elaborated by the research team before running the future-workshops.

– Regional scenario description: a 40–50 page report with qualitative and quantitative descriptions of the pilot region in the year 2020. For this description, selected indicators were used: inhabitants, places of work in the different sectors, commuters, social infrastructure, agricultural businesses, settlements, land use (e.g. built-up land, arable land, pastures, forests, etc.), etc.

– Future pictures: 4–6 photographs with visualizations of anticipated landscape changes corresponding to the above scenario written description. Hereby the researchers used photographs, aerial or landscape from the pilot region and produced photomontages on the computer (see future pictures below).

– Future stories: 3–5 stories (1–3 pages each) describing the everyday life of various actors from the pilot region in the year 2020, conveying aspects, which cannot be expressed either in data or in maps or pictures (e.g. personal behaviour, values, opinions, family life, etc.). The future stories were partly written by members of the research team. The others – as the result of a productive co-operation with local schools – were written by pupils from the pilot region aged between 10 and 18 years.

The three products elaborated in step 3 were the basic material for step 4, the local appraisal and development of local and regional strategies. This step was carried out under the framework of the so called 'Cultural Landscape Future Workshops'.

These workshops were held in each of the seven REGALP pilot regions. They were organized by the research team in close co-operation with local stakeholders. The researchers provided the methodology, the agenda and all the scenario material, the locals provided the room, catering and contacts to the participants. The workshops were attended on average by 15 participants, amongst them mayors, representatives of regional administration and important economic sectors, NGOs and the local population.

---

3 Alpine Convention: http://deutsch.cipra.org/
4 European Landscape Convention (2000): www.nature.coe.int/english/main/ landscape/conv.htm

In each pilot region, the workshops consisted of two parts.

– The first workshop was entitled: 'The Future of the Region XY'. The local participants described their opinions on the future development of their region and were then given the scenarios of the research team. These scenarios were presented through future pictures of cultural landscape (photomontages and future-maps), which enabled visualisation and mental transfer of the participants into the future. In addition, the researchers (and/or the locals) prepared future stories describing everyday life of specific actors (e.g. mayor, farmer, hotel manager, etc). in the year 2020. Some of these stories were read out. The participants' expectations for the future were collected from answers to the question "What changes do you expect to happen in the region xy by 2020?" The expectations of the locals and the researchers were compared and discussed, key topics for further work were determined.

– The second workshop, which took place at most 2 months after the first, was entitled: 'From Future to the Present'. The aim was to bring the expectations for the future, elaborated in the first workshops, to the needs of action today. The key question of the second workshop was: "What do we have to do today in order to promote desired, and to avoid undesired, developments?" The expected outcomes of the second workshop were twofold: a list of project ideas and plans for local/regional actions to be undertaken. These actions should then be developed further and carried out within the framework of regional development programmes like LEADER+, INTERREG and others. Secondly, the needs and requirements (recommendations) for interventions at higher levels (e.g. the State, EU, etc.) were developed.

Results

Results of steps 1 and 2: the Alps in the year 2020; integrated scenarios of the cultural landscape

The results of REGALP as well as other alpine research indicate that the gap between prosperous central regions and marginalized peripheral areas in the Alps will increase in the coming years, and the metropolitan areas outside the Alps will gain more and more influence on the alpine area.

Apart from the macro-trends and general policies which can be considered as external factors in the future development of the alpine area, there are also internal factors for future development of the different regional development types within the Alps. Two main factors have been identified.

– 'Alpine remoteness' of a region determines its situation of marginality; alpine remoteness is a complex factor made up by by the distance from centres, altitude, topographic features and transport facilities.

– 'Presence and use of endogenous resources' determines regional development potential and indicates the physical presence of diverse resources (e.g. wood, water, beautiful landscapes, local products or skills, etc.) as well as local policies and attitudes.

Further zonings are listed in Table 4.4 and Figure 4.5, together with some of their peculiarities in both scenarios:

Table 4.4. General features of the main clusters.

Cluster	General peculiarities	Inertial Scenario	Towards Sustainability Scenario
Local centres	Alpine towns in which most of the population, economic activities and services are concentrated. They are disappearing in border alpine regions with a strong influence from metropolitan areas outside the Alps.	Fewer in number but more important than in 2003.	Not wider than in 2003.
Commuter areas with own activities	They gravitate on the local centres, but they have some local activities (in different sectors). They are areas of in- as well as out-commuting. Quality of life could be better than in the local centres.	Not wider than in 2003.	They are wider than in the Inertial Scenario. Different sectors of economic activities are present ('polifunctional growing areas').
Residential commuter areas	Typical areas of out-commuting, often losing services and local identity.	Wider than in 2003.	Not wider than in 2003. Local services and local transport network are quite good.
Growing peripheral areas	To be in a peripheral position in some cases can be a 'resource'. Soft tourism, multifunctional agriculture, protection of nature and/or other activities give these areas a good chance for development.	Potentials cannot be realized.	Wider than in 2003.
Steady peripheral areas	Areas in which the difficulties due to alpine remoteness are in some way counterbalanced by local activities and economic resources.	Less wide than in the Toward Sustainability Scenario; less wide than in 2003 because of the strong decrease of services in peripheral areas that causes depopulation.	Wider than in 2003; polifunctionality among different sectors of economic activities can be a good chance to maintain steady conditions.
Declining peripheral areas	Areas of out-migration as well as of out-commuting. Services as well as local economic activities are decreasing. Re-naturalization characterizes large areas.	Wider than in 2003.	Less wide than in 2003, as in many areas new local resources can be used for 'sustainable' economic activities.

Cluster	General peculiarities	Inertial Scenario	Towards Sustainability Scenario
Tourist areas	Situated at the highest altitudes and linked to winter sports, most of all. Some of them are growing (due to a massive presence of immigrants, tourists and entrepreneurs), others are facing crises.	There are some important growing areas, but a large number of tourist communities at lower sea levels are facing a serious crisis.	Some strong classical tourist areas, soft tourism is a real alternative. Other economic activities help.

Source: Castiglioni *et al.* (2004)

Results of Step 3: Regional cultural landscape scenarios

Working in selected pilot regions and involving local and regional actors in project activities was a core element of REGALP. The research team co-operated with local stakeholders in the following seven pilot regions:
– Wipp Valley, Austria (Brenner transit traffic route);
– Lower Enns Valley/Lower Tauern, Austria (peripheral agricultural area);
– Visp/Saas Valley, Switzerland (highly developed tourist and strong urban area);
– Le Trièves, France (agricultural area not far from Grenoble);
– Isarwinkel, Germany (agricultural area with tourism, influenced by the Munich urban area;
– But Valley, Carnia, Italy (peripheral area with some tourism);
– Upper Sava Valley, Slovenia (peripheral area with old industries in transition).
The selection of the pilot regions was based on the following criteria:
– all main spatial development types in the Alps (e.g. urban regions, tourist areas, peripheral areas, etc.) and all alpine states had to be covered;
– regional stakeholders had to be willing to co-operate with the research team;
– regional planning and programming activities (like LEADER+) had to be taken into account;
– regions should be covered by EU and national subsidies;
– size of pilot regions in the different countries should be similar.
Regional scenarios were elaborated for each of these seven pilot regions. As mentioned in above, the scenarios consisted of a report, the future pictures and the future stories. This material can be downloaded from the REGALP website.
Typical examples for elements of the regional cultural landscape scenarios, future pictures from Carinthia and Slovenia together with parts of future stories from France are presented below.

Future pictures

The pictures 'Ruttach 1998 and 2020' show the process of reforestation on alpine pastures. Less accessible sites are relinquished whereas the easily accessed land around the farm building is intensified.
The example of Žirovnica in Slovenia shows urban sprawl, a typical problem of alpine valley floors.

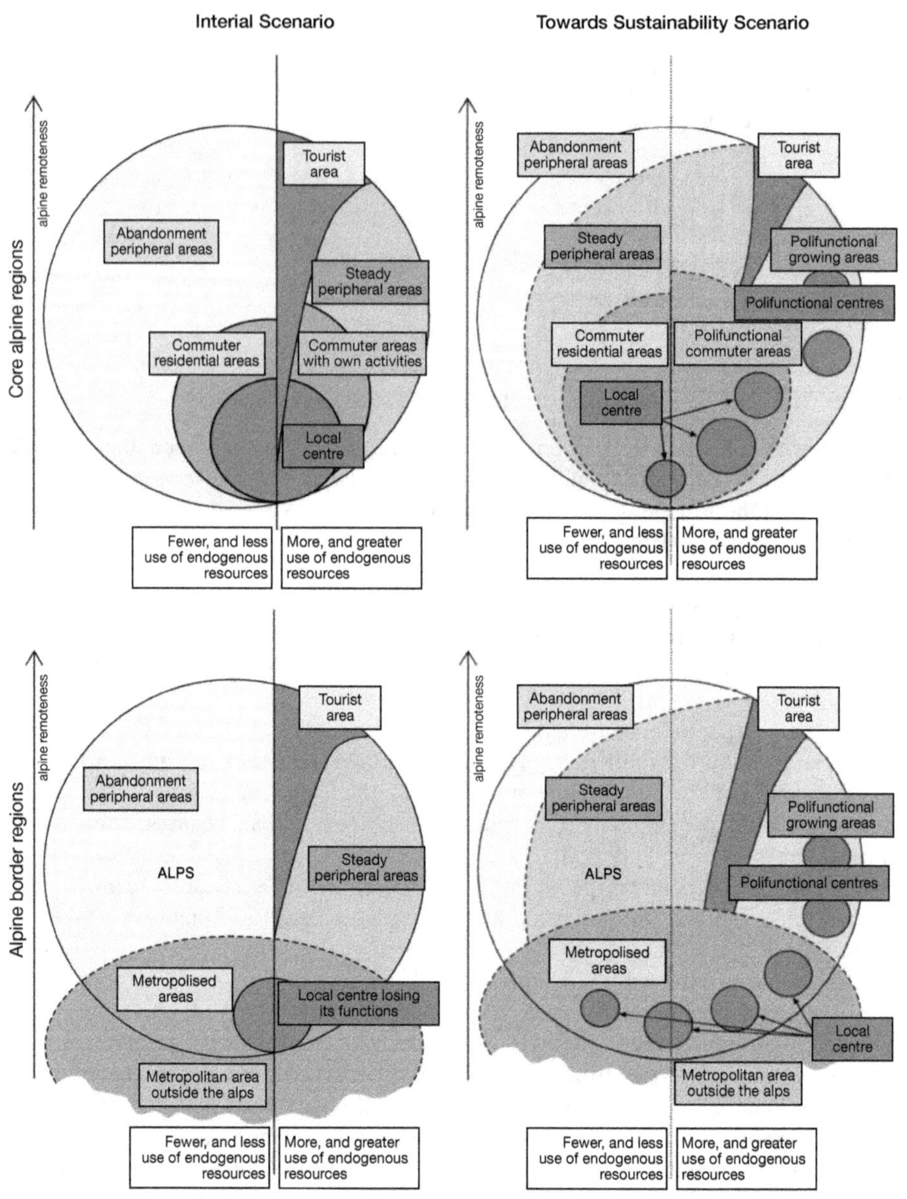

Source: Castiglioni, B., Grosutti, J., Massarutto, A., Troiano, S., Virgilio, T. (2004)

Figure 4.5. Integrated scenarios of the cultural landscape.

Future stories

The future stories, from which the authors have taken some paragraphs, were written by local schoolchildren in the pilot region of Le Trièves, France.

Figure 4.6. Ruttach (Austria) in 1998.

Figure 4.7. Reforestation in Ruttach in 2020?

Un scénario catastrophe pour le Trièves

« Nous sommes en 2023. J'ai 30 ans, je suis mariée et j'ai 3 enfants. Tous les jours, j'entends les voitures, les trains, les camions. Le maire vient une heure par jour et il fait mettre des feux de stop partout. Aux rives, il y a un immeuble de 300 étages. A St Jean d'Hérans, il y a 6 maisons et une ligne droite. La montagne n'a plus d'arbres. Les voitures polluent les campagnes. »

Figure 4.8. Žirovnica (Slovenia) in 2003.

Figure 4.9. Urban sprawl in Žirovnica in 2020?

« Nous sommes en 2023. J'ai 31 ans. Le village de St Jean d'Hérans est extrêmement pollué. Les poubelles sont renversées, il y a des produits chimiques et du caca partout. Aucune loi n'existe plus : cambriolages, vols à main armée, bancs sciés en deux, voici quelques exemples. La police est partout ; on ne peut rien faire sans être contrôlé. »

Un scénario de rêve pour le Trièves

« Nous sommes en 2023. J'ai 30 ans, je suis mariée et j'ai 3 enfants. on est tranquille, on n'entend pas le bruit des voitures, des trains, des camions. Cela sent bon le foin. Il y a une piscine publique, une salle des fêtes. A coté du parc, il y a un parc à chevaux,

une grande école. Les gens sont sympas. Ça ne sent pas la pollution. Il y a des champs d'herbes, des petits villages sympas. »

« Nous sommes en 2023. J'ai 31 ans. Le village de St Jean d'Hérans est devenu génial : deux immeubles, une piscine publique, une piste de quad, le stade de France, un terrain de tennis où Roland Garros va bientôt se passer, un terrain de basket où il se passe des matchs de NBA. On peut aussi faire des balades. »[5]

Results of step 4: appraisal and strategies

As described above, two workshops were held in each pilot region. The first workshop focused on future expectations, the second on actual needs for action in order to promote desired development and avoid undesired developments. Table 4.5 lists the main expectations of participants in the first workshop in Wipptal in Austria.

Table 4.5. The main expectations for the future (including number of points) of local participants in the first workshop in Wipptal, Austria.

Negative expectations		Positive expectations	
Dead villages – lacking infrastructure	10	Recreation potential	15
Increasing environmental impacts (noise and pollution from traffic)	9	Decreasing traffic and environmental impacts because of Brennerbasistunnel	10
Decrease of agriculture and agricultural businesses	9	Good economic structures	8
Dying tourism	4	Wood processing industry: used for energy production and building material	7
No social life (people moving away)	4	Tourism: cultural landscape as a new potential for new target groups	7

Source: Pfefferkorn and Favry (2003)

In the second workshop, researchers and local participants tried to step back from the future to the present by asking: "What do we have to do today in order to promote desired development and avoid undesired development?"

In answering this question, workshop participants and researchers developed the following five ideas for projects and actions.
– Idea 1: Regional trademark 'Wipptal'.
– Idea 2: Direct marketing measures – co-operation between agriculture and tourism.
– Idea 3: Cultural forum Wipptal – calendar of regional events.
– Idea 4: Local recreation strategy for Wipptal.
– Idea 5: Congress on alpine agriculture.

The project ideas developed by participants in the second workshop in Wipptal (documents to be downloaded from the REGALP website), are presented in Table 4.6:

---

[5] In French in the original text.

Table 4.6. Project ideas developed by the participants in the second workshop in Wipptal, Austria.

	Idea 4: Local recreation strategy for Wipptal
Aims	To create an added value from the local recreation tourism (e.g. gastronomy, ski-lifts, tourism, agricultural subsidies) Control of development, less dependence on external factors
Content: what has to be included?	To assess the status quo To find out the interests of the communities: common interests, identify possible conflicts, work on conflicts and look for solutions, define a common strategy Balance of benefit and expenses between the communities Integration of local residents Marketing based on a common strategy, increase of attractiveness and guidance to visitors
Work programme	First step: discussion in the course of the next LEADER+-association management meeting on 1 July 2003 (item on the agenda)
Project organization and partnership	LEADER+ working group tourism, agriculture, communities, regional authority for spatial planning, tourism association, maybe involvement of population
Costs and financing	To be examined in the context of further project steps

Source: Pfefferkorn and Favry (2003)

In all the pilot regions, similar outputs were produced by the participants. In order to ensure further continuity of local action, the ideas were then developed further by the local stakeholders under the framework of LEADER+, INTERREG or other programmes.

Conclusions and discussion

From the perspective of members of the REGALP team, the workshops were a good and useful tool for integrating the perspectives of the local population into the research work. Equally, the locals were also satisfied: most of the local workshop participants gave strong positive feedback at the end of the second workshop (for each workshop, a protocol in the national language has been elaborated by the research team and sent to the participants). The atmosphere was warm and open, the pictures and stories were very helpful in making the complex issues understandable. The opportunity to take part in development projects and to exchange views with other people from their region was rated as positive. Some participants regretted that there were not more such events in their region.

From the point of view of the authors, it was concluded from the experiences gained that landscape is a good communication tool for transdisciplinary research: it appeals to people directly and emotionally, since it is connected to everyday activities and even to individual experiences of childhood. Thus, the landscape topic has proved to be useful for integrating locals into regional planning and decision-making. This could open new perspectives for participatory planning and decision-making in a greater dimension, as the importance of public participation and governance issues has been increasing in the last years. On the one hand, this is due to international strategies, procedures and legal issues like the Aarhuis Convention (UNECE), the Local Agenda 21 (UN), the Convention on Biodiversity (UN), the Flora-Fauna-Habitat Directive (Natura 2000, EC), the Water Framework Directive (EC), the EIA and SEA Regulations (EC) and the European White Book 'Governance' (EC). All these procedures and guidelines make demands for public participation although they do not include detailed regulations.

On the other hand, individuals such as landowners, interest groups like farmers' associations and the civil society in general demand – and are demanded – for more and more involvement in planning and decision-making – and, as we have seen by REGALP and other examples, also into research. Public participation and new ways of decision-making are often experienced as a major challenge – or even as an excessive demand by, for example, decision-makers, planners and persons working in administration at different state levels as well as by economical actors, researchers, interest groups and the citizens themselves. Thus, there is a urgent need for upgrading the culture of co-operation in research and planning: by developing tools and providing room for negotiation (as in the workshops described above) as well as through educating and training stakeholders in 'process and negotiation skills'.

A crucial aspect of the transdisciplinary approach in the REGALP project was expressed by some participants asking the question: 'What will happen after these two workshops? How will communication and project development continue?'

From the positive, and also the negative experiences, gained in the different projects, we can summarize in the following comments. If researchers decide to involve locals in their work, they have to be very careful because they awaken expectations that normally cannot be fulfilled within a *research* project. Researchers come... and after a short time researchers leave, but locals (have to) stay! In order to ensure that the activities initiated (in this case by the workshops) can continue, the whole transdisciplinary procedure has to be embedded into existing local and regional structures: there has to be somebody who cares about the developed ideas and who stays in contact with the local workshop participants. This role can be played by members of local or regional planning boards, LEADER Action Groups, or regional development agencies. If these partners are not closely involved already in the preparation of the process, then the continuation and implementation of ideas will mostly fail. It is only the continuation of these processes and the implementation of ideas that makes the transition from research to action become reality!

Conclusion

Yves LUGINBÜHL

The approach that led to these many contributions to landscape knowledge and their transfer to political action reveals an initial aspect of considerable importance: the diversity of the issues dealt with and, in particular, the development of research on these issues in the majority of European countries. It goes without saying that western Europe is in a privileged position, but we know that other countries in central and eastern Europe are also involved in research on this subject. Nevertheless, it seems obvious that there are gaps in the approaches between the northern and southern countries, not necessarily in the volume of knowledge produced, but rather in the methods used. Research in the southern European countries appears to be more oriented towards the knowledge of those involved in the landscape, their representations and their social practices, which does not mean that research in northern European countries does not deal with these issues. Instead, it approaches it from another angle, with more emphasis on perceptions or preferences, undoubtedly because it is more focused on political action and because northern European research aims at more immediate solutions, and also probably because the institutions responsible for the protection, management and development of landscapes are more receptive than in the south, having been involved in this action for a longer time.

This development and diversity of scientific research on the landscape as well as of research of a more direct nature is certainly indicative of a manifestation of the hope that underlies 'landscape' approaches to take the aspirations of the concerned populations in terms of land-use planning into account, as well as research that is increasingly sensitive at the ecological and aesthetic level. It cannot be denied that the recent European Landscape Convention has played a role in this development; after all, doesn't it actually promote the participation of societies in the development of their living environment and their continuous integration into the decision-making process, as witnessed in the Årrhus convention? This development is obviously a sign of the times, but what exactly does this trend towards the awareness of more 'everyday' landscapes, as described in Article 5 of the European Landscape Convention, really mean? The preceding papers also show that the commitment of research only applies to inhabited landscapes and ones that are not considered to be either remarkable or exceptional. It is the everyday landscape that draws the attention of researchers and land developers today, interested by the dynamics of these landscapes and the problems that they raise, as well as by the practices or representations of the people that interact with them on an everyday level.

All of this being said, we can, nevertheless, observe a remarkable permanence in landscape models. If some researchers imagine that new models appear, particularly in peri-urban spaces or in the new creations of park or landscape designers, they should take a second look. The models that have structured landscape thinking in Europe since the Renaissance are still around and still functional; landscapes continue to be picturesque, sublime, pastoral or bucolic. We would be curious to see what post-modern models look like. If they do exist, we have not seen any yet or, perhaps, they are still just on the drawing board.

The other obvious aspect is the emergence of a landscape market with its own stakeholders, for example, designers, engineers, technicians, companies, managers, etc. Whereas a landscape market arose with the development of tourism, as emphasized by Elisée Reclus in 1866, this market adopted the techniques and practices of landscapers and their institutional partners, in both the public and private spheres. Landscaping pays – it feeds into ecological engineering as well as regional enhancement. The landscape has thus become a means for politicians to cast the spaces in which they exercise their influence in the best possible light: landscape is a symbol of European regions and nations, those selling regional products, wines and other food-related products.

However, beyond this development that can be evaluated in terms of the amount of research or operations that mobilise scientific knowledge within the framework of landscape action and that appear to be extremely numerous[1], we must obviously ask the questions on which landscape researchers focus at this time. What is the effect of this development on landscape theory in Europe? And what are the dimensions that define this research priority?

Lessons learned

Even if these last decades since the 1960s have given rise to multiple discussions on the meaning of the term 'landscape'[2], it seems that most of the researchers that contributed to this work agree on attributing a sense of two-dimensional social construction to the landscape: (1) material, referring to the biophysical aspects of the landscape; and (2) physical, referring to representations and perceptions. The European Landscape Convention's definition of landscape fits perfectly within this framework: "...an area, as perceived by people, whose character is the result of the action and interaction of natural and/or human factors". It thus refers to the two dimensions, material and immaterial, and to the interaction established between the two, i.e. between the biophysical and the social.

---

[1] In the case of France, we estimate that less than 10% of the national territory has not yet been the object of a study or landscape inventory (all types of studies included).

[2] These meanings ranged, on a continuum, from the most material definition (the landscape of first-generation ecologists and physical geographers) to a definition that referred only to the immaterial dimension and gave the landscape the sense of a relationship without a material existence.

The second lesson deals with the social aspects. It includes the formal stakeholders responsible for the implementation of public policies, and the informal stakeholders, that is, civil society whose sensitivities are taken into account within the framework of social representations and perceptions, thus implying that the landscape does not exist without stakeholders. A parallel could be drawn with the art of directing: the landscape is what the stakeholders, formal and informal, direct, to satisfy their needs, and their dreams, of a better future. This also means that landscape does not exist without political action, that is, today, without direct or indirect public policies. This is what leads to the development of 'research action', undisputedly in the process of development, albeit with its ups and downs. This is clearly revealed in the papers included in this book: producing knowledge about landscapes in order to be able to act. Even if research action is still in its beginning stages, even if it advances by trial and error, it is because it casts doubts on the unique relevance of scientific knowledge. Today, it is the sum of knowledge, both scientific and lay, that must be mobilized within the framework of 'transdisciplinarity', the new form of interdisciplinarity that interacts between scientists and political stakeholders, both institutional and informal.

It is probably the issue of the environment that is primarily responsible for changing the meanings of the landscape. Whereas certain conceptual trends attempted to limit the landscape to its aesthetic dimension, analyses of actual experiences made it possible to reinstate environmental issues within the framework of social sensitivities, particularly through the experience of the confrontation of the individual with biophysical materiality, or through a social memory and social relationships. It was necessary to re-establish the lived-in landscape within its spatial, temporal and social context in order to see the signs of ecological concern that could be transposed into aesthetic sensitivities. Admittedly, the landscape is not the environment, but – and here is one of the lessons that we have learned from these papers – the environment/landscape relationship should be looked at more closely in order to contribute to a better understanding of the environment. Just like contributing to the evaluation of biodiversity that was barely touched on here. Even if we now know that certain species require time to adapt to new environments, that it is necessary to ensure the continuity and the connectivity between biophysical structures to enable species to establish themselves in transformed environments, we have still not established a definitive relationship between types of landscapes and biodiversity. We can only say that open landscapes lose part of their species stock when they are closed, that is, in the case of reforestation. These relationships between biodiversity and landscape require long-term research that rarely takes place within the same time-frame as political decisions. We need observations on repeated seasonal events to provide proof of a significant modification of the biodiversity. This can, nevertheless, be approached through the analyses of vegetation in a city where human imports are essential.

We thus arrive at the core of an essential issue concerning the landscape, but that has not yet been analysed in depth, i.e. natural succession. There is no doubt that it contributes to the production of knowledge. However, it is often just skimmed over without being precisely analysed, whereas it is the basis of policies or, in any case, should be. Many papers speak of the processes of densification, desertification, abandonment, fragility and gentrification, without actually defining these terms and analysing them in terms of their relationship to the landscape and without them being the object of an analysis of their representations. However, we know that these representations are the driving force of policy. They, therefore, deserve to be studied and elaborated in great detail. One of the trends in the present production of knowledge effectively lies in the loss of in-depth knowledge of natural succession, in favour of analyses of arguments and representations. This means that we overlook the fact that these successions are the consequence of multiple acts of landscape transformation – individual, collective, institutional, public and private. By not taking these representations into account, we no longer measure the gap that exists between what is said and what actually exists. We, therefore, fall into another excess that is represented by a sort of impressionism that has nothing to do with the artistic movement that was truly dedicated to depicting the natural succession of landscapes and that, in particular, provided an accurate perception of the relationships between social transformations and landscape transformations.

As several authors observed, natural succession processes proceed by superposition of new symbols and new toponyms that are indicative at the same time of a social change and new practices. This is undoubtedly the source of new references or landscape models that, as we have already mentioned, are still on the drawing board and have not yet clearly appeared. The analysis of natural selection processes requires that the factors involved be studied in detail at the same time. Some authors speak of innovations or events that influence these processes, but the economic aspects are rarely mentioned; it is still in the representations of change that new ideas appear, making it possible to shed light on internal contradictions and the disregard for historical time, in particular. Among these factors, legislation is also a consideration with its ability to organize and segregate space and, as a result, landscapes.

It is, therefore, the knowledge of these processes and the factors determining evolution that constitute or should constitute one of the foundations for policy development. Policy is not always the direct result of this knowledge, which is still incomplete. It may also be dependent on the context in which it is developed and where strong pressures in terms of trends, overall measures and slogans exist that do not always take local conditions and their application into account. For example, we know that international agreements to protect nature influence the representations of institutional stakeholders who interpret them in political terms, ignoring the local context. Decentralization, which should bring decision-making closer to the local reality, often leads to a reformulation of policies in terms of regional priorities; the landscape then either disappears or serves the purposes of a new power that transforms it into a local symbol.

Policy development cannot be isolated from changes in the context within countries that do not yet have a solid legislative framework for actions concerning the landscape. It must also take new social and ecological processes into consideration, as we have already mentioned. However, one of the conclusions we can draw from the papers in this collection is undoubtedly the lack of co-ordination of policies directly linked to the landscape with sectorial policies, such as those of water, housing or agriculture, in particular. Once again, we can only regret the disregard for economic factors.

Policy implementation

This phase of the public and voluntary landscape production process is most certainly the one that gives rise to the most contributions, ranging from the stages of knowledge of landscape characteristics to the implementation that brings all the participants together in a participative procedure. These methods are the most detailed and lead us to analyse the relevance of one method over another and the advantages of one tool compared with another.

Questions have arisen, first in the discussions and papers: should holistic methods – those that attempt to take the entire landscape into account – be given priority, or should we instead focus on iterative methods that, on the basis of an object, the tree, for example, attempt to reconstruct the landscape and the complexity of its operation? This issue has not yet been resolved and thus leaves researchers masters of their own ability to analyse their research object.

Some of the papers here make it possible to pinpoint the role of the different categories of stakeholders. Concerning elected officials, the authors examined when interventions take place and problems that arise as a result of the gap between electoral time and the time necessary for understanding, scheduling, negotiating or arranging financing. Decentralization and the importance attributed to the local level may result in the risk of modern 'feudalism', a return to the exercise of local power that is difficult to control.

Landscape policies are always on the lookout for new landscape models, which are long in coming, as we have already observed; landscape values featured in projects are often those of the past, whereas projects are designed for the future. Could sociability, modern urbanity and peri-urbanity or citizen participation serve as a basis for these models? Or should we look for them in the ephemeral or in the expression of the relationship between nature and culture that seems to no longer pit the first against the second in light of our improved everyday knowledge about nature? The development of poverty and multiculturalism, however, does not seem to inspire adapted landscape policies.

Policy development still encounters difficulties in integrating historical data. In spite of the considerable development of landscape inventories and atlases in Europe (UK, Holland, Spain and France), where knowledge about landscape history makes it possible to avoid the disregard for historic time, it is still difficult to design policies that take the past into account from a point of view other than that of patrimony and that consider it an asset for the future. This is perhaps because land-use projects do not create definitive landscapes, as would be the tendency with current designs. However, we know that landscapes will change, as will the stakeholders' representations of them. Furthermore, these representations evolve on their own, independent of any radical transformations in the landscapes themselves.

Methodologies

Corresponding to different objectives, the methodologies aim at defining the status of a landscape, understanding the social demand, imagining the landscape of the future and evaluating policies.

Characterizing the landscape is the usual first phase to which several countries seem to be committed; France, Spain, Holland, UK and the Wallonia region of Belgium have made progress in this area. Using different methods, they attempt to integrate archaeological and historic knowledge that some people feel is essential for defining the future and ensuring the transition between the past and the future. However, knowledge derived from empirical know-how is still poorly understood, whereas it could contribute to projecting the landscape at the local level.

Understanding the social demand. The concern for this objective testifies to the interest that research and research/action displays in terms of the stakeholders' aspirations and their representations of the landscape. Generally speaking, the papers included here first emphasize the question of the diversity of know-how and, in particular, the confrontation between empirical knowledge, scientific knowledge and technical knowledge. It is from this point of view that a collective reading can become a tool of reflection for the analysis of the materiality of a landscape, its representations and the consensus-building process. However, from the beginning, it raises the problem of the relevance of categories of stakeholders to be consulted: local stakeholders, of course, but which ones, people involved in land regulation, such as landowners and farmers, local experts or simple citizens, the 'ordinary' inhabitants of a region. Some worry about the potential abuse of results obtained from consultations with the population. What about minority stakeholders? How can they be integrated into processes where they may feel alienated or not dare to express their opinions? Obviously, to consult does not just mean to listen, but rather to initiate the exchange of information. These methods, which play an important role in the implementation of citizen participation, also have their price: they require time that does not necessarily correspond to 'political time'. And how can we design these consultations to provide the concerned populations with the certainty that their opinion counts? This is undoubtedly one of the most important lessons to be learned: the participation of populations in the formulation of landscape projects involves a deep-rooted need on their part – to be listened to, and to not have the feeling of having been dragged into negotiation procedures that only appear to be so on the surface and that can be used as an alibi by politicians. We can, therefore, understand the gulf that has been created between the world of politics and populations that feel that their aspirations are not understood for what they actually represent.

Imagining the landscape of the future. By examining the papers in this collection in detail, even those based on fundamental research, we can nevertheless see the underlying thread that consists of providing knowledge earmarked for action. It is, therefore, not the fault of research that it is unable to imagine the potential use of its results for the benefit of political action. It is probably because, as yet, there are no effective means of transferring knowledge towards this objective. In any case, it is obvious that the research approach that considers the landscape as an object modifies that of the land-use project, not only because it introduces a qualitative aspect and coherency, but because it tends to give rise to a new situation within the territorial framework. It reveals different representations that can – and must – be adopted by the stakeholders if they are given the time necessary to do so.

The experiences referred to in the papers in this book that deal with policy implementation attempt to innovate in terms of approach, particularly when they introduce the notion of stakeholder participation. They are, perhaps, not yet adequately known, too isolated from normative action to constitute a source of innovative experiences, as we would have expected. A welcome initiative on the part of public institutions would be to take advantage of these innovations by revealing their failures and accomplishments at the same time. These accomplishments obviously depend on the means established; the landscape project can lead to the construction of a territorial identity, to the socialization or naturalization of public space, but it can also contribute to the 'museumification' of landscapes, the production of artificial or segregative landscapes, as well as the loss of environmental or social functionalities.

The landscape is an object of negotiation, but can also be a tool for negotiation by lending itself to a collective reading, either *in situ* or in photographic representations, and thus makes it possible, in relation to the stakeholders' representations, to identify the issues, the categorial projects that may also become landscape projects when they are brought together and made coherent. Thus, the ongoing discussion between researchers, politicians and civil society may produce conceivable future scenarios inspired by adopted and negotiated measures. These debates can be independent of institutional and technical-oriented procedures and take place within the framework of informal knowledge-exchange networks by making it possible to speak the unspeakable. Nevertheless, all landscape projects remain dependent on the future and on transformations that may arise, even during the development and implementation of the project. This is why some people propose an adaptive land development process that conceives of the project as an experience in the process of evolving and that integrates, on the other hand, information that these transformations produce as the implementation of the project advances.

Evaluating policies and their effects. This issue is only lightly touched on in this book, although the question of evaluation sometimes arises in a critical reading of policy implementation. The evaluation of a project or a landscape policy entails a series of questions that are not dealt with here. What are we evaluating? Is the rural landscape easier to evaluate than the urban landscape because it more clearly reveals environmental and development problems? Or is this distinction only the effect of an increased enhancement of the rural landscape in relation to the role it plays in European collective representations? How can we evaluate notions of tranquillity, access or well-being? And finally, at what scale do we evaluate the effects of a policy or a landscape project?

Evaluation also implies that indicators have been developed, but how can we imagine indicators that take population project acceptability into account? Do we not need to totally rethink this question by imaging indicators that can take account of social conflicts, landscape opening and closing processes, and the use of vegetation to evaluate the biodiversity produced by certain landscapes? How can we design indicators that make it possible to monitor economic and political effects, particularly in terms of creating new jobs, or the added value of an economic sector that takes landscape considerations into account?

Through these papers, research in terms of policy implementation thus reveals the issues concerning the transfer of knowledge to political action, and first and foremost, the one illustrated by the sharing of know-how, between expert, empirical and technical knowledge. The scientific community is not yet ready to admit that the knowledge of local stakeholders has the same status as the knowledge produced through rigorous protocols and by production of evidence. However, this is the price we must pay so that research/action can emerge, and the landscape is particularly well adapted in this respect, provided that the social representations of the stakeholders are revealed.

Research/action development must also face the restraints imposed by time: research time that can be long, particularly when it is necessary to verify ecological processes, action time and political time that can diverge and lead to conflicts between politicians and scientists.

Research and the transfer of knowledge that it produces in terms of action encounter, among other things, the complexity of processes, particularly those that interact between nature and social issues, between the material and the immaterial, between expert and lay knowledge. The languages that stakeholders use provide a source of complexity that research draws on to establish power leverage to compensate for its weakness in relation to politics, but it must also demystify its own language, translate it for 'ordinary' stakeholders, so that the dialogue between science and the uninitiated individual is reciprocal and leads to a fertile exchange. Moreover, the social sciences use lay arguments as a way of comprehending the world and as a means to understand the representations of nature and landscapes. They should, therefore, make an effort to be accessible to local stakeholders by granting them access to their knowledge by translating it into easily understood concepts, being careful not to reduce the complexity of the processes in question.

Research can thus contribute to political action and, by contributing to experimental development, to development experiments that put research protocols to the test as well. It must consider that landscape management is a process in constant evolution where experts and 'ordinary' citizens pool their ideas of the world through negotiated instruments of collective participation.

Admittedly, this book is undoubtedly not representative of all the research being done in Europe today, but it is a first approach, limited, but diverse enough to highlight the emergence of a European scientific community that aims to innovate and establish relationships with political action. It asks only to be endowed with the means to address the future of European landscapes and their contribution to social and individual well-being.

Bibliography

ADGER W.N., 2003. Building resilience to promote sustainability. *In IHDP Update 02/2003*, 2003, p.1ff.

AGEYMAN J., 1990. Black people in a white landscape: social and environmental justice. *Built environment*, 16 (3): 232-236.

AGLI N., 1988. *Analyse et extension du centre de la ville de Biskra*. Mémoire de DEA en urbanisme, Paris Villemin.

ALDRED O., FAIRCLOUGH G. J., 2003. *Historic Landscape Characterisation: Taking stock of the Method – The National HLC Method Review 2002*. London, English Heritage.

ALKAMA D., 1995. *Analyse typologique de l'habitat : cas de Biskra*. Thèse de Magister, Université de Biskra, Algérie.

ALLAIRE G., DUPEUBLE T., 2003. De la multifonctionnalité à la multi-évaluation de l'activité agricole. *Économie Rurale*, (275) : 51-65.

ALPHANDÉRY P., 1996. La nature de Disneyland Paris. *Courrier de l'environnement de l'INRA*, n° 28, août.

AMBLARD H., BERNOUX P., HERREROS G., LIVIAN Y-F. 1996. *Les nouvelles approches sociologiques des organisations*. Seuil, 245 p.

ANTROP M., VAN DAMME S., 1995. Landschapszorg in Vlaanderen: onderzoek naar criteria en wenselijkheden voor een ruimtelijk beleid met betrekking tot cultuurhistorische en esthetische waarden van de landschappen in Vlaanderen. Gent (download from http://www.geoweb.ugent.be/services/index.asp [in Dutch]).

ANTROP M., 1997. The concept of traditional landscapes as a base for landscape evaluation and planning. The example of Flanders Region. *Landscape and Urban Planning*, (38): 105-117.

ANTROP M., 2000a. Background concepts for integrated landscape analysis. *Agriculture, Ecosystems and Environment*, 77, 17-28.

ANTROP M., 2000b. Changing patterns in the urbanized countryside of Western Europe. *Landscape Ecology*, 15,3, 257-270.

ANTROP M., 2001. De landschapsatlas, Methode. *In* Hofkens E., Roossens I. (Eds.), 2001. *Nieuwe impulsen voor de landschapszorg. De landschapsatlas, baken voor een verruimd beleid*. Brussel, Ministerie van de Vlaamse Gemeenschap, Afdeling Monumenten en Landschappen, 21-43.

ANTROP, M., 2003. The role of cultural values in modern landscapes. The Flemish example. *In* Palang, H. et Fry, G., *Landscape Interfaces. Cultural heritage in changing landscapes*, Kluwer Academic Publishers, Dordrecht, 91-108.

ANTROP M., BOURGEOIS J., CORDEMANS C., LACHAERT P.-J., ROGGE E., THOEN E., VAN EETVELDE, V., 2004. A transdisciplinary landscape study of the archaeology, history and geography of the Pajottenland (Flanders, Belgium) – the case of Gooik. *In* de Boer I., Carsens G J., Van der Valk A., 2004., *Multiple Landscapes: Merging Past and Present in Landscape Planning*. Fifth Int. Workshop on Sustainable Land Use Planning, Intern. Studygroup On Multiple Use of Land (ISOMUL) and Land Use Planning Group, Wageningen University, 31 p. + CD-ROM.

ANTROP M., 2004. Landscape change and the urbanisation process in Europe. *Landscape and Urban Planning*, 67, 1-4, 9-26.

ARAMBERRI J., 2001. The host should get lost: Paradigms in the Tourism Theory. *Annals of Tourism Research*. 28, 738-761.

AREND M., 2003. Können Stadtplanung und Wohnungsmarktpolitik einen Beitrag zur besseren Durchmischung und zur Integration von MigrantInnen leisten? *In* Wicker H.-R., Fibbi R., Haug W. (Eds.), *Migration und die Schweiz: Ergebnisse des Nationalen Forschungsprogramms "Migration und interkulturelle Beziehungen"*. Seismo, Zürich, 237-255.

ARLER F., 2000. Aspects of landscape or nature quality. *Landscape Ecology*, 15, 291-302.

Association OREE, 2005. *Vade-mecum de la concertation locale*, Paris, Association ORÉE.

AUSTAD I., 2000. The future of traditional agriculture landscapes: retaining desirable qualities. *In* Klijn J., Vos W. (eds.), 2000, *From Landscape Ecology to Landscape Science*. WLO, Kluwer Academic Publ., Wageningen, 43-56.

AZNAR O., 2002. *Services environnementaux et espaces ruraux : une approche par l'économie des services*. Faculté de sciences économiques et de gestion, Dijon, Université de Bourgogne.

BAČNAR D., ČERNIC-MALI B., KERBLER-KEFO B., PRAPER S., 2004. Making Public the View of Locals. WP 5, Task 5.5/Task 5.6 Conference of Regions. D 5.4 Proceedings of the Conference of Regions. Regional Development and Cultural Landscape Change: The Example of the Alps. Evaluating and Adjusting EU and National Policies to Manage a Balanced Change. Ljubljana.

BAGES R., 1998. L'espace de la citoyenneté dans les communes rurales entre ouverture et repli identitaire : les maires face aux résidents secondaires et aux étrangers. *In* Bages R. Granié A.-M. (Eds), *Comment les ruraux vivent-ils et construisent-ils leur(s) territoire(s) aujourd'hui ?*. Toulouse, Université de Toulouse Le Mirail, 201-209.

BAHRDT H-P., 1968. *Die moderne Grossstadt: Soziologische Überlegungen zum Städtebau*. Wegner, Hamburg.

BALDESCHI P. (a cura di), *Il paesaggio agrario del Montalbano*, Firenze, Passigli, pp. 81-122.

BARBAGLI M., CASTIGLIONI M., DALLA ZUANNA G., 2003. *Fare famiglia in Italia, un secolo di cambiamenti,* Bologna, Il Mulino.

BARLOW K., COCKLIN C., 2003. Reconstructing rurality and community: plantation forestry in Victoria, Australia. *Journal of Rural Studies,* 19, 503-519.

BAROUCH G., 1989. *La décision en miettes : système de pensée et d'action à l'œuvre dans la gestion des milieux naturels.* Paris, L'Harmattan.

BARR C. and PETIT P., Eds., 2001. *Hedgerows of the world: their ecological functions in different landscapes.* UK, IALE.

BARREDO J.I., LAVALLE C., DEMICHELI L., KASANKO M., MCCORMICK N., 2003. *Sustainable urban and regional planning: The MOLAND activities on urban scenario modelling and forecast. Luxembourg.* Office for Official Publications of the European Communities, 2003.

BARRET P., BEURET J.-E., BILLETTE DE VILLEMEUR C., 2003. *Guide pratique du dialogue territorial. Concertation et médiation pour l'environnement et le développement local.* Paris, Collection Pratiques.

BARTHÉLÉMY C., JACQUÉ M., 2002. La gestion des espaces protégés en Camargue : la construction sociale de l'homme. *Faire savoir,* (2) : 41-48.

BASQUE GOVERNMENT, 1990. *Mapa de Paisaje de la Comunidad Autónoma del País Vasco.* Vitoria.

BASSAND M., STEIN V., COMPAGNON A., JOYE D., MEURY M., 1999. Les espaces publics urbains. *Revue Économique et Sociale,* (4) : 241-252.

BASSAND M., COMPAGNON A., JOYE D., STEIN V., 2001. *Vivre et créer l'espace public.* Lausanne, Presses polytechniques et universitaires romandes.

BÄTZING W., 1993. *Der sozio-ökonomische Strukturwandel des Alpenraums im 20.* Bern, Jahrhundert.

BÄTZING W., MESSERLI P., PERLIK M., 1995. *Regionale Entwicklungstypen. Analyse und Gliederung des schweizerischen Berggebietes.* Bern.

BÄTZING W., 1998. *Die Alpen im Spannungsfeld der europäischen Raumordnungspolitik. Einführungsreferat zum Expertenseminar 'die aktuelle Situation im Alpenraum und das Projekt* REGIONALP (EFRE Art. X), Salzburg.

BAUDRY J., BUNCE R.G.H., BUREL F., 2000. Hedgerow diversity: an international perspective on their origin, function, and management. *Journal of Environmental Management,* (60): 7-22.

BAUDRY J., JOUIN A, 2003. De la haie aux bocages : organisation, fonctionnement et gestion. *INRA Editions,* Ministère de l'Écologie et du Développement durable, Paris, 1-474.

BAUDRY J., N'DIAYE A., 2003. Évaluation des plantations et de leurs caractéristiques écologiques., *In* H. Lamarche, ed, *Bocagement, reconstitution et protection du bocage. Évaluation des politiques publiques de paysagement du territoire,* p. 119-147.

BAUDRY J., JOUIN A. coord., 2003. *De la haie aux bocages. Organisation, dynamique et gestion.* Versailles, INRA Editions, 435 p.

BECK U., 2001 (1986). *La société du risque. Sur la voie d'une autre modernité.* Paris, Aubier.

BELL S., 1999. *Landscape: Pattern, Perception and Process.* E.&F.N, London. Spon.

BELL S., 2001. Landscape pattern, perception and visualisation in the visual management of forests. *Landscape and Urban Planning* 54, 201-211.

BELL S., WARD THOMPSON C., TRAVLOU P., 2003. Contested views of freedom and control: children, teenagers and urban fringe woodlands in central Scotland. *Urban Forestry and Urban greening 2*, 87-100.

BELL S., MORRIS N., FINDLAY C., TRAVLOU P., MONTARZINO A., GOOCH D., GREGORY G., WARD THOMPSON C., 2004. Nature for people: the importance of green spaces to East Midlands communities. *English Nature Research Report 567*, Peterborough, English Nature.

BENATIA F., 1976. *Du sous-développement au développement urbain ; les bidonvilles d'Alger.* APC Alger. p 32.

BENCHEIKH H.M.F., 1999. *The process of urbanisation and its impact on the urban environment of the oasis, the case of Biskra.* Séminaire institut d'architecture, Biskra.

BENSON J., ROE M., 2000. *Landscape and Sustainability.* Taylor and Francis, 304 p.

BENTON T.G., VICKERY J.A., WILSON J.D., 2003. Farmland biodiversity: is habitat heterogeneity the key? *Trends-in-ecology-and-evolution.* 18 (4): 182-188.

BERGUES M., 1995. Des vaches au marais : de l'élevage traditionnel à l'animal comme outil de gestion paysagère. *In* Voisenat C. (ed.), *Paysage au Pluriel ; pour une approche ethnologique des paysages*, Paris, Éditions de la Maison des Sciences de l'Homme, 151-166.

BERQUE A., 1995 (1991). De paysages en outre-pays. *In* Roger A. (dir.), *La théorie du paysage en France (1974-1994).* Seyssel, Champ Vallon, 346-359.

BERQUE A., 2000. *Écoumène. Introduction à l'étude des milieux humains.* Belin, Paris.

BERRY A.R., BROWN I. W., (Eds), 1995. *Managing Ancient Monuments: An Integrated Approach*, Association of County Archaeology Officers/Clwyd County Council, Mold.

BESSE J-M., 2000. *Voir la Terre. Six essais sur le paysage et la géographie.* Arles, Actes Sud ENSP/Centre du Paysage.

BILLAUD J.P., STEYAERT. Agriculture et conservation de la nature : raisons et conditions d'une nécessaire co-construction entre acteurs. *Actes des journées AFPF* 23-24, mars 2004 Biodiversité des prairies.

BLATRIX C., 1999. Le maire, le commissaire enquêteur et leur « public ». La pratique politique de l'enquête publique. *In* CRAPS, CURAPP, (coll.), *La démocratie locale. Représentation, participation et espace public.* Paris, Presses Universitaires de France, 161-176.

BLONDEL J., 1995 [1979]. *Biogéographie.* Paris, Masson, Coll. Ecologie, 297 p.

BLOWERS A., 1999. Nuclear Waste and Landscapes of Risk. *Landscape Research,* (Vol. 24, n° 3), 241-264.

BOERI S., BASILICO G., 1997. *Sezioni del paesaggio italiano*, Tavagnacco, Art&.

BOERI S., 2003. *Multiplicity, USE-Uncertain states of Europe*, Milano, Skira.

BOLGER D.T., SCOTT T.A., ROTENBERRY J.T., 2001. Use of corridor-like landscape structures by bird and small mammal species. *Biological Conservation*, 102, 213-224.

BOLTANSKI L., THÉVENOT, L., 1991. *De la justification. Les économies de la grandeur.* Paris, Gallimard, 483 p.

BORG I. SHYE S., 1995. *Facet Theory: Form et Conten.* London, Sage.

BOURDIEU P., 1992. Un doute radical. *In Réponses.* Paris, Seuil, 207-217.

BOUTFNOUCHET M., 1985. *Système social et changement social en Algérie.* Alger, OPU, 164 p.

BOUVIER-DACLON N., SÉNÉCAL G., 2001. Les jardins communautaires de Montréal : Un espace social ambigu. *Loisir et société/Society and Leisure 24*, (2) : 507-531.

BRADLEY A., BUCHLI V., FAIRCLOUGH G.J., MILLER J, SCHOFIELD J., 2004. *Change and Creation, Historic landscape Character 1950-2000*, English Heritage/Atkins. www/change-andcreation.org.uk

BREAKELL B. 2002. Missing persons: who doesn't visit the people's parks. *Countryside recreation* 10 (1): 13-17.

BREAKWELL G.M., 1990. *Interviewing*. Exeter, British Psychological Society.

BRIFFAUD S., 2002. Pour une pédagogie de la médiation paysagère : une problématique pour la formation des professionnels du paysage. *Colloque « Gérer les paysages de montagne pour un développement concerté et durable »*. Florac, 7 p.

BRIFFAUD S., 1994. *Naissance d'un paysage : les montagnes pyrénéennes à la croisée des regards, XVIe-XIXe siècles*. Archives départementales des Hautes-Pyrénées, Tarbes.

BRIFFAUD S., 2001. Comment peut-on évaluer les effets d'une politique sur les paysages ? *In* Berlan-Darqué M., Terrasson D. (Eds), *Politiques publiques et paysages. Actes du séminaire d'Albi*. Cemagref-Ministère de l'aménagement du territoire et de l'environnement, Paris, p. 47-52.

BRITISH WATERWAYS, 2002. *Waterways for People*. British Waterways, www.britishwaterways.co.uk/downloads/Final

BRITISH WATERWAYS, 1995. *Birmingham and Black Country Canals Perception Survey*.

BRITISH WATERWAYS. http://217.79.111.204/downloads/IWAAC %20Social %Inclusion

BRUN J.-J., DEBARBIEUX B., DUCOURTIOUX S., PETIT J., LHEUREUX P., 2002. Une recherche interdisciplinaire et exploratoire avec un Parc naturel régional : descriptions de clairières en Chartreuse. *Natures Sciences Sociétés*. 10, 42-51.

BUCHECKER M., HUNZIKER M., KIENAST F., 2003. Participatory landscape development: overcoming social barriers to public involvement. *Landscape and Urban Planning*, (64): 29-46

BUIJS A., PEDROLI B., LUGINBÜHL Y., 2003. From hiking through farmland to farming in a leasure landscape: Changing social perceptions of the European landscape. Submitted to *Landscape Ecology*, Nov. 2003.

BUKOW W.-D., NIKODEM C., SCHULZE E., YILDIZ E., 2001. Die multikulturelle Stadt: Von der Selbstverständlichkeit im städtischen Alltag. *Interkulturelle Studien* 6. Leske + Budrich, Opladen.

BURGESS J., 1995. *Growing in Confidence: a study of perceptions of risk in urban fringe woodlands*, Countryside Commission, Cheltenham, Technical report CCP 457, http://www.nottingham.ac.uk/sbe/planbiblios/bibs/country/09.html

CADIOU N, LUGINBÜHL Y. Modèles paysagers et représentations des paysages en Normandie Maine. *In Paysage au Pluriel, pour une approche ethnologique des paysages*. Coll. Ethnologie de la France, cahier 9, Paris, Édition de la Maison des Sciences de l'Homme, p 19-34, 1995.

CALLON M., LASCOUMES O., BARTHES Y., 2001. *Agir dans un monde incertain. Essai sur la démocratie technique*. Seuil, 357 p.

CALLON M., 1986. Éléments pour une sociologie de la traduction. La domestication des coquilles Saint-Jacques et des marins-pêcheurs dans la baie de Saint-Brieuc. *L'année sociologique*, 169-208.

CANCELA D'ABREU A., PINTO-CORREIA T., OLIVEIRA R., 2004. *Contribuição para a Identificação e Caracterização das Paisagens do Portugal Continental*, Universidade de Évora, Lisboa, Direcção Geral de Ordenamento do Território e Desenvolvimento Urbano.

CANDAU J., RUAULT C., 2002. Discussion pratique et discussion stratégique au nom de l'environnement : différents modes de concertation pour définir des règles de gestion des marais. *Économie rurale*, (270), 19-35.

CANDAU J. (coord.), AZNAR O., GUÉRIN M., LE FLOCH S., MICHELIN Y., MOQUAY P., 2003. *Acteurs locaux et initiatives publiques dans le domaine du paysage : une analyse des processus de construction des interventions publiques localisées.* Cemagref – ENITAC – ENGREF.

CANDAU J., AZNAR O., GUÉRIN M., MICHELIN Y., MOQUAY P., 2003. Normes en conflit et compétences interactionnelles : débats autour d'un paysage rural. *In Normes sociales et processus cognitifs.* (ed SACO), Poitiers, Maison des Sciences de l'Homme et de la Société, 27-31.

CANDAU J., AZNAR O., GUÉRIN M., LE FLOCH S., MICHELIN Y., MOQUAY P., 2003. *Acteurs locaux et initiatives publiques dans le domaine du paysage. Une analyse du processus de construction des interventions publiques localisées.* Rapport de recherche, Programme de recherche MEDD « Politiques publiques et paysages », Cemagref-ENITAC-ENGREF.

CANTER D., 1977. *The Psychology of Place.* London, Architectural Press.

CARPENTER S., WALKER B., ANDERIES J.M., ABEL N., 2001. From Metaphor to Measurement: Resilience of What to What? *Ecosystems* 2001/4, 765-781.

CASIMIRO P., 2002. *Uso do Solo, Teledetecção e Estrutura da Paisagem* – Concelho de Mértola, Dissertação de Doutoramento, Universidade Nova de Lisboa, Lisboa.

CASSATELLA C., 2001. *Iperpaesaggi*, Collana Universale di Architettura, Torino, Testo & Immagine.

CASSI L., MARCACCINI P., 1998. "Toponomastica, beni culturali e ambientali. Gli 'indicatori geografici'per un loro censimento". *Mem. Soc. Geogr. It.*, LVI.

CASSI L., 2004. "Nuovi toponimi". *Italia – Atlante dei tipi geografici*, Firenze, Istituto Geografico Militare, Tav. 152.

CASTIGLIONI B., 2002. *Percorsi nel paesaggio*, Torino, Giappichelli Editore.

CASTIGLIONI B., GROSSUTTI J., MASSARUTTO A., TROIANO S., VIRGILIO T., 2004. *Developing integrated cultural landscape scenarios in the Alps for the year 2020.* WP4 Work Package Report. Regional Development and Cultural Landscape Change: The Example of the Alps. Udine, Evaluating and Adjusting EU and National Policies to Manage a Balanced Change.

CASTIGLIONI B., FERRARIO V., 2005. "Tra Montello e città diffusa. La percezione del paesaggio e delle sue trasformazioni", CASTIGLIONI B. (a cura di), *Montello.* 3KCL – Karstic Cultural Landscape. Architecture of a unique relationship people/territori, Montebelluna, Museo Civico di Storia Naturale e Archeologia.

CAUQUELLIN A., 2002. *Le site et le paysage.* Paris, Quadrige/Puf, Presses Universitaires de France.

CCD, 1997. *Convenção das Nações Unidas de Combate à Desertificação, Programa de Acção Nacional, Organização Nacional para a aplicação da CCD.*

CENTRAL SCOTLAND FOREST (2003). *The Central Scotland Forest* http://www.csct.co.uk/csf/index.htm the Central Scotland Forest Trust, visited 13/07/03.

CHALAS Y., 2005, Quelle ville pour demain ?, *Villes en évolution*. Paris, Documentation française, 11-31.

CHANDERNAGOR A., 1993. *Les maires en France, XIXᵉ-XXᵉ siècle*. Paris, Fayard.

CHESTERS A. 1997. Who's been left out. *In* E. BLAMEY (Ed), *Making Access for All a Reality*. Proceedings, Countryside Recreation Network, Cardiff University, 32-36.

CHEVALLERIE H. de la, 1999. Kleingärten – eine soziale Aufgabe. *Stadt und Grün*, 48 (6): 388-393.

CHIESI L., COSTA P., 2005. "Il Montalbano dal punto di vista dei suoi abitanti. Una ricerca su territorio, identità e senso del paesaggio nella campagna toscana". BALDESCHI P. (a cura di), *Il paesaggio agrario del Montalbano*, Firenze, Passigli, pp. 81-122.

CICOUREL A.V., 1973. *Cognitive sociology*. Middlesex, England, Penguin Education.

CÍLEK V., 2000. Paměová struktura krajiny a památné kameny (memory structure of landscape and memorial stones). *In Kulturní krajina, téma pro 21 století aneb proč ji chránit?*, Ministry of Environment of Czech republic, 69-73.

CÍLEK V., 2002. *Krajiny vnitrní a vnejsí*. Dokorán, Praha.

CLARKE R.V., 1999. Les technologies de la prévention situationnelle. *In les cahiers de la sécurité intérieure*, n° 21, 3ᵉ trimestre, p 101.

CLARK J., DARLINGTON J., FAIRCLOUGH G.J., (Eds) 2003. *Pathways to Europe's Landscape*. Heide, EPCL/EU.

CLARK J., DARLINGTON J., FAIRCLOUGH G.J., 2004. *Using Historic Landscape Characterisation (the national HLC Applications Review 2003)*. Preston, English Heritage/Lancashire County Council.

CLAVAL P.L., 2004. The Languages of Rural Landscapes. *In* Palang H., Sooväli H., Antrop M., Setten S., 2004, *European rural landscapes: persistence and change in a globalising environment*. Kluwer Academic Publishers, 11-40.

COLEY R.L., KUO F.E., SULLIVAN W.C., 1997. Where does community grow? The social context created by nature in urban public housing. *Environment and Behavior 29* (4), 468-494.

COLLECTIF, 2002. *Connaissance et gestion des habitats et des espèces d'intérêt communautaire*. Tome 6. Espèces végétales. Cahiers d'habitats Natura 2000, Paris, La Documentation Française, 271 p.

COMMISSARIAT GÉNÉRAL AU PLAN, 1994. *Les zones humides. Rapport d'évaluation. Comité interministériel de l'évaluation des politiques publiques*. Paris, La documentation française, 391 p.

CONAN M., 1994. L'invention des identités perdues. *In* Berque A. (dir.), *Cinq propositions pour une théorie du paysage*. Champ Vallon, Seyssel, 33-49.

CONSEIL SCIENTIFIQUE DE L'ÉVALUATION, 1997. *L'évaluation en développement*. Paris, La Documentation française.

COOPER MARCUS C., FRANCIS C. (Eds.), 1998. *People places: Design guidelines for urban open space*. New York, Van Nostrand Reinhold.

CORDOVIL F. *et al.*, 2003. *A política Agrícola Comum e a União Europeia*, Lisboa, Centro de Informação Europeia Jacques Delors.

COSGROVE D., 1985. Prospect, perspective and evolution of the landscape idea. *Trans. Inst. Br. Geogr.*, N.S. 10, 44-62.

COSGROVE D., 1993. Landscapes and Myth, Gods and Humans. *In* Bender B. (ed.), *Landscape Politics and Perspectives. Explorations in Anthropology Series*. Oxford, Berg Publishers, 351 p.

COSGROVE D. E., 1984. *Social formation and symbolic landscape*. Wisconsin, Madison, The University of Wisconsin Press.

COUNCIL OF EUROPE, 2000. *The European Landscape Convention*. Florence, 20 October 2002.

COUNCIL OF EUROPE, 2000. European Landscape Convention. Florence, *European Treaty Series* – No. 176; www.coe.int/T/E/cultural-co-operation/Environment/Landscape.

COUNCIL OF EUROPE, 2002. The European Landscape Convention, *Naturopa 98*, Strasbourg, Council of Europe.

COUNCIL OF EUROPE, 2000. European Landscape Convention. *T-Land* (2000) 6. Strasbourg

COUNTRYSIDE AGENCY, 1998. Barriers to enjoying the countryside. *Research Note CRN* 11.The Countryside Agency.

COUNTRYSIDE AGENCY, 2002. Countryside Agency policy on rural social exclusion "Rural Social Exclusion AP01/54".

COUNTRYSIDE AGENCY, 2004. Diversity Review. Options for Implementation. *Research Note CRN 75*, The Countryside Agency.

COUNTRYSIDE AGENCY, 2004. *Countryside Quality Counts – Tracking Change in the English Countryside: Constructing an Indicator of Change in Countryside Quality*. Final Report by Nottingham University Consultants Ltd. www.**countryside-quality-counts**.org.uk

COUNTRYSIDE COMMISSION, 1996. *Views from the Past – Historic Landscape Character in the English Countryside:* CCWP 04, Cheltenham, Countryside Commission.

COUNTRYSIDE COMMISSION AND COUNTRYSIDE AGENCY, 1998-99. *Countryside Character volumes*: (1-8), Cheltenham. www.countryside.gov.uk/cci.

COUNTRYSIDE AGENCY AND SCOTTISH NATURAL HERITAGE, 2002. *Landscape Character Assessment: Guidance for England and Scotland*, Cheltenham. www.countryside.gov. uk/cci/guidance

COURTILLAOT J.P., 1979. Dervaux, *In Croissance urbaines de Biskra*. AMC 48, 28.

COUVREUR M., MENSCHAERT J., SEVENANT M., RONSE A., VAN LANDUYT W., DE BLUST G., ANTROP M., HERMY M., 2004. Ecodistricten en ecoregio's als instrumentvoor natuurstudie en milieubeleid, *Natuur.focus*, 3 (2), 51-58.

CROFT A., 2004. *Thames Gateway – Historic Environment Characterisation Project*. Chris Blandford Associates, EH, Kent and Essex County Councils, Uckfield.

CRSTRA, 2002. *Étude scientifique du parc et jardin 5 juillet*. Rapport d'activité, 23-25.

DALLA ZUANNA G., ROSINA A., ROSSI F. (a cura di.), 2004. Il Veneto. Storia della popolazione dalla caduta di Venezia ad oggi. Venezia, Marsilio.

DANTON P., BAFFRAY M., 1995. *Inventaire des plantes protégées en France*. Paris, Nathan, 294 p.

DARRÉ J.-P., 1999. *La production de connaissance pour l'action. Arguments contre le racisme de l'intelligence*. Éditions de la Maison des sciences de l'homme, Paris, Institut national de la recherche agronomique, 244 p.

DARRÉ J.-P., 2006. *La recherche coactive de solutions entre agents de développement et agriculteurs*. Paris, Editions du GRET.

DARWIN C., 1859. *L'origine des espèces*. Paris, Flammarion, coll. G.F., 608 p.

DAVIS N.-D., 1992. The Providence Neighborhood Planting Program. *American Forests*, (98), January/February.

DE ANGELINI A., 2004. "Popolazione e territorio", DALLA ZUANNA G., ROSINA A., ROSSI F. (a cura di), *Il Veneto. Storia della popolazione dalla caduta di Venezia ad oggi*, Venezia, Marsilio.

DE BORGHER M., 2002. Toelichting bij het decreet van 21 december 2001. tot wijziging van het decreet van 16 april houdende bescherming van landschappen. *Monumenten, & Landschappen*, 21, 1, Binnenkrant, 2-4.

DEBROUX J., 1995. Enquête sur un étrange succès : l'analyse paysagère dans le massif de Belledonne. *In* Voisenat C. (ed.), *Paysage au pluriel, pour une approche ethnologique des paysages*. Cahier 9, Paris, Editions de la Maison des Sciences de l'Homme, 209-218.

DEBUSSCHE M., LEPART J., DERVIEUX A., 1999. Mediterranean landscapes changes: evidence from old postcards. *Global Ecology and Biogeography*, (8): 3-15.

DE LA SOUDIÈRE M., 1991. Paysage et altérité. En quête de « culture paysagère » : réflexions méthodologique. *Études Rurales*, « De l'agricole au paysage », janvier-décembre, 121-(124) : 141-150.

DELAUNOIS H., 1960. *Inventaris van de landschappen: provincie Antwerpen*. Bestuur van de Ruimtelijke Ordening, Brussel.

DELCROS P., 1994. *Écologie du paysage et dynamique végétale post-culturale*. « Études » du CEMAGREF, série Gestion des Territoires (13) : 334 p.

DIAMOND J., 2000. Evolution, consequences and future of plant and animal domestication. *Nature*, (418) : 700-707.

DI-MÉO G., HINNEWINKEL J.-C., 1999. Représentations patrimoniales et recompositions territoriales vécues dans l'entre-deux-mers girondin. *Géographie et Cultures*, (30), 71-94.

DONADIEU P., 1995. Pour une conservation inventive du paysage. *In* A. Roger (ed.), *La théorie du paysage en France (1974-1994)*. Seyssel, Champ Vallon, 400-423.

DONADIEU P., 1999. Entre urbanité et ruralité, la médiation paysagiste. *Les Annales de la Recherche urbaine* 85, 215 p.

DONALD, I., 1994a. The Structure of Office Workers' Experience of Organizational Environments. *Journal of Occupational and Organizational Psychology*, (67): 241-258.

DONALD, I., 1994b. Management and Change in Office Environments. *Journal of Environmental Psychology*, (14): 21-30.

DORANDEU R., 1994. Les métiers avant le métier. Savoirs éclatés et modèle notabiliaire. *Politix* 28 (Le métier d'élu : jeux de rôles), 5-15.

DOUGLAS M., 1986. *How Institutions Think*, Syracuse University Press, New York.

DTZ Pieda Consulting, 2001. *Review of social exclusion activity in the Countryside*.

DÜBENDORFER S., 2001. *Das Integrationspotential von Grünräumen (städtische Wälder und Parks) in der Agglomeration Zürich. Möglichkeiten und Wege der Integration ausländischer Kinder und Jugendlicher*. Unpublished diploma thesis, Chair of Forest Policy and Forest Economics, Zürich, ETHZ.

DUCROT O., 1984. *Le dire et le dit*. Paris, Les Éditions de Minuit.

DÜRRENMATT R., FORNELLS J., HAMADAN A., ZOSSO G., 2001. *Sozialintegration im städtischen Grünraum. Unpublished final report of the interdisciplinary project in forest sciences*, sommer semester term.

ENGLISH HERITAGE, 2005. Characterisation, *Conservation Bulletin*, issue 47 Winter 2004, London, English Heritage, (www.english-heritage.org.uk/characterisation)

EUROPEAN COMMISSION, 1999. *European Spatial Development Perspective*, Brussels.

EUROPEAN ENVIRONMENT AGENCY, 2002. Towards an urban atlas: assessment of spatial data on 25 European cities and urban areas, *Environmental Issue Report* No. 30, Publications of the European Communities, Luxembourg, 131 p.

FAIRCLOUGH G. J., LAMBRICK G., MCNAB A., 1999. *Yesterday's World, Tomorrow's Landscape* – the English Heritage Historic Landscape Project 1992-94, London, English Heritage.

FAIRCLOUGH G. J., 2002. Cultural landscape and spatial planning: England's Historic Landscape Characterisation Programme. *In* Green & Bidwell (Eds) 2002, *Heritage of the North Sea Region : Conservation and Interpretation*. Papers presented at the Historic Environment of the North Sea InterReg IIC Conference 29-31 March 2001, South Shields, Donhead, Shaftesbury, 123-149.

FAIRCLOUGH G.J., RIPPON S.J., (Eds) 2002. Europe's Cultural Landscape: archaeologists and the management of change, EAC Occasional Paper no 2, Europæ Archaeologiæ Concilium and English Heritage, Brussels and London FAO 1991. Second interim report on the state of tropical forests by Forest resources assessment 1990 Project. Tenth World Forestry Congress, September 1991, Paris, France.

FAIRCLOUGH G.J., 2003a. Cultural Landscape, Sustainability and Living with Change? *In* Teutonico J. M. and Matero F., (Eds) 2003, *Managing Change: Sustainable approaches to the Conservation of the Built Environment*. Proceedings of the 4th Annual US/ICOMOS International Symposium, Philadelphia, Pennsylvania, April 2001. Los Angeles: The Getty Conservation Institute, 23-46.

FAIRCLOUGH, G. J., 2003b. The long chain: archaeology, historical landscape characterization and time depth in the landscape. *In* Palang and Fry, *Landscape Interfaces: Cultural Heritage in Changing Landscapes*. Landscape Series 1, Dordrecht, Kluwer Academic Publishers, 295-317.

FAO 1991. *Second Interim Report on the State of Tropical Forests by Forest Resources Assessment 1990 Project. Tenth World Forestry Congress, September 1991, Paris*. Food and Agriculture Organization of the United Nations, Rome.

FASSMANN H., KOHLBACHER J., REEGER U. (Eds.), 2002. *Zuwanderung und Segregation: Europäische Metropolen im Vergleich*. Drava verlag, Klagenfurt/Celovec.

FAURE A., 1992. *Le village et la politique. Essai sur les maires ruraux en action*, Paris, L'Harmattan.

FAURE A., POLLET G., WARIN P. (Eds.), 1995. *La construction du sens dans les politiques publiques. Débats autour de la notion de référentiel*. Paris, L'Harmattan.

FAURE G.O., MERMET L., TOUZARD H. 2000. *La négociation. Situations, problématiques, applications*. Paris.

FERNÁNDEZ MUÑOZ S., 2004. La participación pública en la ordenación del paisaje. Una reflexión a partir de tres proyectos en la Región de Murcia. En Mata (Cord), (2004). *El paisaje y la Gestión del Territorio*. Barcelona, Universidad Internacional Menéndez Pelayo.

FERRET J., 1996. *Paroles d'élus. Le travail politique au quotidien*, Eres.

FIRMINO A., 1999. Agriculture and landscape in Portugal, *Landscape and Urban Planning* 46, *Elsevier*, 83-91.

FLEURY A.M., BROWN R.D., 1997. A framework for the design of wildlife conservation corridors With specific application to southwestern Ontario. *Landscape and Urban Planning*, (37), 163-186.

FLOYD M.F., SHINEW K.J., 1999. Convergence and divergence in leisure style among Whites and African Americans: Toward an interracial contact hypothesis. *Journal of Leisure Research 31* (4): 359-384.

FOLKE C., *et al.*, 2002. Resilience and Sustainable Development: Building Adaptive Capacity in a World of Transformations. Scientific Background Paper commissioned by the Environmental Advisory Council of the Swedish Government in preparation for WSSD. International Council for Science. 2002. ICSU Series on Science for Sustainable Development: Resilience and Sustainable Development No. 3., 37 p.

FONTAINE J., LE BART C. (dir.), 1994. *Le Métier d'élu local*, Paris, L'Harmattan.

FORMAN R.T.T., GODRON M., 1986. *Landscape ecology*. London, Wiley and Sons, 619 p.

FORMAN R.T.T., 1995. Some general principles of landscape and regional ecology. *Landscape Ecology* 10(3): 133-142.

FORTIN M.-J., 2005. *Paysage industriel, lieu de médiation sociale et enjeu de développement durable et de justice environnementale : les cas des complexes d'Alcan (Alma, Québec) et de Péchiney (Dunkerque, France).* Thèse de doctorat présentée à l'Université du Québec à Chicoutimi et à l'Université de Paris 1 – Sorbonne.

FRIEDBERG C., M. COHEN, MATHIEU N., 2000. Faut-il qu'un paysage soit ouvert ou fermé ? L'exemple de la pelouse sèche du Causse Méjan. *Natures Sciences Sociétés* 8(4) : 26-42.

FRY G. L., SARLOV HERLIN I., 1995. Landscape design: how do we incorporate ecological, cultural and aesthetic values in landscape assessment and design principles? *In* Griffiths, G.H. (ed.), 1995. *Landscape ecology: theory and application.* Proceedings IALE-UK Conference, Reading, 51-60.

FRY G., 2001. Multifunctional landscapes – towards transdisciplinary research. *Landscape and Urban Planning*, 57, 159-168.

FRY G., JERPÅSEN G., SKAR B., STABBETORP O., 2001. Combining landscape ecology with archeology to manage cultural heritage interests in the changing countryside. *In* Mander Ü, Printsmann A., Palang H., 2000. *Development of European Landscapes.* Conference proceedings IALE European Conference 2001, Tartu: Publicationes Instituti Geographici Universitatis Tartuensis, 2 Vol., 198-202.

GAUDIN J.-P. (dir.), 1991. *Desseins de ville « Art urbain et urbanisme »*, Paris, L'Harmattan, 174 p.

GEHRKE M., 2001. Der Park als Veranstaltungsort: Über Veranstaltungen im Grünen und deren Verträglichkeit. *Stadt und Grün* 50 (5): 325-332.

GERMANN-CHIARI C., SEELAND K., 2004. Are urban green spaces optimally distributed to act as places for social integration? Results of a geographical information system (GIS) approach for urban forestry research. *Forest Policy and Economics* 6: 3-13.

GEYSER – FONDATION DE FRANCE : *100 expériences de dialogue territorial.* Projets soutenus par la Fondation de France 1997-2001, Paris.

GIDDENS A., 1994 (1990). *Les conséquences de la modernité.* Paris, L'Harmattan.

GIGLER U., TÖTZER T., 2005. Case Studies on revitalising inner city industrial sites in Western Europe sustainably: crucial lessons learned in the redevelopment process. KMG – Kolleg für Management und Gestaltung (ed.): ECOS 2004 – 1st European Networks Conference on Sustainability in Practice – *The Conditions and Requirements for a New European Level of Capacity Building*, 1 – 4 April 2004, Conference Reader, Eigenverlag, Berlin/Dessau 2005, 144-151.

GILMOUR A., WALKERDEN G., SCANDOL J., 1999. Adaptive management of the water cycle on the urban fringe: three Australian case studies. *Conservation Ecology* 3(1): 11. [online] URL: http://www.consecol.org/vol3/iss1/art11/.

GOBSTER P.H., 1998. Urban parks as green walls or green magnets? Interracial relations in neighborhood boundary parks. *Landscape and Urban Planning 41* (1): 43-55.

GOBSTER P., 2004. The Social aspects of landscape change: protecting open space under the pressure of development. *Landscape and Urban Planning 69* (Introduction), 149-151.

GOHEEN P.G., 2003. The assertion of middle-class claims to public space in late Victorian Toronto. *Journal of Historical Geography* 29, 73-92.

GÓMEZ MENDOZA J. (dir.). 1999. *Los paisajes de Madrid. Naturaleza y medio rural*. Madrid, Alianza Universidad-Fundación Caja Madrid.

GORGEU Y. *et al.* (dir.), 1995. *La charte paysagère, outil d'aménagement de l'espace inter-communal*. Paris, La documentation française.

GOUGH C., DARIER E., DE MARCHI B., FUNTOVITZ S., GROVE-WHITE R., GUIMARAES PEREIRA A., SHACKLEY S., WYNNE B., 2003. Contexts of citizen participation. *In* Kasemir B., Jager J., Jaeger C.C., Gardner M.T., *Public participation in sustainability science*. Cambridge University Press. Cambridge, 37-61.

GRABOY-GROBESCO, A., 2002. L'environnement des équipements commerciaux. *Études foncières,* n° 98, 29-32.

GRAS R., BENOÎT M., DEFFONTAINES J.P., DURU M., LAFARGE M., LANGLET A., OSTY P.L., 1989. *Le fait technique en agronomie. Activité agricole, concepts et méthodes d'étude*. Paris, INRA, L'Harmattan.

GREEN D., KIDD S., 2004. *Milton Keynes Urban expansion: Historic Environment Assessment*. Buckinghamshire County Council, Milton Keynes and EH, Aylesbury (and EH web page 2003).

GREIDER T., GARKOVICH L., 1994. Landscapes: The social construction of nature and the environment. *Rural Sociology*, 59, 1-24.

GRENVILLE J. (Ed), 1999. *Managing the Historic Rural Environment*, Routledge/English Heritage, London.

GRIFFITHS R., SHAPLAND JM., 1979. *Designing against vandalism*. London, Design Council. 12-16.

GROTH P., BRESSI T.W. (eds.), 1997. *Understanding Ordinary Landscapes*. New Haven, Yale University Press.

GROUNDWORK BLACKBURN AND MANCHESTER METROPOLITAN UNIVERSITY, 1999. *Involving Black and Minority Ethnic Communities*. Groundwork Blackburn and Manchester Metropolitan University, 123.

GRUMBINE R.E., 1996. Reflections on "What is Ecosystem Management". *Conservation Biology*, (11): no. 1, 41-47.

GUÉRIN M., MOQUAY P., 2002. Intercommunalité, pays : les recompositions territoriales, in Perriet-Cornet P. (dir.), *À qui appartient l'espace rural ?* DATAR/L'aube, La Tour d'Aigues, 105-132.

GUISEPELLI E., FLEURY P., LUGINBÜHL Y., 2000. Paysage et développement dans les Alpes du Nord. Organisation entre acteurs et rôle des experts. Communication au colloque « *Nouvelles urbanités, nouvelles ruralités en Europe* », Strasbourg, 10-12 mai, 7 p.

GUISEPELLI E., 2001. *Le paysage comme objet et outil de négociation des actions de développement dans les Alpes du Nord*. Thèse de doctorat de Géographie humaine, sous la direction de Y. Luginbühl, décembre, 484 p+annexes.

GUISEPELLI E., FLEURY P., 2003. *Paysage et agriculture dans les Alpes du Nord. Représentations et aspirations de la société*. GIS Alpes du Nord, 54 p.

GUISEPELLI E., 2005. Les représentations sociales du paysage comme outils de connaissance préalable à l'action. L'exemple des Alpes du nord. *Cybergéo*, N° 309, 3 May 2005.

GUNDERSON L.H., 1999. Resilience, flexibility, and adaptive management: antidotes for spurious certitude? *Conservation Ecology* 3(1): 7. [online] URL: http://www.consecol.org/vol3/iss1/art7.

GUNDERSON L.H., 2000. Ecological Resilience – in theory and application. *Annual Review of Ecological Systems*, (31): 425-439.

HANHÖRSTER H., MÖLDER M., 2000. VI. Konflikt- und Integrationsräume im Wohnbereich. *In* Heitmeyer W., Anhut R. (Eds.), *Bedrohte Stadtgesellschaft: Soziale Desintegrationsprozesse und ethnisch-kulturelle Konfliktkonstellationen*. Juventa Verlag, Weinheim, München, 347-400.

HANHÖRSTER H., 2001. *Integration von Migrantinnen und Migranten im Wohnbereich. Institut für Landes- und Stadtentwicklungsforschung des Landes Nordrhein-Westfalen*, Dortmund.

HARFF Y., LAMARCHE H., 2003. Les agriculteurs et le bocage, p. 175-256. *In* Lamarche H., ed. *Bocagement, reconstitution et protection du bocage. Évaluation des politiques publiques de paysagement du territoire*. INRA CNRS, Nanterre, Rapport de projet PEVS CNRS.

HATCHUEL A., 2000. Intervention Research ad the Production of Knowledge. *In learn: Cow up a tree, Knowing and Learning for Change in Agriculture. Case studies from industrialised Countries*. Paris, INRA Éditions, 55-68.

HÄUSSERMANN H., (Ed.), 2000. Grossstadt: Soziologische Stichworte. Leske + Budrich, Opladen.

HERRING P., 1998. *Cornwall's Historic Landscape – Presenting a Method of Historic Landscape Character Assessment*, Cornwall Archaeological Unit and English Heritage.

HIESS H. *et al.*, 1999. *Verkehr und Kulturlandschaft*. Endbericht. Im Auftrag des Bundesministriums für Wissenschaft und Verkehr. Wien (Österreichisches Kulturlandschafts-forschungsprogramm).

HIESS H., FAVRY E., PFEFFERKORN W., 1999. *Future Pictures and Future Stories of Cultural Landscape*. Research project financed by the Austrian Federal Ministry for Education, Science and Culture under the Austrian Research Focus Cultural Landscapes. Vienna.

HJORTSO C., 2004. Enhancing public participation in natural resources management using Soft OR – an application of strategic option development and analysis in tactical forest planning. *European Journal of Operational Research*, 152-3, 667-683.

HOFKENS E., ROOSSENS I. (eds.), 2001. *Nieuwe impulsen voor de landschapszorg. De landschapsatlas, baken voor een verruimd beleid*. Ministerie van de Vlaamse Gemeenschap, Afdeling Monumenten en Landschappen, Brussels.

HOLLING C.S., 1973. Resilience and stability of ecological systems. *Annu. Rev. Ecol. Syst.* 4, 1973, 1-23.

HOLLING C.S., ed., 1978. *Adaptive Environmental Assessment and Management*. London, John Wiley and Sons, 377.

HOLLING C.S., 1986. The resilience of terrestrial ecosystems: local surprise and global change. *In* W.C. Clark, R.E. Munn, eds., *Sustainable development of the biosphere*. Cambridge, Cambridge University Press, 192-217.

HOLLING C.S., GUNDERSON L.H., 2002. Panarchy. *Understanding transformations in human and natural systems*, Washington, Covelo, London, Island Press, 507.

HUISSOUD T., STOFER S., CUNHA A., SCHULER M., 1999. *Structures et tendances de la différenciation dans les espaces urbains en Suisse*.

HUNZIKER M., 1995. The spontaneous reafforestation in abandoned agricultural lands : perception and aesthetic assessment by locals and tourists. *Landscape and Urban Planning* 31, *Elsevier*, 399-410.

IDF, 1981. *La réalisation pratique de haies brise-vent et bandes boisées*. I.D.F., 151 p.

IFEN, 1996. *25 % de prairies ont disparu depuis 1970*. Paris, Les collections de l'environnement, N° 25, 4 p.

IFN, 2000. *Indicateurs de gestion durable des forêts françaises*. Paris, Ministère de l'Agriculture et de la Pêche, 129 p.

INDOVINA F., 1990. *La città diffusa*, Venezia, DAEST.

INDOVINA F., FREGOLENT L., SAVINO M. (a cura di), 2005. *L'esplosione della città: Barcellona, Bologna, Donosti-Bayonne, Genova, Lisbona, Madrid, Marsiglia, Milano, Montpellier, Napoli, Porto, Valencia, Veneto centrale*, Bologna, Editrice Compositori.

INLAND WATERWAYS AMENITY ADVISORY COUNCIL, 2001. *The inland waterways: towards greater social inclusion*. Final report of the working group on social inclusion. IWAAC.

INRA-SAD ARMORIQUE, CNRS (UMR LADYSS, ECOBIO et COSTEL), ENSP, 2003. *Bocagement, reconstitution et protection du bocage. Évaluation des politiques publiques de paysagement du territoire*. Programme de Recherche Politiques publiques et paysages, 354 p. + annexes.

IVERSON L.R., COOK E.A., 2000. Urban forest cover of the Chicago region and its relation to household density and income. *Urban Ecosystems* 4 (2): 105-124.

JACOBS J., 1961. The death and life of great American cities. *Vintage books*, New-York. 371-372.

JACOBS P., 1985. Achieving Sustainable Development. *Landscape Planning*, 12, Amsterdam, *Elsevier Science Publishers*, 203-209.

JACOBS J., 1993. *Tod und Leben grosser amerikanischer Städte*. Bauwelt Fundamente 4. Vieweg, Braunschweig.

JOBERT B., MULLER P., 1987. *L'État en action : politiques publiques et corporatismes*. Paris, PUF.

KAPLAN S., 1995. The Restorative Benefits of Nature: Toward an Integrative Framework. *Journal of Environmental Psychology*, 15, 169-182.

KAUFMAN S. 1991. Decision Making and Conflict Management Processes in Local Government. *In* R.D., Bowen W. M., Chandler M. O. *et al.*, *Managing Local Government, Public Administration in Practice.* Bingham, Newbury Park (Cal.), Sage, 115-134.

KAUR E., PALANG H., SOOVALI H., 2004. Landscapes in change-opposing attitudes in Saaremaa, Estonia. *Landscape and Urban Planning.* 67, Elsevier 109-120.

KEATING M., 1993. Les interventions économiques des collectivités locales aux États-Unis, en Grande-Bretagne et en France. Les effets politiques et économiques. *In* Biarez S. Nevers J.-Y. (dir.), *Gouvernement local et politiques urbaines.* Grenoble, CERAT, 463-474.

KELLY G.A., 1955. *The Psychology of Personal Constructs*, New York: W.W. Norton

KERGREIS S., 2002. Pratiques des agriculteurs sur leurs bordures de champ : une approche anthropologique et régulationiste., 405-4322002. "Sociologie, économie et environnement" (70ᵉ Congrès de l'ACFAS), Québec. 13-17 mai 2002. *Cahiers du CRISES*, Centre de recherche sur les innovations sociales.

KESSELMAN M., 1967. *The ambiguous consensus. A study of local government in France.* Knopf, New York.

KINGDON J., 1984. *Agendas, Alternatives and Public Policies.* Boston, Little Brown.

KINZIG A.P., GROVE J.M., 2001. Urban-suburban ecology. *Encyclopedia of Biodiversity*, Academic Press, (5): 733-745.

KNEAFSEY M., 2000. Tourism, Place Identities and Social Relations in the European Rural Periphery. *European Urban and Regional Studies,* vol.7, n° 1, pp. 35-50.

KOLLMANN G., LEUTHOLD M., PFEFFERKORN W., SCHREFEL C., 2003. Partizipation. Ein Reiseführer für Grenzüberschreitungen in *Wissenschaft und Planung*. Profil Verlag. Wien, München.

KORPELA K.M., 1992. Adolescents' Favourite Places and Environmental Self-Regulation. *Journal of Environmental Psychology*, (12) 249-258.

KUHNHOLTZ-LORDAT G., 1945. La *silva* le *saltus* et l'*ager* de garrigue. *Annales de l'École nationale d'agriculture de Montpellier*, XXVI, 1-78.

KUO F.E., SULLIVAN W.C., COLEY R.L., BRUNSON L., 1998. Fertile ground for community: Inner-city neighborhood common spaces. *American Journal of Community Psychology*, 26 (6): 823-851.

KUO F.E., SULLIVAN W.C., 2001. Environment and crime in the inner city: Does vegetation reduce crime? *Environment & Behavior*, 33, 343-367.

KUO F.E., SULLIVAN W.C., 2001. Aggression and violence in the inner city: Impacts of environment *via* mental fatigue. *Environment & Behavior*, 33(4), 543-571.

KWEON B.S., SULLIVAN W.C., WILEY A.R., 1998. Green common spaces and the social integration of inner-city older adults. *Environment and Behavior*, 30 (6): 832-858.

LÆSSØE J., IVERSEN T., 2003. *Naturen i et hverdagslivs-perspektiv: En kvalitativ interviewundersøgelse af forskellige danskeres forhold til naturen, Danmarks Miljøundersøgelser* (DMU, National Environmental Research Institute, Denmark). Faglig rapport n° 437.

LAMARCHE H., LUGINBÜHL Y., TOUBLANC M., BAUDRY J., THENAIL C., BUREL F. *et al.*, 2003. *Bocagement, reconstitution et protection du bocage : évaluation des politiques publiques de paysagement du territoire*. UMR LADYSS, ENSP, INRA-SAD, UMR ECOBIO Rennes, Université de Rennes/COSTEL, 11 p.

LASCOUMES P., 1996. Rendre gouvernable : de la traduction au transcodage. L'analyse des processus de changement dans les réseaux d'action publique. *In* CURAPP, *La gouvernabilité*. Paris, PUF.

LASSUS B., 1991. *Les continuités du paysage*. Urbanisme. N° 250.

LATOUR B., 1999. *Politiques de la nature*. Paris, La Découverte, 384 p.

LAURENT C., 1994. L'agriculture paysagiste : du discours aux réalités. *Natures Sciences Sociétés*, 2, 231-242.

LE BART C., 1992. *La rhétorique du maire-entrepreneur*. Paris, Pédone.

LE BART C., 1999. Le savoir-faire politique comme bricolage. *In* Poirmeur Y., Mazet P. (dir.), *Le métier politique en représentations*. Paris, L'Harmattan.

LE BART C., 2003. *Les maires : sociologie d'un rôle*. Villeneuve d'Asq, Presses Universitaires du Septentrion.

LE CŒUR D., BAUDRY J. *et al.*, 1997. Field margins plant assemblages: variation partitioning between local and landscape factors. *Landscape and Urban Planning*, 37(1-2): 57-71.

LE DU L., 1995. *Images du paysage. Télédétection, intervisibilité et perception, l'exemple des Côtes d'Armor*. Thèse de doctorat en géographie, université de Haute Bretagne, 334 p.

LE DU-BLAYO, MORANT P., ROZÉ F., SALIOU P., BUREL F., BUTET A., MILLAN N., RANTIER Y., WOLF A., BAUDRY J., VOLANT D., 2000. *Cartographie et évaluation de la qualité biologique du bocage du département des Côtes d'Armor*. Rapport de recherche dans le cadre du Plan départemental de l'Environnement, 333 p.

LE DU L., 2000. Unités de paysage et télédétection. *In Action paysagère et acteurs territoriaux*, Poitiers, GESTE n° 1, 109-119.

LE DU-BLAYO L., 2004. L'évaluation des paysages bocagers : articulation entre les méthodes des scientifiques et les moyens des gestionnaires. Actes du Colloque UMR 5045 CNRS « *L'évaluation du paysage : une utopie nécessaire ? À la recherche d'indicateurs, de marqueurs pluridisciplinaires* », université Paul Valéry Montpellier.

LE FLOCH S., DEVANNE A.S., DEFFONTAINES J.P. La « fermeture du paysage » : au-delà du phénomène, petite chronique d'une construction sociale. *L'espace géographique* (à paraître).

LE FLOCH S., 2003. *Qu'entend-on par « fermeture du paysage » ?* Paris, ministère de l'Écologie et du Développement durable, Direction de la nature et des paysages, 26 p.

LE FLOCH S., DEVANNE A.S., 2004. *D'espace public en espaces ouverts. Exploration bibliographique sur le thème des interrelations entre personnes et entre personnes et environnement physique*. Cemagref, Bordeaux.

LE FLOCH S., DEFFONTAINES J.P., TERRASSON D., RIBÉREAU-GAYON M.D., 2004. *Retour sur le « paradoxe du paysage » : dix ans de « paysage »*. Séminaire Natures Sciences et Sociétés Dialogues (26 mai 2004), Paris.

LÉON L'AFRICAIN, 1977. *Histoire des villes Africaines*. Alger, SNED.

LEPART J., ESCARRÉ J., 1983. La succession végétale, mécanismes et modèles : analyse bibliographique. *Bulletin d'écologie*, 14 : 133-178.

LEPART J. ET DEBUSSCHE M., 1992. Human impact on landscape patterning: Mediterranean examples. *In* A.J. Hansen et F. di Castri, *Landscapes boundaries. Consequences for biotic diversity and ecological flows*. Ecological Studies n° 92. Springer-Verlag, Berlin. pp. 76-105.

LEPART J., DERVIEUX A., DEBUSSCHE M. (1996). Photographie diachronique et changement des paysages. Un siècle de dynamique naturelle de la forêt à Saint-Bauzille-de-Putois, vallée de l'Hérault. *Forêt Méditerranéenne,* 17 : 63-80.

LEPART, J., P. MARTY, ET Rousset O., 2000. Les conceptions normatives du paysage. Le cas des Grands Causses. *Nature Sciences Sociétés* 8(4) : 16-25.

LEPART J., MARTY P. (2004). L'objet et son image ? Science des représentations ou science des paysages : les enjeux du transfert vers les gestionnaires. *In* Rivière Honegger A. & Puech D., eds.), *L'évaluation du paysage : une utopie nécessaire ? : à la recherche d'indicateurs/ marqueurs pluridisciplinaires.* Communications au colloque de Montpellier, 15-16 janvier 2004, Montpellier, Presses de l'université Paul Valéry, 519-526.

LESSARD G., 1998. An adaptive approach to planning and decision-making. *Landscape and Urban Planning,* v. 44, 81-87.

LEVEQUE C., 2001, *Écologie. De l'écosystème à la biosphère.* Paris, Dunod, 502 p.

LISCHNER K.R.,1994. *Leben zwischen den Häusern: Gestaltung und Nutzung der Aussenräume.* Bericht 25 des NFP 25 ,Stadt + Verkehr', Zürich.

LIZET B., WOLF A-E, CELECIA J. (Dir.)., 1997, Sauvages dans la ville. JATBA, *Revue d'ethnobiologie,* MNHN, 606 p.

LLEWELLYN M., 2003. Polyvocalism and the public : "doing" a critical historical geography of architecture. *Area.* 35, 264-270.

LOIBL W. *et al.,* 2002. *STAU-Wien – Stadt-Umland-Beziehungen in der Region Wien: Siedlungsentwicklung, Interaktion und Stoffflüsse.* Final Report, September 02, ARC Seibersdorf Research Reports ARC-S-0181.

LOIBL W., TÖTZER T., 2003. Modelling Growth and Densification Processes in Sub-urban Regions – Simulation of Landscape Transition with Spatial Agents. *Environmental Modelling and Software,* v. 18, n° 6, 485-593.

LORENZ K.,1997. *Odumírání lidskosti,* Mladá Fronta, Prah.

LOWENTHAL D., (1994). Identity, Heritage, and History. *In* J. R. Gillis (ed.), *Commemorations. The Politics of National Identity.* Princeton (N.J.), Princeton University Press, 41-57.

LUCIANI D., "Insediamento e mobilità nel Nord Est: appunti su una nebulosa senza centro".

LUGINBUHL Y., 2003. *Jardins de tous les désirs d'Europe centrale.* Les Cahiers du paysage, Acte-sud/ENSP 9 & 10, 228-258.

LUGINBÜHL Y., 2003. Temps social et temps naturel dans la dynamique des paysages. *In* Poullaouec-Gonidec P., Paquette S., Domon G. (eds), *Les temps du paysage.* Montréal, Les Presses de l'université de Montréal, 85-104.

LUGINBÜHL Y., 1998. Symbolique et matérialité du paysage. *Revue de l'économie méridionale.* Vol. 46, n° 183, 235-245.

LUGINDÜHL Y., 1990. *Paysages. Textes et représentations du siècle des Lumières à nos jours.* Lyon, La Manufacture.

LUGINBÜHL Y., 1991. Le paysage rural. La couleur de l'agricole, la saveur de l'agricole, mais que reste-t-il de l'agricole ? *Études rurales,* 121-124, pp. 27-44.

LUGINBÜHL Y., 1996. Représentation du paysage rural, représentation de la société : une lecture historique. *In* Jollivet M., Eizner N., *L'Europe et ses Campagnes*. Chap. 9, Presses de Sciences Po, 217-244.

LUGINBÜHL Y., TOUBLANC M., 1998. *De l'utilité de l'émondage à la contemplation du paysage*. Rapport de recherche pour le ministère de l'Environnement et l'INRA-SAD Armorique, 120 p., illustrations.

LUGINBÜHL Y., 2001. Paysage modèle et modèles de paysages. *In* Éditions Odile Jacob, 305 p, *L'Environnement, question sociale*. Paris, 49-56.

LUGINBÜHL Y., 2002. Rural tradition and Landscape Innovation in the eighteenth Century. *In* dir. John DIXON HUNT, *Tradition and Innovation in French Garden Art*. Philadelphia, University of Pennsylvannia Press, 82-92.

LUGINBÜHL Y., TOUBLANC M., 2004. Des talus arborés aux haies bocagères : des dynamiques de pensée du paysage inspiratrices de politiques publiques. 12p. *In actes du colloque international « De la connaissance des paysages à l'action paysagère »*, 2-4 décembre 2004, format CD-Rom.

LUGINBÜHL Y., TOUBLANC M, 2006. Des talus arborés aux haies bocagères : des dynamiques de pensées du paysage inspiratrices de politiques publiques. *In* actes du colloque Mabileau A., 1992, *L'élu local : nouveau professionnel de la République*. Pouvoirs, n° 60, 67-78.

LUZ F., 2000. Participatory landscape ecology – A basis for acceptance and implementation. *Landscape and Urban Planning* 50, 157-166.

LYNCH K., 1982. *Voir et planifier l'aménagement qualitatif de l'espace*. Paris, Édition Bordas, 109 p.

MABRY K.E., BARRETT G.W., 2002. Effects of corridors on home range sizes and interpatch movements of three small mammal species. *Landscape Ecology*, 17: 629-636.

MACINNES L., 2004. Historic Landscape Characterization. In Bishop and Phillips (Eds), *Countryside Planning: New approaches to Management and Conservation*. London, Earthscan. 155-169.

MACINNES L., WICKHAM-JONES C.R., 1992. Time-depth in the countryside: archaeology and the environment. *In* Macinnes L, Wickham-Jones C.R., (Eds), *All Natural Things: Archaeology and the Green Debate*. Oxford, Oxbow Monograph 21, 1-13.

MC LAIN R.J., LEE R.G., 1996. Adaptive Management, Promises and Pitfalls. *Environmental Management*. v. 20, no. 4, 437-448.

MADGE C., 1997. Public parks and the geography of fear, *Tijdschrift voor Economische en Sociale Geografi*, 88 (3), 237-250.

MAIRIE DE PARIS, 2004. *Charte régionale de la biodiversité et des milieux naturels, dossier de presse 18 mars 2004*. Direction générale de l'information et de la communication, 11 p.

MANGIN D, 2005. *La ville franchisée, formes et structures de la ville contemporaine*. Paris, La Villette, 398 p.

MARGUERIE D., ANTOINE A., THENAIL C., BAUDRY J., BERNARD V., BUREL F., CATTEDU I., DIRE MY, GAUTIER M., GEGHARDT A., GUIBAL F., KERGREIS S., LANOS P., LE CŒUR D., LE DU L., MÉROT P., NAAS P., OUIN A., PICHOT D., VISSET L., 2003. Bocages armoricains et sociétés, genèse, évolution et interactions. *In Des milieux et des hommes : fragments d'histoires croisées*. Elsevier, collection environnement, 115-131.

MARINI D. (a cura di), 2002. *Nord Est 2002. Rapporto sulla società e l'economia*, Venezia, Fondazione Nord Est.

MATA OLMO R., SANZ HERRÁIZ (dir.), 2003. *Atlas de los paisajes de España*. Madrid, Ministerio de Medio Ambiente.

MATA R., GÓMEZ J. AND FERNÁNDEZ F., 2001. Paisaje, calidad de vida y territorio. *Análisis Local*, 37, 27-40.

MATLESS D., 2003. The Properties of Landscape. *In* Anderson K., Domosh M., Pile S., Thrift N. (eds.), *Handbook of Cultural geography*. Londres, Sage, 227-232.

MEINDL P., JEDELSKY B., 2003. LandwirtschafT in Wien – Zwischen Stadtplanung und Ökonomie. *Werkstattbericht,* no. 52.

MENGUY B., PERNET A., 2004. *Zones d'activités et paysage. Du constat partagé à de nouvelles mises en œuvre*. Lavoûte-Chilhac, Centre du paysage.

MEYER-CECH K., SEHER W., 2003. Flächenschutz in der Stadtentwicklung von Wien. ELSA e.V. (ed.) : *Local land & soil news*, no. 5/I/03, 13-14.

MICHELIN Y., 2000. De la gestion de l'espace au développement local. *Revue d'Auvergne*, n° hors série, Pays : de l'aménagement au développement des territoires, AUA, Clermont-Ferrand, 47-58.

MICOUD A., 1999. Patrimoine et légitimité des territoires. De la construction d'un autre espace et d'un autre temps commun. *In* Gerbaux F. (ed.), *Utopie pour le territoire : cohérence ou complexité ?* La Tour d'Aigues, Éditions de l'Aube, 53-63.

MICOUD A. 2002. La biodiversité, un objet social certes, mais quel objet sociologique ? *In* VIVIEN F. D. (ed.), *Biodiversité et appropriation : les droits de propriété en question*. Paris, Eleesevier.

MILCHERT J., 1998. Der Park als multikultureller Ort. *Stadt und Grün*, 47 (9): 667-671.

MITCHELL D., 2003. Dead Labor and the Political Economy of Landscape – California Living, California Dying. *In* Anderson K., Domosh M., Pile S., Thrift N. (eds.); *Handbook of Cultural geography*. Sage, Londres, 233-248.

MITSCHERLICH A., 1965. *Die Unwirtlichkeit unserer Städte: Anstiftung zum Unfrieden*. Suhrkamp Verlag, Frankfurt am Main.

MOLLARD A., RAMBONILAZA M., VOLLET D., 2006. Aménités environnementales et rente territoriale sur le marché de services différenciés : le cas du marché des gîtes ruraux labellisés en France. *Revue d'économie politique*, vol. 116, n° 1, 251-277.

MONTANARELLA L., 1999. *Soil at the interface between Agriculture and Environment*. [online] URL : http://europa.eu.int/comm/agriculture/envir/report/en/inter_en/report.htm.

MOQUAY P., 1998. *Coopération intercommunale et société locale*. L'Harmattan, Paris-Montréal (Québec).

MOQUAY P., 2002. Quelle fonction pour les élus demain ? *In* Mairie-conseils, *Et devinez sur qui ça retombe ? Ou la vie quotidienne des maires dans 32 000 communes*. Paris, La Documentation française.

MORAND-DEVILLER J., 1994. Environnement et paysage. *L'actualité juridique, Droit administratif,* 20 septembre, 588-595.

MORET J., 2004. La biodiversité à Paris. *In* Michaud Y. (Dir.), Paris, Odile Jacob, coll. université de tous les savoirs, 217-238.

MOUGENOT C., 2003. *Prendre soin de la nature ordinaire.* Éd. Maison des sciences de l'homme-INRA, 230 p.

MULLER P., SUREL Y., 2000. *L'analyse des politiques publiques*, Paris, Montchrestien.

MULLER P., 2005. Esquisse d'une théorie du changement dans l'action publique. Structures, acteurs et cadres cognitifs. *Revue française de science politique*, vol. 55, n° 1, 155-187.

MÜLLER C., 2002. *Wurzeln schlagen in der Fremde : Die Internationalen Gärten und ihre Bedeutung für Integrationsprozesse.* München, Ökom Verlag.

MUNARIN S., TOSI M.C., 2001. *Tracce di città. Esplorazioni di un territorio abitato: l'area veneta,* Milano, Franco Angeli.

NACEUR F., 1997. Le vandalisme urbain et la population infantile et juvénile. *In actes du séminaire national en architecture,* Biskra, 177-186.

NACEUR F., 2004. La problématique de la dominance masculine au niveau des espaces urbains : cas des villes algériennes. *Collection perspectives*, 8, 240-247.

NATIONAL COMMUNITY FOREST, 2003. *The National Community Forest Partnership: Environmental Regeneration-shaping the future.* Disponible sur : http://www.community-forest.org.uk/, the National Community Forest Partnership, visited 2/10/03

NOWOTNY H., SCOTT P., GIBBONS M., 2002. *Re-Thinking Science – Knowledge and the Public in an Age of Uncertainty.* Policy Press, Bristol.

ÖBIG (Österreichisches Bundesinstitut für Gesundheitswesen), 2002. *Erfassung des Stadtgrüns durch Biotopmonitoring.* Commissioned by MA 22 – Umweltschutz.

O.C. GIS-VLAANDEREN, 2001. *Landschapsatlas.* Brussel: Ondersteunend Centrum GIS-Vlaanderen, CD-ROM.

OLIVEIRA C., 2001. *Lugar e memória (testemunhos megalíticos e leituras do passado).* Lisboa, Colibri.

OLIVEIRA R., 1998. Causas para a desflorestação e degradação da floresta – Estudo de Caso para o concelho de Mértola – Portugal, Mediterrâneo – Desertificação. *Revista semestral*, n°s 12/13, Lisboa, Instituto Mediterrânico, Universidade Nova de Lisboa, 75-93.

OLIVEIRA R., 2001. Política Agrícola Comum – Uma Reflexão Aplicada ao concelho de Mértola. *Revista Municipal* n°3, 1° semestre de 2001, Câmara Municipal de Mértola, 20-33.

OLIVIER L., GALLAND J.-P., MAURIN H., 1995. *Livre rouge de la flore menacée de France.* Tome 1 : espèces prioritaires. Collection Patrimoines Naturels N° 20. Institut d'écologie et de gestion de la biodiversité, Paris, Muséum national d'Histoire naturelle.

OLSSON P., FOLKE C., HAHN T., 2004. Social-Ecological Transformation for Ecosystem Management: the Development of Adaptive Co-management of a Wetland Landscape in Southern Sweden. *Ecology and Society*, Vol. 9, No.4.

OLWIG K.R., 2004. "This Is Not A Landscape": Circulating Reference And Land Shaping. *In* Palang H., Sooväli H., Antrop M., Setten S., 2004, *European rural landscapes: persistence and change in a globalising environment.* Dordrecht, Kluwer Academic Publishers, 41-66.

O.N.S., 1998. *Bilan annuel 1998 statistiques.*

OPENSPACE RESEARCH CENTRE, 2003. *Diversity Review, Options for Implementation.* Cheltenham, Countryside Agency.

ORESZCZYN S., 2000. A systems approach to the research of people's relationships with English hedgerows. *Landscape and Urban Planning*, 50, 107-117.

O'ROURKE E., 1999. Changing identities, changing landscapes: Human-land relations in transition in the Aspre, Roussillon. *Ecumene*, 6, 29-50.

PADIOLEAU J.-G., 1982. *L'État au concret*. Paris, PUF.

PALANG H., FRY G., (Eds), 2003. *Landscape Interfaces: Cultural Heritage in Changing Landscapes*. Landscape Series 1, Dordrecht, Kluwer Academic Publishers.

PALMER J.F., 1997. Stability of landscape perceptions in the face of landscape change. *Landscape and Urban Planning* 37, 109-113.

PALMER J.F., 2004. Using spatial metrics to predict scenic perception in a changing landscape: Dennis. Massachusetts, *Landscape and Urban Planning* 69, 201-218.

PANIGA M., 2003. *Gli spazi verdi luganesi del 21 secolo sotto la percezione della popolazione locale*.

PERLIK M., 2001. Alpenstädte zwichen Metropolisation und neuer Eigenständigkeit. *Geographica Bernensia*, Bern, p 38.

PERROT M., LA SOUDIÈRE (de) M., 1998. La résidence secondaire : un nouveau mode d'habiter la campagne ? *Ruralia*, vol. 02, 137-150.

PFEFFERKORN W., FAVRY E., 2003. Local workshops, Pilot region Wipptal. Annex 2 to Work Package 5 Report. *Regional Development and Cultural Landscape Change: The Example of the Alps. Evaluating and Adjusting EU and National Policies to Manage a Balanced Change*. Vienna.

PFEFFERKORN W., EGLI H. R., MASSARUTTO A., 2005. Regional Development and Cultural Landscape Change. In the Alps. The Challenge of Polarisation. From analysis and scenarios to policy recommendations. *Geographica Bernensia*, G 74. Switzerland, Institute of Geography, University of Berne.

PFEFFERKORN W., FAVRY E., 2005. Futurs Alpins / The Future of the Alps. *La revue de géographie alpine / Journal of alpine research*. Grenoble, Tome 93 No 2.

PFISTER S., 2003. *Stadtnahe Grünräume für Ältere: Befragungen von Zürichs Seniorinnen und Senioren zu den Grünflächen ihrer Stadt*.

PICKETT S.T.A., CADENASSO M.L., GROVE J.M., 2004. Resilient cities: meaning, models, and metaphor for integrating the ecological, socio-economic, and planning realms. *Landscape and Planning* 69 (2004), 369-384.

POUSIN F. (dir.), 2000. Pouvoir des figures. *Cahiers de la recherche architecturale et urbaine*, 8.

PRÄSIDIALDEPARTEMENT DER STADT ZÜRICH, 2002. *Gute Beziehungen schaffen: Integrationsprojekte der Stadt Zürich*. Zürich, Präsidialdepartement der Stadt Zürich, Fachstelle für Stadtentwicklung.

PRIEUR M. ET DUROUSSEAU S., 2004. *Étude de droit comparé sur la participation du public en matière de paysage dans le contexte de la mise en œuvre de la convention européenne du paysage*. Strasbourg, Conseil de l'Europe, T-FLOR 3 (2004) 6, 47 p.

QUINLAN A., 2003. Resilience and adaptive capacity. Key components of sustainable social-ecological systems. *In IHDP, Update 02/2003*, p 4f.

REBOLLO O., 2002. *Bases político-metodológicas para la participación*. Ciudades para un futuro más sostenible, 24.

REGIÓN DE MURCIA, 2002. *Análisis, diagnóstico y propuestas sobre el paisaje de la Huerta de Murcia y la Vega Media*. Dirección General de Ordenación del Territorio. 3 vols. (Mimeo).

REGIÓN DE MURCIA, 2003. *Análisis, diagnóstico y propuesta de directrices del paisaje del noroeste de la Región de Murcia*. Dirección General de Ordenación del Territorio. 6 vols. (Mimeo).

REGIÓN DE MURCIA, 2004. *Análisis, diagnóstico y propuesta de directrices del paisaje del Altiplano de la Región de Murcia*. Dirección General de Ordenación del Territorio. 4 vols. (Mimeo).

REGIONAL CONSULTING, BIOTECHNICAL FACULTY OF THE UNIVERSITY OF LJUBLJANA; URBAN PLANNING INSTITUTE OF THE REPUBLIC OF SLOVENIA, E.C.O., 2001. *Sustainable development of Alpine cultural landscapes in the Austria – Slovenia border region*. Vienna, Klagenfurt, Ljubljana.

REGIONAL CONSULTING, L&R SOCIAL RESEARCH; URBAN PLANNING INSTITUTE OF THE REPUBLIC OF SLOVENIA, E.C.O., 2002. *EU-Extension to the East: Chances and Risks for Sustainable Development of Cultural Landscapes in the Styria-Slovenia Border Region*. Research project financed by the Austrian Federal Ministry for Education, Science and Culture under the Austrian Research Focus "Cultural Landscapes". Vienna, Klagenfurt, Ljubljana.

ROBERT M., 2001. Localité et changement social, le petit bout de la lorgnette ? *Observatoire des rapports entre rural et urbain*, vol. 3, 1-6.

ROBINSON R. A, SUTHERLAND W. J., 2002. Post-war changes in arable farming and biodiversity in Great Britain. *Journal of Applied Ecology*, 39 (1): 157-176.

ROBSON C., 1993. *Real World Research*. Oxford, Blackwell.

ROGER A., 1978. *Nus et paysages. Essai sur la fonction de l'art*. Paris, Aubier.

ROGER A., 1997. *Court traité du paysage*. Paris, Gallimard.

RÖLING N.G., WAGEMAKERS M.A.E, ed., 1998. *Facilitating Sustainable Agriculture: Participatory Learning and Adaptive Management in Times of Environmental Uncertainty*. Cambridge University Press, Cambridge, UK and New York, New York, USA.

ROMÃO C., 1997. *Manuel d'interprétation des habitats de l'Union européenne*. DG Environnement, Sécurité Nucléaire et Protection Civile, Ecosphèr, 109 p.

ROTMANS J., KEMP R., 2003. *Managing Societal Transitions: Dilemmas and Uncertainties: the Dutch energy case-study*. OECD Workshop on the Benefits of Climate Policy: Improving Information for Policy Makers. [online] URL: http://www.oecd.org/dataoecd/6/31/2483769.pdf.

RUDOLF F., 1998. *L'environnement, une construction sociale. Pratiques et discours sur l'environnement en Allemagne et en France*. Strasbourg, Presses universitaires de Strasbourg, 184 p.

SABLET M., 1988 – Des espaces urbains agréables à vivre. Places, rues, squares, jardins. Paris, *Le moniteur*, p. 49

SAOULI A. Z., 1989. *The revival of traditional Housing, case of Biskra*. Thèse de Magistère, Grande-Bretagne, université de York.

SAURÍ D., DEL MORAL L., 2001. Recent developments in Spanish water policy. Alternatives and conflicts at the end of the hydraulic age. *Geoforum*, 32, 351-362.

SAUVEZ M., 2001. *La ville et l'enjeu du « développement durable »*. Rapport au ministre de l'aménagement du territoire et de l'environnement, Paris, Documentation française, 436 p.

SCOTT A., 2002. Assessing Public Perception of Landscape: The LANDMAP experience. *Landscape Research* 27-3, 271-295.

SCOTT M.J., CANTER D.V., 1997. Picture or Place? A Multiple Sorting Study of Landscape. *Journal of Environmental Psychology* 17, 263-281

SCOTT M.J., 1998. *The Environmental Correlates of Creativity and Innovation in Industrial Research Laboratories*. PhD Thesis University of Liverpool.

SEGGERN H., VON TESSIN W., 2002. Einen Ort begreifen: Der Ernst-August-Platz in Hannover : Beobachtungen – Experimente – Gespräche – Fotos. *In* Riege M, Schubert H. (Eds.), *Sozialraumanalyse : Grundlagen – Methoden – Praxis*. Leske + Budrich, Opladen 265-280.

SHEETS V. L., MANZER C. D., 1991. Affect, Cognition, and Urban Vegetation: Some Effects of Adding Trees Along City Streets. *Environment and Behavior*, 23(3): 285-305.

SHERLOCK K., 2001. Revisiting the concept of hosts and guests. *Tourist Studies*. 1, 271-295.

SHYE S., ELIZUR D., HOFFMAN M., 1994. *Introduction to Facet Theory; Content Design and Intrinsic Data Analysis in Behavioural Research*. Thousand Oaks: Sage.

SLEE B., CURRY N., JOSEPH D., 2002. *Social exclusion in countryside leisure in the United Kingdom: the role of the countryside in addressing social inclusion*. A report for the Countryside Recreation Network. Countryside Recreation Network 122.

SIEVERTS T., 2003. *Cities without cities, an Interpretationv of the* Zwischenstadt, Londra/New York, Spon/Routledge.

SIMON P., 1997. Les usages sociaux de la rue dans un quartier cosmopolite. *Espaces et Sociétés* 90/91 : 43-68.

SIRIANNI C., FRIEDLAND L., 2001. *Civic innovation in America*. London, California press, 371 p.

SOLECKI W.D., WELCH J.M., 1995. Urban parks: green space or green walls? *Landscape and Urban Planning* 32 (2): 93-106.

SOLTNER D., 1978. *L'arbre et la haie*. Angers, Collection Sciences et Techniques Agricoles, 5e édition, 208 p.

SOLTNER D., 1994. *Planter des haies*. Coll Sciences et Techniques Agricoles. 7e édition, 104 p.

SOZIALDEPARTEMENT DER STADT ZÜRICH, 2001. *Projekt „Sicherheit / Intervention / Prävention" sip : Bericht über die Pilotphase*. Zürich, Sozialdepartement der Stadt Zürich.

SPITTHÖVER M., 2003. Integration oder Segregation? Öffentliche Freiräume und ihre Besucher in Kassel-Nordstadt. *Stadt und Grün* 52 (2): 24-30.

STEFULESCO C., 1993. *L'urbanisme végétal*. Paris, IDF, 323 p.

STEINNOCHER K. *et al.*, 1999. *Monitoring Urban Dynamics. Metropolitan Area of Vienna, final report*, August 1999, ARC Seibersdorf Research Reports OEFZS-S-0038.

STEINNOCHER, K., TÖTZER, T., 2001. Analyse von Siedlungsdynamik durch Verknüpfung von Fernerkundung und demographischen Daten. Federal Environmental Agency Ltd. – Austria (ed.): *Versiegelt Österreich? Der Flächenverbrauch und seine Eignung als Indikator für Umweltbeeinträchtigungen*. Wien, Conference Papers, CP-030, 39 47.

STEINNOCHER K., KÖSTL M., 2002. Verdichtung oder Zersiedelung? Eine Analyse des Flächenverbrauchs im Umland von Wien. In: M. Schrenk (ed.): *CORP 2002: Beiträge zum 7*. Symposion zur Rolle der Informationstechnologie in der und für die Raumplanung, 193-200.

STEWART WP., LIEBERT D., LARKIN K.W., 2004. Community identities as visions for landscape change. *Landscape and Urban Planning* 69, 315-334

STOOL-KLEEMANN S., 2001. Barriers to nature conservation in Germany: A model explaining opposition to protected areas. *Journal of Environmental Psychology*, Academic Press, 369-385

STONE D., 1989. Causal Stories and the Formation of Policy Agendas, *Political Science Quarterly*, n° 2, 281-300.

SUSTALP, 2000. *Evaluation of Instruments of the European Union regarding their Contribution to Sustainable Environment and Agriculture in the Alps*. Bozen, Brussels.

SWAFFIELD S., 1998. Contextual meanings in policy discourse: a case study of language use concerning resource policy in the New Zealand high country. *Policy Sciences*, vol. 31, 99-224.

TACK G., VAN DEN BREMPT P., 2001. De Landschapsatlas: Resultaten. *In* Hofkens E., Roossens I. (eds.), 2001, *Nieuwe impulsen voor de landschapszorg. De landschapsatlas, baken voor een verruimd beleid*. Brussel, Ministerie van de Vlaamse Gemeenschap, Afdeling Monumenten en Landschappen, 44-62.

TESSIN W., 2003. Anonymität und Kommunikation im öffentlichen Freiraum. *Stadt und Grün* 52 (2) : 19-23.

THENAIL C., CODET C., 2003. Systèmes techniques de gestion des bordures de champs en exploitation agricole, et intégration des haies nouvelles., 150-171. *In* Lamarche H. (ed.), *Bocagement, reconstitution et protection du bocage. Évaluation des politiques publiques de paysagement du territoire*. Inra-CNRS, Rapport de projet, Nanterre.

THE SHELTAIR GROUP, 2003. *A sustainable urban system: the long-term plan for Greater Vancouver*. [online] URL: www.sheltairgroup.com.

THIEBAUT L., 2002. Commentaire : le paysage : marchandise ou symbole d'identité ? *Nature, sciences, sociétés*, 10, 54-55.

THUM K.P., 1980. *Es grünt so grau: 25 alternative Gestaltungsvorschläge zur Grünflächenplanung*. Paperpress, Wien.

TOUBLANC M., 2004. Un dispositif d'évaluation sommaire au service d'une action publique incertaine : l'exemple la politique de reconstitution du bocage dans le département des Côtes-d'Armor. *In* Puech D., Rivière Honegger A. (eds), *L'évaluation du paysage : une utopie nécessaire ? À la recherche d'indicateurs/ marqueurs pluridisciplinaires*, université Paul Valéry Montpellier III, Gap, Éditions Louis Jean, 465-485.

TOMPKINS E. L., ADGER W. N., 2004. Does adaptive management of natural resources enhance resilience to climate change? *Ecology and Society* 9(2): 10. [online] URL : http://www.ecologyandsociety.org/vol9/iss2/art10.

TÖTZER T., GIGLER U., 2005. Managing change: Lessons learned in case studies on revitalising old industrial sites in European cities. *In* Schrenk M. (ed.), *Proceedings CORP 2005 (Competence Center of Urban and Regional Planning) "Real Models – Unreal World – Planning and dealing with the unpredictable"*. Vienna University of Technology, [online] http://mmp-tk1.kosnet.com/corp/archiv/papers/2005/CORP2005_TOETZER_GIGLER.pdf.

TRAVLOU P., 2003. *People and Public open Space in Edinburgh City Centre: Aspects of Exclusion*. Presented at the 38th International Making Cities Liveable Conference, Carmel, California, 19-23 October.

TRESS B., TREES G., 2001. Capitalising on multiplicity: a transdisciplinary systems approach to landscape research. *Landscape and Urban Planning* 57, 143-157.

TRESS B., TRESS G., VAN DER VALK A., FRY G. (eds.), 2003. Interdisciplinary and Transdisciplinary Landscape Studies: Potential and Limitations. Wageningen, *Delta Series* 2.

TRIVELLY E., 2004. *Quand les moutons s'en vont... Histoire et représentations sociales du boisement des pelouses sèches du sud-est de la France.* Aix-en-Provence, université de Provence, 263 p.

TROSA S. (1992). Le rôle de la méthode dans l'évaluation, à travers l'expérience du conseil scientifique de l'évaluation en France. *Politiques et management public* 10 (3) : 83-102.

TURCO A., 2003. "Abitare l'avvenire. Configurazioni territoriali e dinamiche identitarie nell'età della globalizzazione", *Boll. Soc. Geogr. It.*, XII, VIII, 3-20.

TURRI E., 1974. *Antropologia del paesaggio*, Milano, Edizioni di Comunità.

TURRI E., 1998. *Il paesaggio come teatro. Dal territorio vissuto al territorio rappresentato,* Venezia, Marsilio.

TWIGGER-ROSS C.L., UZZELL D.L, 1996. Place and Identity Processes. *Journal of Environmental Psychology* 16, 205-220.

ULRICH R.S., 1981. Natural versus urban scenes: some psychophysiological effects. *Environment and Behaviour* 13, 523-556.

ULRICH R.S., SIMONS R. F., LOSITO B. D., FIORITO E., MILES M. A., ZELSON M.,1991. Stress Recovery During Exposure to Natural and Urban Environments. *Journal of Environmental Psychology*, 11, 201-230.

URBAIN J.D., 2002. *Paradis verts. Désirs de campagne et passions résidentielles.* Paris, Payot.

VALLERANI F., 2000. "Il Veneto e le seduzioni palladiane tra senso del luogo e postmoderno". COSGROVE D., *Il paesaggio palladiano: la trasformazione geografica e le sue rappresentazioni culturali nell'Italia del XVI secolo*, Verona, Cierre, pp. 9-30.

VALLERANI F., VAROTTO M., 2005. *Il grigio oltre le siepi. Geografie smarrite e racconti del disagio in Veneto*, Portogruaro, Nuova Dimensione.

VAN DER LEEUW S.E., ASCHAN-LEYGONIE C., 2000. *A long-term perspective on resilience in socio-natural systems.* Paper presented at the workshop on "System shocks – system resilience" held in Abisko, Sweden, May 22-26, 2000.

VAN EETVELDE V. ANTROP M., 2005. The significance of landscape relic zones in relation to soil conditions, settlement pattern and territories in Flanders. *Landscape and Urban Planning*, 70 (2005) 127-141.

VAN HOORICK G., 2000. *Juridische aspecten van het natuurbehoud en de landschapszorg, Intersentia Rechtswetenschappen*, Antwerpen – Groningen.

VAN-TILBEURGH V., 2002. L'institution de la mer d'Iroise comme nature remarquable. *Natures, sciences, sociétés* 10 (4) : 31-37.

VERA F.W.N., 2000. *Grazing ecology and forest history.* CABI Publishing, Oxon et New York, 506 p.

VOISENAT C. (éd.), 1995. *Paysage au pluriel. Pour une approche ethnologique des paysages*, Paris, Éditions de la Maison des Sciences de l'Homme.

VOS W., MEEKES H., 1999. Trends in European cultural landscape development: perspectives for a sustainable future. Landscape and Urban Planning 46, *Elsevier*, 3-14.

VOS W., KLIJN J., 2000. Trends in European landscape development: prospects for a sustainable future. *In* Klijn J., Vos W. (eds.), 2000, *From Landscape Ecology to Landscape Science.* Wageningen, WLO, Kluwer Academic Publ., 13-30.

WALTERS C.J., HOLLING C.S., 1990. Large-scale Management Experiments and Learning by Doing. *Ecology*, v. 71, no. 6, 2060-2068.

WARD THOMPSON C., 2002. Urban open space in the 21st century. *Landscape and Urban Planning* 60, 59-72.

WARD THOMPSON C., 2002. Urban open space in the 21st century. *Landscape and Urban Planning* 60 (2): 59-72.

WARD THOMPSON C., ASPINALL P., BELL S., FINDLAY C., WHERRETT J, TRAVLOU, P., 2004. *Open Space and Social Inclusion: Local Woodland Use in Central Scotland.* Edinburgh: Forestry Commission.

WENT D., DYSON-BRUCE L., VINDEDAL K., 2003. *Historic Environment Issues in the proposed London – Stansted – Cambridge growth area: an indicative study of the Harlow – Stansted area).* London, English Heritage (www/english-heritage.org.uk).

WIEL M., 1999. *La transition urbaine ou le passage de la ville pédestre à la ville motorisée.* Liège, Pierre Mardaga.

WILSON J.-Q., KELLING G., 1982. Broken Windows. *Atlantic Monthly*, n° 211, 29-38.

ZANCHINI E., 2002. Paessaggi e partecippazione. *In* Clementi A. (de.), *Interpretación di paesaggio.* Roma, Meltemi, 292-310.

ZERBI M. C., 1993. *Paesaggi della geografia,* Torino, Giappichelli Editore.

ZONNEVELD I.S., 1995. *Land Ecology.* SPB Academic Publishing, Amsterdam.

List of authors

Marc Antrop, Ghent University, Geography Department, Belgium
marc.antrop@ugent.be

Olivier Aznar, Cemagref Clermont, Aubière, France
olivier.aznar@clermont.cemagref.fr

Nicolas Ballesteros, Swiss Federal Institute of Technology (ETH),
Human-Environment-Systems Institute, Zurich, Switzerland
nicolas.ballesteros@env.ethz.ch

Jacques Baudry, INRA-SAD Armorique, Rennes, France
jacques.baudry@rennes.inra.fr

Simon Bell, Edinburgh College of Art, Edinburgh, United Kingdom
s.bell@eca.ac.uk

Nathalie Blanc, CNRS, UMR LADYSS 7533, Institut de Géographie, Paris, France
nathali.blanc@wanadoo.fr

Françoise Burel, Université de Rennes 1, ECOBIO, UMR 6553 CNRS, Rennes, France
francoise.burel@univ-rennes1.fr

Jacqueline Candau, Cemagref Bordeaux, France
jacqueline.candau@cemagref.fr

Benedetta Castiglioni, Università di Padova, Dipartimento di Geografia, Italy
etta.castiglioni@unipd.it

Barbara Černič Mali, Rosinak & Partner ZT GmbH, Austria

Marianne Cohen, Université Paris 7, UFR GHSS, Paris, France
cohen@paris7.jussieu.fr

Anne-Sophie Devanne, Cemagref Bordeaux, France
anne-sophie.devanne@cemagref.fr

Milena Dneboská, University of Évora, Portugal

Veerle Van Eetvelde, Ghent University, Geography Department, Belgium
veerle.vaneetvelde@ugent.be

Graham Fairclough, English Heritage, London, United Kingdom
graham.fairclough@english-heritage.org.uk

Santiago Fernández Muñoz, Universidad Carlos III,
Departemento de Humanidasdes, Madrid, Spain
scfernan@hum.uc3m.es

Viviana Ferrario, IUAV Venezia, Dottorato di Ricerca in Urbanistica, Italy
viviana.ferrario@inwind.it

Philippe Fleury, SUACI Montagne - GIS 'Alpes du Nord', Chambéry, France
fleury.gis@wanadoo.fr

Marie-José Fortin, Université du Québec, Chicoutimi (UQAR), Canada
marie-jose_fortin@uqar.ca

Ute Gigler, ARC Systems Research GmbH, Seibersdorf, Austria
ute.gigler@arcs.ac.at

Sandrine Glatron, Image et Ville, Strasbourg, France
sandrine.glatron@lorraine.u-strasbg.fr

Marc Guérin, Cemagref Antony, France
marc.guerin@cemagref.fr

Emmanuel Guisepelli, SUACI Montagne – GIS 'Alpes du Nord', Chambéry, France
eguisepelli@suacigis.com

Mario Klesczewski, Conservatoire des Espaces Naturels du Languedoc Roussillon,
Montpellier, France
cen-lr@wanadoo.fr

Didier Le Cœur, INRA-SAD Armorique, Rennes, France
didier.lecoeur@agrocampus-rennes.fr

Laurence Le Du-Blayo, CNRS Université de Rennes 2, COSTEL UMR 6554, France
laurence.ledu@uhb.fr

Sophie Le Floch, Cemagref Bordeaux, France
sophie.le-floch@cemagref.fr

Jacques Lepart, CNRS, UMR 5175, Montpellier, France
lepart@cefe.cnrs-mop.fr

Yves Luginbühl, CNRS, UMR LADYSS, Paris, France
luginbuh@univ-paris1.fr

Pascal Marty, CNRS, UMR 5175, Montpellier, France
marty@cefe.cnrs-mop.fr

Rafael Mata Olmo, Universidad Autónoma, Madrid, Spain
   rafael.mata@uam.es

Yves Michelin, ENITAC, Lempdes, France
   michelin@gentiane.enitac.fr

Patrick Moquay, Cemagref Bordeaux, France
   patrick.moquay@cemagref.fr

Farida Naceur, Université de Biskra, Département d'Architecture, Algeria
   naceur.farida@caramail.com

Rosário Oliveira, University of Évora, Portugal
   mrgo@uevora.pt

Wolfgang Pfefferkorn, Rosinak & Partner ZT GmbH, Austria
   pfefferkorn@rosinak.at

Klaus Seeland, Swiss Federal Institute of Technology (ETH),
   Human-Environment-Systems Institute, Zurich, Switzerland
   klaus.seeland@env.ethz.ch

Daniel Terrasson, Cemagref Bordeaux, France
   daniel.terrasson@cemagref.fr

Claudine Thenail, INRA-SAD Armorique, Rennes, France
   claudine.thenail@rennes.inra.fr

Catharine Ward Thompson, Edinburgh College of Art, Edinburgh, United Kingdom
   c.ward-thompson@eca.ac.uk

Tanja Tötzer, ARC Systems Research GmbH, Seibersdorf, Austria
   tanja.toetzer@arcs.ac.at

Monique Toublanc, ENSP Versailles, France
   m.toublanc@versailles.ecole-paysage.fr

The coordinators would like to thank Stéphanie Touvron and Julienne Baudel for the logistical and technical support throughout the project in discussions with authors, desktop publishing, proofreading and computer graphics.

Editorial board

Martine Berlan-Darque (Ministry of Ecology and Sustainable Develoment, France)
m.berlan@free.fr

Françoise Burel (CNRS Rennes, France)
francoise.burel@univ-rennes1.fr

Graham Fairclough (English Heritage, London, United Kingdom)
graham.fairclough@english-heritage.org.uk

Josefina Gomez Mendoza (Universidad Autonoma de Madrid, Spain)
Josefina.gomez@uam.es

Yves Luginbühl (CNRS Paris, France)
luginbuh@univ-paris1.fr

Raffaele Milani (University of Bologna, Italy)
milani@philo.unibo.it

Bas Pedroli (ALTERRA-Wageningen, The Netherlands)
bas.pedroli@wur.nl

Jürgen Peters (University of Eberswalde, Germany)
jpeters@fh-eberswalde.de

Teresa Pinto Correia (University of Evora, Portugal)
mtpc@uevora.pt

Daniel Terrasson (Cemagref Bordeaux, France)
daniel.terrasson@cemagref.fr

Hans Karl Wytrzens (University of Natural Resources and Applied Life Sciences,
Vienna, Austria)
wytrzens@boku.ac.at

Imprimé pour vous par Books on Demand (Allemagne)

www.ingramcontent.com/pod-product-compliance
Lightning Source LLC
LaVergne TN
LVHW010815200726
843507LV00003B/606